In-Vitro and In-Vivo Tools in Drug Delivery Research for Optimum Clinical Outcomes

In-Vitro and In-Vivo Tools in Drug Delivery Research for Optimum Clinical Outcomes

Edited by
Ambikanandan Misra
Aliasgar Shahiwala

CRC Press
Taylor & Francis Group
Boca Raton London New York

CRC Press is an imprint of the
Taylor & Francis Group, an **informa** business

CRC Press
Taylor & Francis Group
6000 Broken Sound Parkway NW, Suite 300
Boca Raton, FL 33487-2742

Printed on acid-free paper

International Standard Book Number-13: 978-1-138-55560-0 (Hardback)

Library of Congress Cataloging-in-Publication Data

Names: Misra, Ambikanandan, editor. | Shahiwala, Aliasgar, editor.
Title: In-vitro and in-vivo tools in drug delivery research for optimum clinical outcomes / [edited by] Ambikanandan Misra and Aliasgar Shahiwala.
Description: Boca Raton : Taylor & Francis, 2018. | Includes bibliographical references.
Identifiers: LCCN 2018004185 | ISBN 9781138555600 (hardback : alk. paper)
Subjects: | MESH: Drug Delivery Systems--methods | In Vitro Techniques | Drug Administration Routes
Classification: LCC RM301.25 | NLM QV 785 | DDC 615.1/9--dc23
LC record available at https://lccn.loc.gov/2018004185

Visit the Taylor & Francis Web site at
http://www.taylorandfrancis.com

and the CRC Press Web site at
http://www.crcpress.com

Contents

Preface

The pathway for the fabrication of a clinically viable formulation starts with the physicochemical characterization of the drug candidate, the selection of a strategy for formulation, the evaluation of the formulation for in-vitro and in-vivo performance, and by preclinical and clinical evaluation. Characterizing the physicochemical properties of a drug like solubility, stability, salt and polymorph selection, solid-state properties, and partition coefficient and ionization constant(s) are key elements in establishing the initial feasibility of any formulation development project, and in the subsequent interpretation of the observed in-vivo profile of a given dosage form. A high proportion of experimental drugs have low solubility, which creates challenges in the development of parenteral formulations, and cause poor dissolution and low and variable bioavailability in oral formulations. The meaningful systemic exposure is not only dependent on the solubility, but also on the permeability. Integrating the physicochemical knowledge with biopharmaceutical considerations helps to make informed decisions in the formulation development. This knowledge also provides valuable insight into process development and manufacturing.

The development of novel drug delivery systems involves the optimization of a considerable number of parameters that ultimately impact therapeutic performance. Therefore, proper design and selection of in-vitro characterization tools are of prime importance in gaining an insight into the prospective in-vivo behavior, while proper selection of in-vivo studies help to predict the performance of drug delivery systems in preclinical and clinical studies. However, regulatory bodies only provide general requirements that are not sufficient to predict product clinical viability, superiority, and commercial success. The information available in the literature is also scarce and scattered.

This book, *In-Vitro and In-Vivo Tools in Drug Delivery Research for Optimum Clinical Outcomes* is our sincere attempt to compile the information regarding in-vitro and in-vivo tools in a meaningful logical manner to provide drug delivery research maximum predictability for optimum clinical outcomes. Since the number of drug delivery systems is vast, and in-vivo methods are usually route-specific, the chapters are designed to cover in-vitro and in-vivo tools specific to a particular route of drug delivery. Chapter 1 discusses challenges associated with in-vitro and in-vivo characterizations and common tools that are used for characterizing nanocarriers. Chapter 2 to Chapter 9 provide a comprehensive review of novel approaches and techniques used for characterization and evaluations of drug delivery systems delivered through oral and intraoral, transdermal, nasal–pulmonary, ocular, gastroretentive, colon-targeted, brain-targeted, and parenteral routes of administration.

In the case of an oral route of drug delivery, oral, paroral, gastroretentive, and colon-targeted products with different characterizing parameters are provided as separate sections or chapters. In case of transdermal, nasal, and ocular drug delivery, drug diffusion is the key parameter, whereas in case of pulmonary delivery, drug deposition pattern within lung is of importance. Similarly, in brain delivery, drug transfer across the blood–brain barrier is one of the most important parameters to evaluate and so on. The rate and extent of drug dissolution or release is a key in-vitro property that affects the in-vivo response, and hence is covered in detail under each chapter. Additionally, special emphasis is given to the what, why, and how of different characterization tools to comprehend clinical outcomes of developed delivery systems. Evidence of in-vitro and in-vivo correlations are also provided from the published literature. The chapter references can further be used to supplement the content, since it is difficult to provide accounts of all in-vitro and in-vivo tools used in drug delivery research.

This book is intended to serve as a latest and comprehensive resource for researchers, students, academics, and scientists working in the field of drug delivery research. We believe that after reading this book, the reader will have an enriched knowledge regarding various in-vitro and in-vivo testing methods for different routes of drug delivery, and different types of drug delivery systems. Most importantly, this book will make it possible for readers to appreciate the strengths and weaknesses of each test method which, in turn, will assist them in selecting specific methods that suit their scientific needs.

It gives us immense pleasure to extend our gratitude to the contributors in this book who have brought together their collective experience and expertise in compiling the available information in the literature to produce this book.

Ambikanandan Misra
Aliasgar Shahiwala

List of Abbreviations

5-ASA	5-aminosalicylic acid
AFM	Atomic force microscopy
API	Active pharmaceutical ingredient
ATP	Adenosine triphosphate
AUC	Area under curve
BDM	Biorelevant dissolution media
Caco-2	Cancer coli-2
CC-NPs	curcumin nanoparticles
CDDS	Colon-targeted drug delivery systems
CDER	Center for Drug Evaluation and Research
CFTR	Cystic fibrosis transmembrane conductor regulator
Cmax	Maximum plasma concentration
CO_2	Carbon dioxide
CQAs	Critical quality attributes
DCFH	Dichloro-dihydro-fluoroscein
DCFH-DA	Dichloro-dihydro-fluorescein diacetate
DDS	Drug delivery system
DMSO	Dimethyl sulfoxide
DNA	Deoxyribonucleic acid
DSS	Dextran sulphate sodium
EC	Ethyl cellulose
ES100	Eudragit S 100
FaSSIF	Fasted-state simulated intestinal fluid
FCS	Fetal calf serum
FDA	Food and Drug Administration
FDDS	Floating drug delivery system
FeSSIF	Fed state simulated intestinal fluid
FITC	Fluorescein isothiocyanate
FSC	Forward scatter
FTDC	Flow-through diffusion cell
GC	Gas chromatography
GI	Gastrointestinal
GIT	Gastrointestinal tract
GR	Gastroretentive
GRDF	Gastroretentive dosage forms
HA	Hyaluronic acid
HAART	Highly active antiretroviral therapy
HBS	Hydrodynamically balanced systems
HBSS	Hank's balanced salt solution
HCl	Hydrochloric acid

HCO_3	Bicarbonate ion
HDL	High-density lipoproteins
HLB	Hydrophilic lipophilic balance
HPLC	High-performance liquid chromatography
HPMA-N	(2-Hydroxypropyl) methacrylamide
HPMC	Hydroxypropyl methyl cellulose
IBD	Inflammatory bowel disease
IM	Intramuscular
IMMC	Interdigestive migrating myoelectric complex
IR	Infrared
IRMS	Isotope-Ratio Mass Spectrometry
IV	Intravenous
IVIS	In-vivo imaging system
IVIVC	In-vitro–in-vivo correlation
Ke	Terminal elimination rate constant
LBDS	Lipid-based delivery systems
LDL	Low-density lipoproteins
LSCM	Laser scanning confocal microscope
MCG	Medicated chewing gum
MEM	Minimum essential medium
MPP	Mucus-penetrating particles
MTT	3-(4,5-dimethylthiazol-2-yl)-2,5-diphenyltetrazolium bromide
NCE	New chemical entity
NDDS	Novel drug delivery systems
NLCs	Nanostructured lipid carriers
NMR	Nuclear magnetic resonance
NPs	Nanoparticles
OBT	Octanoic acid breath test
ODTs	Orally disintegrating tablets
PBS	Phosphate buffered saline
PCDCs	Pressure-controlled colon delivery capsules
pDMAEMA	Poly(2-(dimethylamino-ethyl) methacrylate
P-GP	P-Glycoprotein
PHEA	Poly(N-(2-hydroxyethyl)-DL-aspartamide)
PI	Propidium iodide
PK	Pharmacokinetic
PLGA	Polylactic-co-glycolic acid
pSLN	PEGylated solid lipid NPs
PVA	Poly vinyl alcohol
PVP-MA	Poly(1-vinyl-2-pyrrolidone-co-maleic anhydride)
RES	Reticuloendothelial system
ROS	Reactive oxygen species
RPMI	Roswell Park Memorial Institute medium
SC	Subcutaneous

SCFA	Short-chain fatty acid
SD rat	Sprague Dawley rat
SEDDS	Self-emulsifying drug delivery systems
SGF	Simulated gastric fluid
SHIME	Simulator of the human intestinal microbial ecosystem
SIF	Simulated intestine fluid
SiRNA	Small interfering ribose nucleic acid
SLNs	Solid lipid nanoparticles
SMEDDS	Self-microemulsified drug delivery system
SRB	Sulphorhodamine B
SSC	Side scatter
t½	Elimination half-life
TBARS	Thiobarbituric acid reactive substances
TCA	Trichloro acetic acid
TDT	Traditional dissolution test
Tmax	Time to reach Cmax
TNBS	Trinitrobenzene sulfonic acid
UC	Ulcerative colitis
USFDA	United States Food and Drug Administration
USP	United States Pharmacopoeia
UTI	Urinary tract infection
XRD	X-ray diffraction

Editors

Dr. Ambikanandan Misra is a professor of pharmacy, former dean of the faculty of technology and engineering, and head of the department of pharmacy at the Maharaja Sayajirao University of Baroda, Vadodara, India. He has authored 7 books, 41 book chapters, and 166 peer-reviewed publications in reputed international journals. He has on the editorial board of 15 international journals, and a referee to more than 50 journals. He is currently serving as an associate editor of the journal, *Drug Development and Industrial Pharmacy*. He has participated as member and chairman in several academic bodies of national and international stature. Under his supervision, 43 PhD and 132 master's students have completed their dissertations. He has also completed 22 research projects funded by the government and pharmaceutical industry. He is also an advisor to many pharmaceutical and biotechnology industries in India and abroad. He has filed 30 national and international patents, out of which 9 have been granted so far. He is the coordinator of TIFAC CORE in New Drug Delivery Systems, a prestigious project for the Indian government, and recently, he has also initiated an Indo-German project that is funded under DST-DAAD. In the past, he has served as a coordinator and principle investigator of various projects of government of India such as the PURSE program; Department of Science and Technology; Small Business Innovation Research Initiative (SBIRI) project by the Department of Biotechnology, India (DBT); and is the subcoordinator of the Interdisciplinary Life Science program of Advance Research and Education (ILSPARE), DBT. As a recipient of many prestigious awards, including the national award for a research project titled *Liposomal Dry Powder Inhalation* awarded by the vice president of India, "Best Scientist of the Year" by the Association of Pharmaceutical Teachers of India, "Outstanding Manuscript Award by American Association of Pharmaceutical Scientists," "Best Reviewer in Pharmaceutical Sciences for Elsevier Journals" by Elsevier, "Distinguished Alumnus Award" by MSU Pharmacy Alumni Association, and the Illustrious alumnus award by Dr. H.S. Gaur of the University of Sagar, to acknowledge his contribution to the pharmacy profession.

Aliasgar Shahiwala, With more than 15 years of teaching and research experience, Dr. Shahiwala is currently serving as a professor in the department of pharmaceutics and as program director of postgraduate studies at Dubai Pharmacy College, Dubai. Dr. Shahiwala received his master's and PhD in pharmaceutics and pharmaceutical technology from The Maharaja Sayajirao University of Baroda, India with a high research output in the area of novel drug delivery. His postdoctoral research was conducted at Northeastern University, United States, and was specifically focused on applications of nanotechnology in the field of drug delivery and drug targeting. Dr. Shahiwala published several international publications in high impact peer-reviewed journals, four book chapters with internationally renowned publishers, and one patent. He is an editor of the book entitled *Applications of Polymers in Drug Delivery* published by Smithers Rapra Technology, a UK-based publisher. Prof. Shahiwala's research credentials have established him as a reviewer, a member of editorial board, a speaker, and an invited author for various pharmaceutical scientific journals and conferences. Dr. Shahiwala also has more than three years of rich research experience in the formulation and development division of large-scale manufacturers of pharmaceuticals in India as an added technical expertise in his field.

Contributors

Jitendra Amrutiya
Department of Pharmacy
The Maharaja Sayajirao University of Baroda
Vadodara, India

Meena Bansal
Mehr Chand Polytechnic College
Jalandhar, India

Sanjay Bansal
Mehr Chand Polytechnic College
Jalandhar City, India

Priyanka Bhatt
Department of Pharmacy
The Maharaja Sayajirao University of Baroda
Vadodara, India

Mahavir Chougule
School of Pharmacy
Department of Pharmaceutics and Drug Delivery
Pii Center for Pharmaceutical Technology
National Center for Natural Products Research
Research Institute of Pharmaceutical Sciences
The University of Mississippi
Oxford, Mississippi

Priya B. Dubey
Maliba Pharmacy College
Uka Tarsadia University
Bardoli, India

Jaimini K. Gandhi
Maliba Pharmacy College
Uka Tarsadia University
Bardoli, India

Anuj Garg
Institute of Pharmaceutical Research
GLA University
Mathura, India

Keerti Jain
Department of Pharmacy
The Maharaja Sayajirao University of Baroda
Vadodara, India

Ankit Javia
Department of Pharmacy
The Maharaja Sayajirao University of Baroda
Vadodara, India

Rohit Joshi
School of Pharmacy
Department of Pharmaceutics and Drug Delivery
The University of Mississippi
Oxford, Mississippi

Jovita Kanoujia
Department of Pharmaceutical Sciences
Babasaheb Bhimrao Ambedkar University
Lucknow, India

Mohammed Shuaib Khan
Formulation Department-Pharma
Himalaya Global Research Center
Dubai, United Arab Emirates

Manisha Lalan
Babaria Institute of Pharmacy
BITS Edu Campus
Vadodara, India

Rohan Lalani
Department of Pharmacy
The Maharaja Sayajirao University of Baroda
Vadodara, India

Tejal A. Mehta
Institute of Pharmacy
Department of Pharmaceutics
Nirma University
Ahmedabad, India

Ambikanandan Misra
Faculty of Pharmacy
The Maharaja Sayajirao University of Baroda
Vadodara, India

Munira M. Momin
College of Pharmacy
Department of Pharmaceutics
Mumbai, India

Tarul Mulay
School of Pharmacy
Department of Pharmaceutics and Drug Delivery
The University of Mississippi
Oxford, Mississippi

Vivek Patel
Department of Pharmacy
The Maharaja Sayajirao University of Baroda
Vadodara, India

Pranav Ponkshe
School of Pharmacy
Department of Pharmaceutics and Drug Delivery
The University of Mississippi
Oxford, Mississippi

Sumant Saini
University Institute of Pharmaceutical Sciences
Panjab University
Chandigarh, India

Shailendra K. Saraf
Babu Banarasi Das Northern India Institute of Technology
Lucknow, India

Shubhini A. Saraf
Babasaheb Bhimrao Ambedkar University
Lucknow, India

Pranav J. Shah
Maliba Pharmacy College
Uka Tarsadia University
Bardoli, India

Aliasgar Shahiwala
Dubai Pharmacy College
Dubai, United Arab Emirates

Bhupinder Singh
University Insititute of Pharmaceutical Sciences
UGC Centre of Excellence of Nanomaterials, Nanoparticles and Nanocomposites (Biomedical Sciences)
Panjab University
Chandigarh, India

Samipta Singh
Babasaheb Bhimrao Ambedkar University
Lucknow, India

Sanjay Singh
Department of Pharmaceutical Engineering and Technology
Indian Institute of Technology (BHU)
Varanasi, India

Praveen K. Srivastava
Department of Pharmacy
The Maharaja Sayajirao University of Baroda
Vadodara, India

Naazneen Surti
Babaria Institute of Pharmacy
BITS Edu Campus
Vadodara, India

Hemal Tandel
Department of Pharmacy
The Maharaja Sayajirao University of Baroda
Vadodara, India

Hetal P. Thakkar
Department of Pharmacy
The Maharaja Sayajirao University of Baroda
Vadodara, India

Ruchi Amit Thakkar
Department of Pharmaceutics and Drug Delivery
School of Pharmacy
The University of Mississippi
Oxford, Mississippi

1

In-Vitro and In-Vivo Tools in Emerging Drug Delivery Scenario: Challenges and Updates

Hemal Tandel, Priyanka Bhatt, Keerti Jain, Aliasgar Shahiwala, and Ambikanandan Misra

CONTENTS

1.1 Introduction

Administration of drug molecules with conventional dosage forms results in uniform distribution throughout the body without preference to the diseased site, and hence to achieve the desired therapeutic concentration at the site of the disease or at the target site, usually a larger dose of a drug is required to administer. But, administration of larger doses results in the undesirable exposure to organs/cells/tissues and unavoidable side effects/toxic effects. Hence, in order to ensure the safety and efficacy of formulation, scientists are continuously investigating the novel drug delivery system (NDDS) that is specific to the target site where the drug is required to deliver selectively. Although the development of such types of target specific delivery systems is a difficult task to achieve, the discovery, design, and development of novel materials (including microcarriers and nanocarriers like liposomes, microparticles, polymeric nanoparticles, polymeric micelles, dendrimers, solid lipid nanoparticles, hyperbranched polymer drug conjugates, nanostructure lipid carriers, lipid drug conjugates, carbon-based nanomaterials, and quantum dots) has enabled a targeted drug delivery system with potential safety and efficacy.

NDDSs have not only provided us with the optimum dose at the right time and right location by reduction in side effects, they have also provided the pharmaceutical field with the efficient use of expensive drugs, excipients, and reduction in production cost. Moreover, they have proven beneficial to patients in myriad ways, by giving them better therapy, as well as improved comfort and an increased standard of living. The potential advantages revealed by NDDSs are enormous and some of them are summarized as follows:

1. Many of them are biocompatible and biodegradable. They also provide site specificity and thus minimize toxic side effects on other organs and tissues.
2. They are active as well as passive; targeting can be achieved through particle size and surface modifications as novel systems have flexibility to bind with site-specific ligands.
3. They provide controlled drug release rates and particle degradation characteristics, which can be readily varied by the choice of matrix components.
4. High drug loading and drugs can be incorporated into the systems without any chemical reaction, which is a crucial factor for conserving the drug activity.
5. They offer improved therapeutic efficacy and overall pharmacological responses per unit dose.

6. Novel systems have covered almost all the routes of administration including oral, nasal, parenteral, intraocular, etc. (Hnawate and Deore 2017).
7. They offer uniform drug delivery, with enhanced bioavailability because of small size and higher surface area.
8. Low drug dose is required compared to conventional drug dosage form.
9. A wide variety of chemicals or drugs can be formulated into novel systems, including biological components like cytokines, proteins, and peptides.
10. They prevent drug degradation and resulting inactivation occurring due to endogenous chemicals (Kotturi 2015).
11. Functions like active drug targeting, on-command delivery, intelligent drug release, bioresponsive triggered systems, self-regulated delivery systems, systems interacting with the body, and smart delivery are easily achieved with these novel systems.
12. General formulation schemes can be used for making novel systems such as intravenous, intramuscular, or peroral drugs.
13. Cell and gene targeting has become possible.
14. They are superior disease markers in terms of sensitivity and specificity (Costas Kaparissides 2006).

1.2 In-Vitro and In-Vivo Assessment: Why Is It Needed Particularly for NDDS?

It is crucial to find and translate realistic prospects offered by understanding the biological processes for the developing and engineering of effective and safe nanoscale drug delivery systems and medicine that can effectively enhance benefit-to-risk ratio. However, exaggerated in-vivo applications of nanomaterials for a broad range of newly developing and poorly characterized multifunctional molecules and often nonbiodegradable nanoscale materials (e.g., carbon nanotubes, quantum dots, and certain polymeric and metallic nanoparticles), have upraised its harmful effect and possible safety. Researchers rarely address these issues; their attention is simply to establish NDDS for defining the proof-of-concept and usually in animal models which is not relevant to the human disorder/disease in question. In many cases, a slight specificity in pharmacological influence of a NDDS is indicated as successful in the targeting/therapeutic effect (Bawa, Audette, and Reese 2016). On the other hand, NDDS may also exert unexpected side effects and toxicities and hypothetically create a threat to public health and the environment.

The development of these new formulations and drug delivery system is not an easy task and many challenges are faced by scientists during this process. The site-specific delivery of the novel dosage form is challenging. Also after reaching the specific site, parameters like residence time and dosing accuracy are crucial, as they differentiate novel formulation from the conventional one. With nanotechnology, the fabrication of a delivery system with reproducibility and stability is a big challenge (Jain, Mehra, and Jain 2015; Deodhar, Adams, and Trewyn 2017; Sun, Zhang, and Wu 2017; Zhao et al. 2013). Therefore, preclinical testing and performance assessment parameters are very important before clinical studies of novel drug formulation. A key consideration, both in expressions of economics and pragmatism, is planning in-vitro studies that link to in-vivo testing. If the in-vitro data correlate to the in-vivo parameters, it can reasonably be extrapolated that the same results would be obtained in-vivo. Thus, the right selection of related in-vitro models for evaluation of a parameter is crucial part for successfulness of in-vivo parameter testing (Omar and Nadworny 2017). Following subsections discusses different challenges associated with the in-vitro and in-vivo testing of NDDS.

1.2.1 Challenges—In-Vitro Studies

1.2.1.1 Particle Size, Zeta Potential and Stability

The change in particle size of an active pharmaceutical ingredient, formulations or an excipient from macro to micro and nanosize range potentially influence the quality, safety, and efficacy of drug or drug product (Table 1.1). The significance of establishing the relationship between particle size, zeta potential, and stability of product with in-vitro dissolution, the release of drug at targeted site, and bioavailability is essential to emphasize during production with the ultimate goal to efficient and clinically relevant product

TABLE 1.1

Effects on API Properties by Changes in Particle Size

Changes	Effect on Physicochemical and Biological Properties
Reduction in particle size	Enhancement of solubility of API
	Increase in stability of API
	Alteration of pharmacokinetic behavior of drug
	Enabling fast drug release on the other hand also facilitating sustained drug release and thereby absorption of drug
	Increase in bioavailability by specific targeting either by active targeting or passive targeting of API
	Prolongation of retention time in systemic circulation by reduction of reticuloendothelial uptake and also by reduction of glomerular filtration rate
	Reduction in dose
	Improved efficacy and safety

specifications. Some of these challenges of the aforementioned in-vitro studies are as follows:

1. Precise mechanism of production to control various parameters like particle size, zeta potential, and stability for production of NDDSs are the major problems.
2. Designing protocols to study in-vivo kinetic parameter that are appropriate (e.g., dose, sampling times and the route of administration of drug) to identify whether the nanoparticles may release for extended period of time at specific site with the effect of parameters like size and zeta potential. Moreover, stability imparts a greater effect during in-vivo behavior.
3. To observe the changes in systemic absorption of active moiety due to change in nanomaterial size is challenging.
4. Uniformity of particle size data and stability of excipients are important for NDDS formulation. Specifications for particle size of some excipients are not available and may sometimes generate problems.
5. Nanoparticles could have diverse biodistribution profiles with particle size change; therefore, it is necessary to study the biodistribution of biomaterials.
6. The effects of a particle size change on sustained-release nanoformulations are important, and should be necessarily taken into consideration.
7. Normally animal toxicity studies are not always available/performed with an applicable particle size change.

1.2.1.2 Dissolution/Release Studies

The release of the drug from the formulation and biodegradation of polymer used are crucial factors to get into consideration while developing the nanoparticulate system. The release rate of drug candidates depends on solubility of the drug; drug diffusion through the nanoparticle matrix, nanoparticle matrix erosion/degradation, and a combination of the erosion/diffusion process. For conventional formulations, a United States Pharmacopoeia (USP)–recommended dissolution test apparatus is commonly used to study the in-vitro drug release rate. However, in the case of micro and nanoparticulate systems, the use of these apparatus is not recommended.

This is because of the following challenges:

1. It is difficult to achieve sink conditions in the case of nanoparticulate systems which have a high surface area.
2. The separation of dissolved drugs and undissolved particles is also difficult while sampling and analyzing.

3. Sometimes the biodegradation of polymeric matrix of nanoparticulate systems is stimuli sensitive; for instance, when there is a need for a specific enzyme, pH, or temperature-sensitive degradation which, in turn, leads to the release of the drug. Thus, specific conditions are required in such cases.

1.2.2 Challenges—In-Vivo Studies

Novel formulations that survive the preliminary analysis, can be optimized further or rehabilitated to make them safer and/or more effective. With optimized novel formulations, researchers turn their attention to preclinical analysis by the support of a pharmacologist or biologist. To perform the clinical trial on human subjects, a safe starting dose, route of administration and rate of administration, if any, must be established for NDDS.

Because the nanomaterial often affects pharmacokinetic parameters of the API, traditional PK endpoints along with biodistribution study may not be enough to accurately evaluate the safety and efficacy of API and the rate and extent of absorption at the site of action. The dosing accuracy and site specificity with minimum toxicity is achievable with NDDS systems. However, some parameters, like the ability to cross various biological barriers like tumor penetration, nanoproduct and tissue interaction, ability to prevent drug release before reaching the target site and mechanisms that trigger drug release at the target site, should be well characterized (Jain, Mehra, and Jain 2015; Lee et al. 2017). Moreover, plasma and tissue distribution of novel particles may be uneven, and dose dumping and multiple dosing sometimes leads to dose accumulation and may result in toxicity. The degradation of some bioactives like peptides, hormones, monoclonal antibodies, genetic materials, recombinant proteins, etc. impose obstacles in the development of drug delivery systems. Therefore, biodegradation and biocompatibility plays an important role in determining the safety of various drug delivery systems. Developing nanocarriers with optimal clearance characteristics that can minimize toxicity risks and decrease concerns of nanoparticles interference to normal physiological processes should also be considered. It is necessary that the drug delivery system itself and its components should not have any detrimental effect on the normal biological and physiological processes, as well on the vital components of biological systems.

1.3 In-Vitro Characterization of NDDS

In developing commercially viable NDDS products, a deep understanding of product attributes and process parameters are necessary. Product attributes are mainly classified as related to product quality (physicochemical characterizations) and product performance (in-vitro release studies). The drug content, drug loading, reproducibility of the formulations, particle size and shape, distribution, and zeta potential analysis of the produced novel formulation are important parameters in in-vitro characterization. In-vitro drug release and permeation studies, tissue/cell viability, cell uptake mechanisms are the other parameters to evaluate in-vitro studies. All these studies give the idea about the mechanical empathetics about drugs and their mechanisms relative to carriers used in new drug formulation development.

1.3.1 Product Quality Attributes

Critical product quality attributes vary from product to product. As an example, transdermal patches have different product attributes compared to modified release tablets. Product-specific attributes are discussed in detail in the further individual chapters of the book. In the following sections, common approaches in design and development of NDDS and parameters used for their characterizations are discussed.

1.3.1.1 Complexes

The complexes of a drug with a polymer or a carrier has been an area of wide investigation in designing a NDDS. These complexes are formed by electrostatic interaction between the drug and polymer or polymer–polymer. The use of such interactions has also been done for complexing DNA with cationic polymers, such as polyethyleneimine and chitosan for gene therapy and vaccination (Patil et al. 2016). The formulation of complexes involves a preliminary interaction based on charge followed by a secondary intra-complex formation with new bonds, and finally the hydrophobic interaction of the secondary complexes. The formation of complexes is influenced by the solvent used, temperature of processing, electrolyte amount, pH of the media, charge density, and ionic strength (Lankalapalli and Kolapalli 2009). Inclusion complexes of hydrophobic drugs with various derivatives of cyclodextrins have also been reported to improve the solubility and permeation through the biological membranes of poorly water soluble drugs (Frömming and Szejtli 1994). Solid dispersion and solid solution are other approaches used to improve their solubility and dissolution rate. This leads to an improvement in the oral absorption rate of the drug, which is due to the solubilizing effect of the carriers employed during the preparation, decrease in the particle size of the drug to molecular level and improved dispersion and wetting of the drug due to the carrier. The improvement in dissolution rate is observed due to an increase in surface area of the particle in contact with the dissolution media (Serajuddin 1999). The characterization method involves the measurement of several physicochemical properties, such as turbidity, ionic strength, and pH as a function of a weight ratio of a polymer. Others include thermal analysis, IR, NMR, viscosity measurement, XRD, ionization constant determination, etc. (Van Leeuwen, Cleven, and Valenta 1991; Dautzenberg 2000; Pogodina and Tsvetkov 1997; Anlar et al. 1994).

1.3.1.2 Conjugates

To maximize the amount of drugs that become available at only the particular site of action, thus minimizing the unwanted exposure to healthy tissues, an alteration of pharmacokinetics of the drug is required (Peer et al. 2007; Torchilin 2005). Such alteration can be done by formulation to a novel drug delivery vehicle or by a covalent modification of the drug. In the latter approach, either a prodrug is prepared or the formulation is surface modified using suitable ligands for achieving superior pharmacokinetics and pharmacodynamics. Prodrugs can consist of either antibody drug conjugates (Bhatt, Vhora, et al. 2016) or peptide drug conjugates (Vhora et al. 2015; Chari, Miller, and Widdison 2014). Whereas for surface modification, peptides, protein, antibodies, aptamers or organic molecules can be employed (Kratz 2008). The essential difference between the two strategies is that in case of prodrugs, conversion occurs to the active form after being processed by the cellular enzymes and the modification is for improving the stability of the drug in-vivo. In the case of surface modification using ligands, the ligand is not metabolized, or the conjugate is not converted to a free form, rather the ligand assists in targeting of the drug to the desired site, impacting the biodistribution of the drug (Khandare and Minko 2006). Antibodies are used because they have selective binding capability. For peptide conjugations, a linker is employed for bond formation with drug molecule. The characterization tools used are similar to those for complexes.

1.3.1.3 Encapsulated Systems

Encapsulation is one of the promising methods for the preparation of nanocarriers like nanospheres, nanocapsules, nanoparticles, nanogels, and polymeric micelles. The encapsulation method is based on the special affinity of polymeric systems to encapsulate the therapeutic agents and also the ability of these systems to release the loaded active drug molecules, including genetic materials, intracellularly. This method has been used for the loading of various drugs (including doxorubicin, amphotericin B, cisplatin, etc.) into nanocarriers (including nanoparticles, dendrimers, carbon nanotubes, etc.) to achieve controlled and targeted release. In-vitro and in-vivo characterization, challenges in this method includes various parameters (including in-vitro parameters like drug loading content, drug loading efficiency, drug release kinetics, etc.), and in-vivo characterization challenges are in-vivo drug release kinetics, with the ability to elicit sustained release and maintain therapeutic concentration at target site, interaction with biological components, biodistribution, and drug targeting efficiency (Barouti, Jaffredo, and

Guillaume 2017; Crivelli et al. 2017; Li et al. 2017; Bae et al. 2017; Kaur et al. 2017; Bajwa et al. 2016).

1.3.1.4 Lipid-Based Systems

Lipid-based delivery systems (LBDS) are one of the most promising delivery systems for drugs, proteins, and peptides, which are known to improve the solubility and bioavailability of poorly water-soluble drugs, minimize the rate of degradation, control the release, and reduce adverse effects of therapeutic drugs and enhance their accumulation at a target site. LBDS can be tailored based on the requirement of final formulation considering the disease indication, administration route, stability, efficacy, and toxicity of the product, as well as the cost. Because of their proven efficacy and safety, they are applicable as potential carrier systems for pharmaceuticals, nutraceuticals, diagnostics, and vaccines. LBDS are broadly categorized as:

1. Emulsions, including microemulsion, self-emulsifying drug delivery systems (SEDDS), self-microemulsifying drug delivery systems (SMEDDS), and nanoemulsion (Patel et al. 2017; Tandel et al. 2017).
2. Vesicular systems, such as liposomes, transferosomes, colloidosomes, etc. (Gandhi et al. 2015).
3. Lipid particulate systems, which includes solid lipid nanoparticles (SLNs) (Pandya et al. 2018), lipid drug conjugates, nanostructured lipid carriers (NLCs), and lipospheres (Shrestha, Bala, and Arora 2014). Formulation parameters affecting the bioavailability of drugs from LBDDS are lipid digestion, mean emulsion droplet size/particle size and lipophilicity of the API and nature of lipids. Some of the challenges in formulating LBDS are their complex physicochemical properties, challenges in stability, limited solubility of some drugs in lipids, preabsorption of formulation in GIT specifically for oral route and lack of in-vitro and in-vivo predicative characterization methods.

The systematic method of formulation development of LBDS begins with the preselection of excipients depending on their composition of fatty acids, melting point of lipids, HLB values of surfactants, digestibility and disposability followed by screening of particular excipients for their solubility, stability and compatibility, as well as for their dispersion or dissolution properties. The next step is the identification of a method of preparation suitable for the intended dosage form, formulation optimization considering the loading of the drug, particle size and release profile of the drug. The prepared formulations are then evaluated for physicochemical characterization (appearance, color, odor, taste, density, pH value, viscosity measurement, % transmittance, zeta potential, globule/particle size, and transmission electron microscopy), in-vitro characterization and in-vivo studies using the appropriate animal models. In order to evaluate the performance of an excipient during formulation development and to predict in-vivo performance, it is necessary to design an in-vitro testing method.

1.3.2 Product Performance Attributes

1.3.2.1 In-Vitro Drug Release Studies

With the emergence of controlled and targeted drug delivery systems, the drug release studies are conducted analogously to drug dissolution studies for conventional dosage forms. In contrast to conventional dosage forms where the in-vitro drug dissolution test is employed as a quality control tool, the purpose of carrying out drug release study for NDDS to know is the mechanism and rate of drug release. In-vitro release testing is important for both quality control purposes as well as to predict in-vivo performance. Nowadays, it is recommended as part of the demonstration of bioequivalence between the test and reference products in the approval of most generic drugs.

Compendial methods are mainly developed for oral (USP type 1 and 2 apparatus) and transdermal (USP 5, 6 and 7 apparatus) formulations. The in-vitro release of a drug from extended, sustained, and pulsatile release tablets are generally carried out using paddle-type tablet dissolution apparatuses maintained at 37°C containing 900 ml of dissolution medium at an agitation rate of 50–75 rpm. To mimic gastrointestinal (GIT) conditions, 0.1 N HCl buffer solution with a pH around 1.2 is used as dissolution medium that mimics the pH conditions of the stomach for the first 2 hours is followed by the change of dissolution medium with phosphate buffer pH 6.8 for the next 6 hours to mimic in-vitro pH conditions of the intestines. The solution is withdrawn at a prefixed time interval and analyzed spectrophotometrically after a suitable dilution. Similarly, the in-vitro characterization of tablet floating behavior, such as floating lag time (the time the tablets took to emerge on the water surface) and the duration of floating, are evaluated in a paddle-type dissolution apparatus mimicking the GIT conditions (Varshosaz et al. 2006). However, compendial methods pose different problems for their suitability as drug release methods for NDDS, such as:

1. For USP 1 and 2 apparatus, large volumes of media are required, which is not practical for the estimation of low-dose NDDS. Also, particulate systems need a dialysis bag/tube to place the sample. Evaporation of the media due to the longer duration of testing of some of the controlled-release products, such as parenteral depots, may be another issue.
2. A flow-through cell apparatus is a widely accepted dissolution test by USP, *European Pharmacopoeia* as Apparatus 4 and Apparatus 3 by JP for testing poorly soluble products or low-dose formulations

with sustained release. Different types of cells are available for testing tablets, powders, suppositories, hard and soft gelatin capsules, implants, semisolids, suppositories, and drug-eluting stents (Fotaki 2011). Originally designed for poorly soluble compounds, different flow-through cell systems have been evolved for novel dosage forms, such as drug-eluting stents and depot preparations.

In recent years, considerable progress has been made for the in-vitro drug release assessment of micro/nanoparticulate systems which include the dialysis and reverse dialysis bag diffusion technique, side-by-side diffusion cells with artificial or biological membranes, agitation followed by ultracentrifugation/centrifugation, and ultrafiltration, or centrifugal ultrafiltration techniques.

1.3.2.2 Hemolytic Assay

For evaluating the hemocompatibility of the delivery system, the hemolytic assay serves as an important tool for determining the ex vivo, pH-dependent behavior. The designed drug delivery system should have an essential property of having endosomolytic behavior, which is essential for facilitating the cytoplasmic delivery of the cargo. Erythrocytes provide a model membrane for evaluating this behavior. Herein, erythrocytes are collected by centrifugation from the blood and diluted to a specific concentration. The incubation of the formulation at different concentrations or weight ratio at a different pH (i.e., pH 7.4, pH 6.8 and pH < 6.8) are carried out. Comparison of the results are done with control untreated samples and positive untreated samples to obtain the hemolytic potential of the test samples (Kichler et al. 2001; Behr 1997; Dawson and Dawson 1986). For cytosolic delivery, it is preferred that the test samples exhibit no hemolysis at pH 7.4, thus substantiating inertness during circulation, and be hemolytic at endosomal pH (i.e., < 6.5) (Dobrovolskaia et al. 2008). The results expressed as percentage of hemolysis are used to evaluate the acute in-vitro hemolytic activity of the nanoparticles at 37°C.

1.3.2.3 Electrolyte Stability Study

Various drug delivery systems intended for systemic administration encounter blood as the primary circulation medium for their transport throughout the body. Blood consists of components such as plasma, red blood cells each of which may get involved in interaction with the formulation. Further, the formulation also may interact with the plasma proteins that may affect their biodistribution (Aggarwal et al. 2009; McElnay 1996). Electrolytes in the blood may lead to charged-based interactions, resulting in flocculation of the system leading to destabilization. Such interactions are particularly encountered with colloidal formulation, wherein electrostatic forces predominate for stabilization. However, for minimizing the impact, steric stabilization by PEGylation can be carried out which prevent particles from coming close enough to allow van der Waals attractive forces between the particles to dominate, and thus create steric barriers resulting in steric stabilization (Morgenstern et al. 2017; Parrott and DeSimone 2012).

1.3.2.4 Cell Culture Studies

Since the knowledge of intracellular delivery and understanding of intracellular trafficking of the drug of chief importance in the development of drug delivery systems, various in-vitro cell line studies can be used to help in predicting pharmacogenomic hypotheses and clinical response. In-vitro cell line studies are essential as they offer the methods for primary evaluation of the direct effects of drugs and formulations on cells and tissues so as to form a basis for in-vivo animal studies and clinical studies. These studies are important, which can give ideas on clinical applicability in pathological conditions and understand molecular mechanisms, as well as can screen the test samples for their efficacy and toxicity.

Primary cell cultures are sometimes not used for experimental studies due to their poor stability, as they undergo constant adaptive alterations and it is challenging to select a time period of when the total cell population is homogenous or stable. After confluence, some cells may transform and become insensitive to contact inhibition and overgrow, therefore it is necessary to keep the cell density low to maintain the original phenotype. After the first subculture or a passage, the culture is called cell line. In each subsequent subculture, a population of cells with the capacity to rapidly grow will predominate while slow growing cells dilute out. In most cases, the culture becomes stable after three passages.

The propagation and growth of cell lines requires a culture media with distinct chemical composition to confirm consistent quality and reproducibility. Mostly, the cells grow efficiently well at pH 7.4 and in a 5% CO_2 environment, as the CO_2 gas phase after dissolution into a culture medium can establish an equilibrium with HCO_3^- ions present in the medium to maintain the pH. Besides, HCO_3^- ingredients such as pyruvate, high concentration of amino acids are used as buffering agents in culture media. The requirements of temperature rely on body temperature of the animal, from which cells are obtained and thus kept at 37°C.

In the following are some of the key cell culture studies those are employed on regular basis for in-vitro characterization of various novel drug delivery formulations.

1.3.2.5 Cellular Uptake and Cell Binding Studies

To study intracellular uptake of delivery systems, in-vitro cell uptake studies are generally carried out either using confocal microscopy or flow-cytometry tools. Formulations are previously loaded with fluorescent dye or tagged with fluorescent moiety and allowed to come into contact with the cell lines of interest for a particular period of time. Then, those cells are observed for cellular uptake of formulation using a confocal microscope or flow cytometer.

Confocal microscopy is a very powerful tool to study cellular binding and the intracellular delivery of drugs. This method is commonly used to study the binding of targeted drug delivery systems to specific receptors or protein ligands (ligand binding assay) and their cellular uptake (Vhora et al. 2014). The laser scanning confocal microscope (LSCM) is an essential component of modern day biomedical research applications. It uses a laser as light source, a sensitive photomultiplier tube detector, and a computer to control the scanning mirrors and build images. The specimens are labelled with one or more fluorescent probes. The confocal microscopy also offers the advantage of greater resolution, due to the use of highly sensitive photomultiplier tube detectors. In cellular biology, the confocal microscopy has been used for visualizing intracellular organelles, cellular uptake, intracellular localization of drugs, and drug delivery systems using fluorescent probes.

Flow-cytometry is a powerful technique for characterizing cells in clinical diagnosis and biomedical research for quantifying aspects about their size, internal complexity, and surface markers. In a flow cytometer, the suspension of cells is hydrodynamically focused in a single-cell wide stream of fluid containing a fast-moving sheath fluid around the slow-moving cell suspension. This laminar stream of particles is subsequently interrogated by one or more laser beams placed perpendicular to it, and will only illuminate single cell at a time. The light scattered in forward direction (FSC) is proportional to the size of the cells, while the light scattered in perpendicular direction (SSC) correlates with intracellular granularity or complexity. Thus scattering itself gives information about the size and composition of the cells (Ibrahim and van den Engh 2007). The second technique of detection relies on the use of fluorescent probes attached to cells, which gives fluorescence after an interaction with lasers at an interrogation point and emits light at longer wavelengths. Here, the nonfluorescent cells will be counted as negative while the fluorescent cells will be counted as positive cells. Further, the intensity of emission gives information about the number of fluorescent probes.

1.3.2.6 In-Vitro Cytotoxicity Studies

Cytotoxicity or viability assays are important, as they can be used to check the susceptibility of cells, toxicity of a developed formulation, and potentially give an idea on in-vivo cell injury or the toxicity of the sample substance or formulation. Cell viability of adherent cell lines can be evaluated by the methods described in Table 1.2 (Mickuviene, Kirveliene, and Juodka 2004).

TABLE 1.2

Methods to Study In-vitro Cytotoxicity in Cell Lines

No.	Categories	Method and Specifications
1	Loss of membrane integrity	• Important in estimating the measure of cellular damage • Some membrane-extracting cyclodextrin carriers and cationic particles, i.e., cationic liposomes, polyplexes, and amine-terminated dendrimers, exhibit their toxic effects by membrane disruption (Hong et al. 2004; Vaidyanathan et al. 2015). • Measurement includes trypan blue exclusion assay, lactate dehydrogenase leakage study, use of fluorescent dyes, and flow-cytometry (Decker and Lohmann-Matthes 1988; Korzeniewski and Callewaert 1983; Aeschbacher, Reinhardt, and Zbinden 1986).
2	Loss of metabolic activity	• Assays that measure metabolic activity include tetrazolium dye reduction, ATP, and 3H-thymidine incorporation assay • In this assay, MTT—a yellow, water-soluble tetrazolium dye—is metabolized by the live cells to purple, water-soluble, formazan crystals. Formazans can be dissolved in DMSO and quantified by measuring the absorbance of the solution at 550 nm. Comparison between spectra of samples of untreated and nanoparticle-treated cells can provide a relative estimate of cytotoxicity (Bhatt et al. 2015; Alley et al. 1988).
3	Loss of monolayer adherence	• Monolayer adherence is frequently assessed by staining the total protein, following the fixation of adherent protein (Lam et al. 2004). • The sulforhodamine B total protein-staining assay is also used for the determination of monolayer adherence.
4	Cell cycle analysis	• Cell cycle analysis study includes staining of DNA with propidium iodide and flow cytometric analysis (Tuschl and Schwab 2004). The method can determine the effect of treated formulations on the progression of cell cycles and on cell death (Bhatt, Lalani et al. 2016).

1.3.2.7 Oxidative Stress

Nanoparticles, due to their large surface area, as well as cationic contaminants and redox active materials, can expedite reactive oxygen species (ROS) generation. The measurement of ROS is mostly conducted by the fluorescent dichlorodihydrofluoroscein (DCFH) assay (Black and Brandt 1974). An ROS probe, DCFH-DA undergoes deacetylation intracellularly followed by ROS-mediated oxidation to a fluorescent species, the excitation wavelength is 485 nm and emission wavelength is 530 nm, respectively. Another method is thiobarbituric acid reactive substances (TBARS) assay which is usually is used for the measurement of lipid peroxidation products.

1.3.2.8 Apoptosis and Mitochondrial Dysfunction

Necrosis and apoptosis are main two mechanisms describing nanoformulation-induced cell death. Flow-cytometry is an important tool for the simultaneous analysis of apoptosis and necrosis in a single type of cell population. DNA-binding dye propidium iodide (PI), which is a fluorescent dye that is impermeable to viable plasma membrane, is used to study necrosis by measuring the permeability of PI through the plasma membrane. The staining of cells with Annexin V in conjunction with propidium iodide (PI) is used to assess early and late phase apoptotic cells (Lalani et al. 2016). The method is based on the principle that viable cells with intact membranes would not allow any of the dye and thus they are similar as unstained cells. Whereas the dead cell membranes or damaged cells are permeable to both the dyes, thus the cells which are in late apoptosis or already dead are both Annexin V and PI positive, the cells that are in early apoptosis are Annexin V positive and PI negative. Various techniques are employed for the detection of mitochondrial dysfunction, including oxygen consumption assessment (via a polarographic technique), measurement of ATPase activity (via a luciferin–luciferase reaction), membrane potential (via a fluorescent probe analysis) and morphology (via electron microscopy) (Gogvadze, Orrenius, and Zhivotovsky 2003).

1.4 In-Vivo Performance Evaluation of NDDS

The most commonly used models for preclinical pharmacokinetic testing are rodents (including rats, mice, and guinea pigs) and non-rodents (including dogs, non-human primates, pigs and mini pigs, rabbits, goats, sheep, etc.).

All pharmacokinetic studies should be evaluated according to validated procedures. The animal species, breed, number of animal/subject in group, weight, age (adult, young, neonate), and gender should be specified and justified. Primary studies should be performed in an adequate number of healthy animals/subjects. Preliminary study shows a safety profile after that the study performed on specific model.

NDDSs evaluate by various methods, such as intravenous, intraperitoneal, buccal, intraperitoneal, ocular, oral, sublingual, topical, subcutaneous, intratracheal, intranasal, etc. The route of administration is the specific focus, as this influence more on the absorption of the API from the novel product.

For individual animals, a dose based on body weight should be expressed; if the dose is intended to be on a body area basis, dose should be expressed both on a body area and body weight basis. For fixed formulations (e.g., tablet/capsule) that are difficult to administer precisely (i.e., on mg/kg body weight basis), dose should be adjusted to individual body weight. When the formulation is administered via drinking water or food, the daily dose of the drug in mg/kg body weight should be calculated, moreover the animal's behavior should be closely observed (daily consumption of food and drinking water, delay between intake of drug and sampling, etc.). For single or fixed combinations, bioavailability studies should be performed for each API, as well as for the combination of product-possible interactions. Exceptions may be considered for certain topical or local treatments, but omission of data should be justified. The appropriate statistical analysis should be performed to determine linearity. For established drug molecules, where dose linearity exists in the animal, single-dose studies, corresponding to the highest intended therapeutic dose, are generally sufficient. Where there is lack of dose-linearity or a very steep dose/effect curve, studies with three different dose levels may be essential and if a range of doses is suggested, the central dose being the median of the therapeutic dose range. If the drug is used only at one dose level, a single-dose study may be adequate, but should be justified. Multiple-dose studies and steady-state investigations should be led with the recommended dosage regimens (dose, number of administrations, frequency of dosing, etc.); multiple-dose testing should give insight into the linearity, accumulation, distribution, steady-state levels and induced effects (e.g. altered metabolism rate and altered disposition). The comparison of plasma concentration profiles after the administration of the first and last dose of novel formulation, as well with the standard drug or with conventional drug formulation, is highly desirable. Biological fluids (blood, plasma, serum, urine, etc.) and organs/tissues are selected for the pharmacokinetic profile of the investigation of novel formulations. Plasma is the most commonly employed biological fluid for kinetic studies. For blood sampling, the following attention should be required: sampling procedure, site of blood collection, material used for sampling, anticoagulant, and conditions of centrifugation to obtain plasma. The stability of the substance must be evaluated. The number of blood samples and the timing of sample collection should be proper to allow for an adequate determination of absorption, distribution, and elimination. Blood samples in the post-absorption phase, as well for sustained drug formulations, should be collected and observed over a long period, as it is necessary for the purpose of the analysis. In some cases, the collection of other biological fluids and tissues may be evaluated (e.g., urine). Collection of some of these fluids may require special attention (e.g., immediate pH measurement, storage conditions till analysis, etc.). The use of a chemical assay method is generally preferable (e.g., gas chromatography, high performance liquid chromatography methods) to evaluate the amount of API in subject after administration of novel formulation. (2010, use 2000; Toomula 2011).

The following parameters should be found out to observe the effect of novel drug delivery vs. conventional drug delivery:

- Cmax: Maximum plasma concentration obtained graphically from the plasma concentration versus time profile
- Tmax: Time to reach Cmax following drug administration, obtained graphically from the plasma concentration versus time profile

- AUC0–t: Area under the plasma concentration-time curve from time 0 (administration) to time t (last sampling time), calculated through the trapezoidal method
- AUC0–inf: Area under the plasma concentration-time curve from time 0 (administration) extrapolated to infinity
- Ke: Terminal elimination rate constant
- t½: Elimination half-life, calculated as 0.693/Ke

The appropriate mathematical procedure should be employed to generate fundamental parameters (compartmental and/or "non-compartmental" analysis). Appropriate kinetic computer programs should be used (regression methods, weighting factor, etc.). Pharmacokinetic parameters should be calculated using the time concentration data from individual animals, and these parameters should be expressed with the indication of the mean and variation values. Standard equations or equivalent calculations should be used to calculate the pharmacokinetic parameters and the interpretation provided of the values obtained (use 2000; Toomula 2011).

1.5 Regulatory Guidance

In October 2015, the Center for Drug Evaluation and Research (CDER) of the US Department of Health and Human Services, Food and Drug Administration published an article titled "Nonclinical Safety Evaluation of Reformulated Drug Products and Products Intended for Administration by an Alternate Route" addressed to "Guidance for Industry and Review Staff." The guidance states that much of the available nonclinical information used to support the approval of an initial formulation can be used to support the safety of new formulations, but in some cases these data may not be sufficient to support additional approval because changes in the formulation could produce a new toxicity, as the new formulation could be used in a different way. This is particularly true if the drug's route of administration is different or the duration of use changes markedly. In those cases, additional nonclinical studies might be recommended to ensure that the toxicity of a new formulation and use is fully characterized. If the new formulation is used in a manner similar to previous formulations, the need for additional nonclinical data generally will be limited. However, if the new formulation is used in a substantially different way (e.g., new route, longer duration), the need for additional nonclinical data becomes greater, and additional nonclinical information may be needed even if no change is made in the composition of the formulation. The guideline discloses the various parameters to be considered before developing a new formulation or a new drug delivery system (guideline).

The United States Food and Drug Administration (FDA) has established CDER's Nanotechnology Risk Assessment Working Group (Nano Group) to assess the potential impact of nanotechnology on drug products. One of its major initiatives has been to conduct a comprehensive risk management exercise to find out the potential impact of pharmaceutical ingredients, nanomaterials and excipients on drug product safety, quality, and efficacy. This exercise concluded that current review practices and regulatory guidance are capable of detecting and managing the potential risks to quality, safety, and efficacy when a drug product incorporates a nanomaterial. The drug products containing nanomaterials should be analyzed for the four risk management areas which have been identified by the CDER working group: (1) potential consideration of how to perform the characterization of nanomaterial properties and improvement in analytical method used for this characterization; (2) the suitability of in-vitro tests to estimate drug performance for drugs with nanomaterials; (3) the understanding of properties arising from nanomaterials that may result in varied toxicity and biodistribution profiles for drug products containing nanomaterials; and (4) enhanced understanding of how the particle size can vary, which can affect the product quality and performance. Also, CDER has been supporting regulatory research in the area of NDDSs, which are especially focused on in-vitro and in-vivo characterization, safety, and equivalence (between reference and new product) considerations. (Tyner et al. 2017). Although the impact of nanomaterials on pharmaceutical products is still unclear, guidelines on risk assessment in food products and cosmetics are available and offer a screening of future developments in pharma products (Wacker 2014).

1.6 Conclusions

Drug delivery research is a translational process, and many formulations failed to be translated into clinical product after promising results in either in-vitro conditions or in-vivo animal experiments. Chapter 1 discussed the various problems which are encountered during the formulation of a drug delivery products, its evaluation and what are the major hurdles in successful clinical translation. In-vitro an in-vivo study of NDDS products focus on assisting the journey of preclinical research into clinical, usually via a faster, easier, cheaper, and more effective route. From the reports of various scientists, we could conclude that a thorough and systematic in-vitro and in-vivo assessment of any novel drug delivery products is required for development of a formulation with substantial potential to reach to the clinical practices.

References

2010. "Guidance for Industry M3(R2) Nonclinical Safety Studies for the Conduct of Human Clinical Trials and Marketing Authorization for Pharmaceuticals, ICH, Revision 1, 2010."

Aeschbacher, Martin, Christoph A. Reinhardt, and Gerhard Zbinden. 1986. "A rapid cell membrane permeability test using fluorescent dyes and flow cytometry." *Cell Biol Toxicol* no. 2 (2):247–55.

Aggarwal, Parag, Jennifer B. Hall, Christopher B. McLeland, Marina A. Dobrovolskaia, and Scott E. McNeil. 2009. "Nanoparticle interaction with plasma proteins as it relates to particle biodistribution, biocompatibility and therapeutic efficacy." *Advanced Drug Delivery Reviews* no. 61 (6):428–37.

Alley, Michael C., Dominic A. Scudiero, Anne Monks, Miriam L. Hursey, Maciej J. Czerwinski, Donald L. Fine, Betty J. Abbott, Joseph G. Mayo, Robert H. Shoemaker, and Michael R. Boyd. 1988. "Feasibility of drug screening with panels of human tumor cell lines using a microculture tetrazolium assay." *Cancer Res* no. 48 (3):589–601.

Anlar, Şule, Yilmaz Çapan, Olgun Güven, Ahmet Göğüş, Turgay Dalkara, and Atillâ Hincal. 1994. "Formulation and *in vitro–in vivo* evaluation of buccoadhesive morphine sulfate tablets." *Pharmaceutical Research* no. 11 (2):231–6.

Bae, Ki Hyun, Susi Tan, Atsushi Yamashita, Wei Xia Ang, Shu Jun Gao, Shu Wang, Joo Eun Chung, and Motoichi Kurisawa. 2017. "Hyaluronic acid-green tea catechin micellar nanocomplexes: Fail-safe cisplatin nanomedicine for the treatment of ovarian cancer without off-target toxicity." *Biomaterials* no. 148:41–53. doi: 10.1016/j.biomaterials.2017.09.027.

Bajwa, Neha, Neelesh K. Mehra, Keerti Jain, and Narendra K. Jain. 2016. 2016. "Targeted anticancer drug delivery through anthracycline antibiotic bearing functionalized quantum dots." *Artif Cells Nanomed Biotechnol* no. 44 (7):1774–82. doi: 10.3109/21691401.2015.1102740.

Barouti, Ghislaine, Cédric G. Jaffredo, and Sophie M. Guillaume. 2017. "Advances in drug delivery systems based on synthetic poly(hydroxybutyrate) (co)polymers." *Progress in Polymer Science* no. 73 (Supplement C):1-31. doi: https://doi.org/10.1016/j.progpolymsci.2017.05.002.

Bawa, Raj et al. "Translational Challenge in Medicine at the Nanoscale." In *Handbook of Clinical Nanomedicine: Law, Business, Regulation, Safety, and Risk*, edited by Raj Bawa; S.R. Bawa; R.N. Mehra, 1291-346. Singapore: Pan Stanford Publishing, 2016.

Behr, Jean-Paul. 1997. "The proton sponge: a trick to enter cells the viruses did not exploit." *CHIMIA International Journal for Chemistry* no. 51 (1–2):34–6.

Bhatt, Priyanka, Nirav Khatri, Mukesh Kumar, Dipesh Baradia, and Ambikanandan Misra. 2015. "Microbeads mediated oral plasmid DNA delivery using polymethacrylate vectors: An effectual groundwork for colorectal cancer." *Drug Deliv* no. 22 (6):849–61. doi: 10.3109/10717544.2014.898348.

Bhatt, Priyanka, Imran Vhora, Sushilkumar Patil, Jitendra Amrutiya, Chandrali Bhattacharya, Ambikanandan Misra, and Rajashree Mashru. 2016. "Role of antibodies in diagnosis and treatment of ovarian cancer: Basic approach and clinical status." *J Control Release* no. 226:148-67. doi: 10.1016/j.jconrel.2016.02.008.

Bhatt, Priyanka, Rohan Lalani, Rajashree Mashru, and Ambikanandan Misra. 2016. "Abstract 2065: Anti-FSHR antibody Fab' fragment conjugated immunoliposomes loaded with cyclodextrin-paclitaxel complex for improved *in vitro* efficacy on ovarian cancer cells." *Cancer Research* no. 76 (14 Supplement): 2065.

Black, Michael J., and Richard B. Brandt. 1974. "Spectrofluorometric analysis of hydrogen peroxide." *Anal Biochem* no. 58 (1):246–54.

Chari, Ravi V.J., Michael L. Miller, and Wayne C. Widdison. 2014. "Antibody–drug conjugates: An emerging concept in cancer therapy." *Angewandte Chemie International Edition* no. 53 (15):3796–827.

Costas Kaparissides, Sofia Alexandridou, Katerina Kotti and Sotira Chaitidou. 2006. "Recent Advances in Novel Drug Delivery Systems." *Azojomo Journal of Materials.* doi: 10.2240/azojono0111.

Crivelli, Barbara, Theodora Chlapanidas, Sara Perteghella, Enrico Lucarelli, Luisa Pascucci, Anna Teresa Brini, Ivana Ferrero, Mario Marazzi, Augusto Pessina, and Maria Luisa Torre 2017. "Mesenchymal stem/stromal cell extracellular vesicles: From active principle to next generation drug delivery system." *J Control Release* no. 262:104–17. doi: 10.1016/j.jconrel.2017.07.023.

Dautzenberg, Herbert. 2000. Light scattering studies on polyelectrolyte complexes. Paper read at Macromolecular Symposia.

Dawson, Rex, and M.C. Rex M.C. Dawson. 1986. Data for biochemical research.

Decker, Thomas, and Marie-Luise Lohmann-Matthes. 1988. "A quick and simple method for the quantitation of lactate dehydrogenase release in measurements of cellular cytotoxicity and tumor necrosis factor (TNF) activity." *Journal of Immunological Methods* no. 115 (1):61–9. doi: https://doi.org/10.1016/0022-1759(88)90310-9.

Deodhar, Gauri V., Marisa L. Adams, and Brian G. Trewyn. 2017. "Controlled release and intracellular protein delivery from mesoporous silica nanoparticles." *Biotechnol J* no. 12 (1). doi: 10.1002/biot.201600408.

Dobrovolskaia, Marina A., Jeffrey D. Clogston, Barry W. Neun, Jennifer B. Hall, Anil K. Patri, and Scott E. McNeil. 2008. "Method for Analysis of Nanoparticle Hemolytic Properties *In Vitro*." *Nano Letters* no. 8 (8):2180-2187. doi: 10.1021/nl0805615.

Fotaki, N. 2011. "Flow-through cell apparatus (USP apparatus 4): Operation and features." *Dissolution Technologies* no. 18 (4):46–9.

Frömming, Karl-Heinz, and József Szejtli. "Cyclodextrin Inclusion Complexes." In *Cyclodextrins in Pharmacy,* edited by Karl-Heinz Frömming and József Szejtli, 45–82. Dordrecht: Springer Netherlands, 1994.

Gandhi, Manit, Tosha Pandya, Ravi Gandhi, Sagar Patel, Rajashree Mashru, Ambikanandan Misra, Hemal Tandel, 2015. Inhalable liposomal dry powder of gemcitabine-HCl: Formulation, *in vitro* characterization and *in vivo* studies, *International Journal of Pharmaceutics* no. 496, 886–95. doi.org/10.1016/j.ijpharm.2015.10.020.

Gogvadze, Vladimir, Sten Orrenius, and Boris Zhivotovsky. 2003. "Analysis of mitochondrial dysfunction during cell death." *Curr Protoc Cell Biol* no. Chapter 18: Unit 18.5. doi: 10.1002/0471143030.cb1805s19.

Guideline, FDA. *FDA Guideline.* Available from http://www.fda.gov/Drugs/GuidanceComplianceRegulatoryInformation/Guidances/default.htm.

Hnawate, Ravi M., and Deore P. 2017. "Nanoparticle-novel drug delivery system: A Review." *Nanoparticle-novel drug delivery system: A Review* no. 5 (5):9–23.

Hong, Seungpyo, Anna U. Bielinska, Almut Mecke, Balazs Keszler, James L. Beals, Xiangyang Shi, Lajos Balogh, Bradford G. Orr, James R. Baker Jr, and Mark M. Banaszak Holl. 2004. "Interaction of poly(amidoamine) dendrimers with supported lipid bilayers and cells: Hole formation and the relation to transport." *Bioconjug Chem* no. 15 (4):774–82. doi: 10.1021/bc049962b.

Ibrahim, Sherrif F., and Ger van den Engh. 2007. "Flow cytometry and cell sorting." *Adv Biochem Eng Biotechnol* no. 106:19-39. doi: 10.1007/10_2007_073.

Jain, Keerti, Neelesh K. Mehra, and Narendra K. Jain. 2015. "Nanotechnology in Drug Delivery: Safety and Toxicity Issues." *Curr Pharm Des* no. 21 (29):4252–61.

Kaur, Avleen, Keerti Jain, Neelesh K. Mehra, and Narendra. K. Jain. 2017. "Development and evaluation of targeting ligand-anchored CNTs as prospective targeted drug delivery system." *Artif Cells Nanomed Biotechnol* no. 45 (2):242-250. doi: 10.3109/21691401.2016.1146728.

Khandare, Jayant, and Tamara Minko. 2006. "Polymer–drug conjugates: Progress in polymeric prodrugs." *Progress in Polymer Science* no. 31 (4):359–97.

Kichler, Antoine, Christian Leborgne, Emmanuel Coeytaux, and Olivier Danos. 2001. "Polyethylenimine-mediated gene delivery: A mechanistic study." *Journal of Gene Medicine* no. 3 (2):135–44.

Korzeniewski, Carol, and Denis M. Callewaert. 1983. "An enzyme-release assay for natural cytotoxicity." *J Immunol Methods* no. 64 (3):313–20.

Kotturi, Nagashree. 2015. "Novel Drug Delivery System." *Research and Reviews: Journal of Pharmaceutics and Nanotechnology* no. 3 (2):33–6.

Kratz, Felix. 2008. "Albumin as a drug carrier: Design of prodrugs, drug conjugates and nanoparticles." *Journal of Controlled Release* no. 132 (3):171–83.

Lalani, Rohan A., Priyanka Bhatt, Mohan Rathi, and Ambikanandan Misra. 2016. "Abstract 2063: Improved sensitivity and *in vitro* efficacy of RGD grafted PEGylated gemcitabine liposomes in RRM1 siRNA pretreated cancer cells." *Cancer Research* no. 76 (14 Supplement):2063.

Lam, Chiu-Wing, John T. James, Richard McCluskey, and Robert L. Hunter. 2004. "Pulmonary toxicity of single-wall carbon nanotubes in mice 7 and 90 days after intratracheal instillation." *Toxicol Sci* no. 77 (1):126–34. doi: 10.1093/toxsci/kfg243.

Lankalapalli, S., and V.R.M. Kolapalli. 2009. "Polyelectrolyte complexes: A review of their applicability in drug delivery technology." *Indian Journal of Pharmaceutical Sciences* no. 71 (5):481.

Lee, Kate Y.J., Gee Young Lee, Lucas A. Lane, Bin Li, Jianquan Wang, Qian Lu, Yiqing Wang, and Shuming Nie. 2017. "Functionalized, Long-Circulating, and Ultrasmall Gold Nanocarriers for Overcoming the Barriers of Low Nanoparticle Delivery Efficiency and Poor Tumor Penetration." *Bioconjug Chem* no. 28 (1):244–52. doi: 10.1021/acs.bioconjchem.6b00224.

Li, Weichang, Siqi Liu, Hang Yao, Guoxing Liao, Ziwei Si, Xiangjun Gong, Li Ren, and Linge Wang. 2017. "Microparticle templating as a route to nanoscale polymer vesicles with controlled size distribution for anticancer drug delivery." *J Colloid Interface Sci* no. 508:145–153. doi: 10.1016/j.jcis.2017.08.049.

McElnay, James, C. "Drug interactions at plasma and tissue binding sites." In *Mechanisms of Drug Interactions*, 125–49. Springer, 1996.

Mickuviene, Ingrida, Vida Kirveliene, and Benediktas Juodka. 2004. "Experimental survey of non-clonogenic viability assays for adherent cells *in vitro*." *Toxicol In Vitro* no. 18 (5):639-48. doi: 10.1016/j.tiv.2004.02.001.

Morgenstern, Josefine, Pascal Baumann, Carina Brunner, and Jürgen Hubbuch. 2017. "Effect of PEG molecular weight and PEGylation degree on the physical stability of PEGylated lysozyme." *International Journal of Pharmaceutics* no. 519 (1):408–17.

Omar, Amin, and Patricia Nadworny. 2017. "Review: Antimicrobial efficacy validation using *in vitro* and *in vivo* testing methods." *Advanced Drug Delivery Reviews* no. 112 (Supplement C):61-68. doi: https://doi.org/10.1016/j.addr.2016.09.003.

Pandya, Nilima T., Parva Jani, Jigar Vanza, and Hemal Tandel. 2018 "Solid lipid nanoparticles as an efficient drug delivery system of olmesartan medoxomil for the treatment of hypertension." *Colloids and Surfaces B: Biointerfaces* no. 165:37–44.

Parrott, Matthew C., and Joseph M. DeSimone. 2012. "Drug delivery: Relieving PEGylation." *Nature Chemistry* no. 4 (1):13–14.

Patel, Ankit, Nilima Pandya, Hemal Tandel. 2017. "Development and Optimization of Self-Microemulsifying Drug Delivery System for Bosentan Monohydrate by D-Optimal Mixture Design: *In-Vitro* and *In-Vivo* Evaluation." *Inventi Rapid: NDDS* no. 2018 (1):1–15.

Patil, Sushilkumar, Priyanka Bhatt, Rohan Lalani, Jitendra Amrutiya, Imran Vhora, Atul Kolte, and Ambikanandan Misra. 2016. "Low molecular weight chitosan-protamine conjugate for siRNA delivery with enhanced stability and transfection efficiency." *RSC Advances* no. 6 (112):110951–63. doi: 10.1039/C6RA24058E.

Peer, Dan, Jeffrey M. Karp, Seungpyo Hong, Omid C. Farokhzad, Rimona Margalit, and Robert Langer. 2007. "Nanocarriers as an emerging platform for cancer therapy." *Nature Nanotechnology* no. 2 (12):751–60.

Pogodina, Natalia V., and Nickolay V. Tsvetkov. 1997. "Structure and dynamics of the polyelectrolyte complex formation." *Macromolecules* no. 30 (17):4897–904.

Serajuddin, Abu. 1999. "Solid dispersion of poorly water-soluble drugs: Early promises, subsequent problems, and recent breakthroughs." *Journal of Pharmaceutical Sciences* no. 88 (10):1058–66. doi: 10.1021/js980403l.

Shrestha, Hina, Rajni Bala, and Sandeep Arora. 2014. "Lipid-Based Drug Delivery Systems." *Journal of Pharmaceutics* no. 2014:10. doi: 10.1155/2014/801820.

Sun, Xiao, Guilong Zhang, and Zhengyan Wu. 2017. "Nanostructures for pH-sensitive drug delivery and magnetic resonance contrast enhancement systems." *Curr Med Chem*. doi: 10.2174/0929867324666170406110642.

Tandel, Hemal, Dhaval Shah, Jigar Vanza, and Ambikanandan Misra. 2017. "Lipid based formulation approach for BCS class-II drug: Modafinil in the treatment of ADHD." *Journal of Drug Delivery Science and Technology*. no. 37 (Supplement C):166–83. doi.org/10.1016/j.jddst.2016.12.012.

Toomula, Nishant, Sathish D. Kumar, Phaneendra M. Arun. 2011. " Role of Pharmacokinetic Studies in Drug Discovery." *Journal of Bioequivalence & Bioavailability* no. 3 (11).

Torchilin, Vladimir P. 2005. "Recent advances with liposomes as pharmaceutical carriers." *Nature Reviews Drug Discovery* no. 4 (2):145–60.

Tuschl, Helga, and Christina E. Schwab. 2004. "Flow cytometric methods used as screening tests for basal toxicity of chemicals." *Toxicol In Vitro* no. 18 (4):483–91. doi: 10.1016/j.tiv.2003.12.004.

Tyner, Katherine M., Nan Zheng, Stephanie Choi, Xiaoming Xu, Peng Zou, Wenlei Jiang, Changning Guo, and Celia N. Cruz. 2017. "How Has CDER Prepared for the Nano Revolution? A Review of Risk Assessment, Regulatory Research, and Guidance Activities." *The AAPS Journal*, 1–13.

The european agency of the evaluation of medicinal producs evaluation of medicines for veterinary. 2000. Guidelines for the conduct of pharmacokinetic studies in target animal species. edited by The european agency of the evaluation of medicinal producs evaluation of medicines for veterinary use.

Vaidyanathan, Sriram, Kevin B. Anderson, Rachel L. Merzel, Binyamin Jacobovitz, Milan P. Kaushik, Christina N. Kelly, Mallory A. van Dongen, Casey A. Dougherty, Bradford G. Orr, and Mark M. Banaszak Holl. 2015. "Quantitative Measurement of Cationic Polymer Vector and Polymer/pDNA Polyplex Intercalation into the Cell Plasma Membrane." *ACS Nano* no. 9 (6):6097–109. doi: 10.1021/acsnano.5b01263.

Van Leeuwen, Herman P., Rob F. M. J. Cleven, and Pavel Valenta. 1991. "Conductometric analysis of polyelectrolytes in solution." *Pure and Applied Chemistry* no. 63 (9):1251–68.

Varshosaz, Jaleh, N. Tavakoli, and F. Roozbahani. 2006. "Formulation and *in vitro* characterization of ciprofloxacin floating and bioadhesive extended-release tablets." *Drug Delivery* no. 13 (4): 277–85.

Vhora, Imran, Sandip Patil, Priyanka Bhatt, Ravi Gandhi, Dipesh Baradia, and Ambikanandan Misra. 2014. "Receptor-targeted drug delivery: Current perspective and challenges." *Ther Deliv* no. 5 (9):1007–24. doi: 10.4155/tde.14.63.

Vhora, Imran, Sandip Patil, Priyanka Bhatt, and Ambikanandan Misra. 2015. "Protein- and Peptide-drug conjugates: An emerging drug delivery technology." *Adv Protein Chem Struct Biol* no. 98:1–55. doi: 10.1016/bs.apcsb.2014.11.001.

Wacker, Matthias G. 2014. "Nanotherapeutics—Product Development Along the "Nanomaterial" Discussion." *Journal of Pharmaceutical Sciences* no. 103 (3):777–84. doi: 10.1002/jps.23879.

Zhao, B., Xue Q. Wang, Xion Y. Wang, Hua Zhang, Wen B. Dai, Jun Wang, Zhen L. Zhong, Hou N. Wu, and Qiang Zhang. 2013. "Nanotoxicity comparison of four amphiphilic polymeric micelles with similar hydrophilic or hydrophobic structure." *Part Fibre Toxicol* no. 10:47. doi: 10.1186/1743-8977-10-47.

2

Intraoral and Peroral Drug Delivery Systems

Mohammed Shuaib Khan, Pranav J. Shah, Priya B. Dubey, and Jaimini K. Gandhi

CONTENTS

2.1 Introduction

Oral drug delivery has been branded for decades as the "most widely used and utilized" route of administration owing its advantages over other routes of administration. Oral drug delivery is widely accepted for delivering drugs and active pharmaceutical ingredients (APIs), macromolecules, peptides, and nanostructured drug deliveries, which offer numerous advantages like ease of administration, a maximum active surface area compared to other delivery systems, and provision of uniform drug delivery (Shah et al. 2014). A reputed British pharmaceutical journal predicted the future of oral solid dosage forms a way back in 1895, stating that "Tablets have had their day and will pass away to make room for something else" (Patel and Patel 2010). Oral drug delivery usually involves three primary routes of administration:

1. Intraoral sublingual, in which the drug formulation is placed under the tongue
2. Intraoral buccal, in which dosage form is placed in contact with buccal mucosa
3. Peroral gastrointestinal, in which drug formulation is passed through the mouth into the gastrointestinal tract

Oral mucosal drug delivery is a technique of systemic drug delivery that offers several advantages over both injectable and enteral methods. Due to the high vascularity of oral mucosa, drugs that are absorbed through it directly get entered into systemic circulation, and bypass the gastrointestinal tract, as well as first-pass metabolism. For drugs that require a rapid onset of action, it is a more comfortable and convenient delivery route than the intravenous route. Not all drugs, however, can be administered through the oral mucosa because of the characteristics of the oral mucosa and the physicochemical properties of the drug (Kulkarni et al. 2009). Different drug delivery system designs and characterizations are required due to anatomical and permeability differences between the sublingual and buccal mucosa. The sublingual route provides fast absorption, rapid onset of action, and high bioavailability, since it is highly permeable and has a rich blood supply. In contrast, the buccal mucosa is less permeable and does not provide the rapid absorption, which results in less bioavailability compared to the sublingual route. The following are the challenges associated with intraoral (buccal and sublingual) drug delivery.

- Not all drugs can be efficiently absorbed through the oral mucosa due to a smaller surface area (170 cm^2, of which ~50 cm^2 represents non-keratinized tissues), charge on constituents, enzymes that break down peptides, continuous secretion of saliva that leads to dilution of drugs, and instability at buccal pH.

- The loss of a dissolved or suspended drug and, ultimately, the involuntary removal of the dosage form which takes place due to the swallowing of saliva.
- Overhydration may lead to a slippery surface; the structural integrity of the formulation may get disrupted by swelling and hydration of the bioadhesive polymers. A constant flowing down of saliva within the oral cavity makes it very difficult for drugs to be retained for a significant amount of time in order to facilitate absorption in this site.
- Drugs with a bitter or unpleasant taste or an obnoxious odor have the tendency to irritate mucosa and therefore cannot be administered.
- For uncooperative or unconscious patients, intraoral drug delivery cannot be useful as they cannot receive anything by mouth. Thus, it necessitates the usefulness of other routes of drug delivery.
- Smoking should be avoided during medication because it will alter the normal physiology of buccal cavity and may affect the buccal absorption of drugs.
- In certain pathological conditions that affect the integrity of the mucosa, such as blisters or mucositis, intraoral drug delivery is not possible.
- In order for a drug to pass through the oral mucosa, it must first diffuse through the lipophilic cell membrane, and then pass through the hydrophilic interior of the cells of the oral epithelium. Thus, the oral mucosa provides both hydrophilic and hydrophobic barriers that must be overcome for efficient mucosal delivery. In the case of binding with other components, bioavailability will be greatly reduced (Patel et al. 2014).
- The use of permeation enhancers, however, must consider issues such as local tissue irritation, long-term tissue toxicity, and enhanced permeability to pathological microorganisms.
- Permeabilities between different regions of the oral cavity vary greatly because of the diverse structures and functions. In general, the permeability is based on the relative thickness and degree of keratinization of these tissues in the order of sublingual > buccal > palatal (Zhang et al. 2002; Sayani and Chien 1996; Patel et al. 2013; Patel et al. 2011).

There is no doubt that the peroral route is most commonly employed route of drug administration. The major focus of drug delivery research through peroral route is on controlled drug delivery systems. As a result, different types of controlled drug delivery systems, such as delayed release (mostly enteric coating), continuous release (matrix, reservoir, osmotic, ion exchange resins), gastroretentive systems (low density, high density, modified shape, mucoadhesive systems), colon-targeted systems, and Peyer's patch targeted systems are invented. Mouth dissolving and dispersible systems are also in the market to avoid swallowing difficulties and improved patient compliance.

2.2 Evaluation of Intraoral Drug Delivery Systems

Different evaluation tools used for the characterization of intraoral dosage forms depends on the type of dosage form. For example, the disintegration time is important for fast dissolving and dispersible tablets and films, while mucoadhesion is important for buccal tablets, gels, and patches. It is not possible to describe all the methods used to determine product quality attributes, only those with parameters that affect product performance are discussed further.

2.2.1 Evaluation of Taste Masking

For oral formulations, especially orally disintegrating tablets (ODTs), palatability of the formulation is a very crucial parameter that can affect acceptance of therapy and patient compliance (Sohi, Sultana, and Khar 2004a). Along with all the merits offered by solid dosage form, ODTs also provide accuracy in dosing, enhanced stability, ease of administration, and rapid disintegration of tablets that results in quick dissolution and rapid absorption, which provides the rapid onset of action. Hence, taste-masking strategies are useful to mask bitter or unpleasant tastes of APIs.

Commonly used strategies of taste-masking include polymer coating, organoleptic methods, hot-melt extrusion, microencapsulation, complexation, and spray-drying (Sohi, Sultana, and Khar 2004b). Taste is a very subjective perception. To evaluate the taste of formulation, there are two in-vivo tests; the first is human panel testing and second is frog taste nerve responses. In the first method, large groups of healthy human volunteers are asked to taste bitter drugs in the formulation of study. Then they are asked to relatively rate the taste of formulation based on its organoleptic characteristics (Anand et al. 2007; Miyanaga et al. 2003). In the second method, an alternating current (AC) amplifier is connected to glossopharyngeal nerve of bull frogs and responses to bitter drugs and test formulations are taken. The height of the peak obtained is used to evaluate taste-masking (Katsuragi, Kashiwayanagi, and Kurihara 1997). However, in-vivo palatability testing of taste-masked formulations by these two methods are unwillingly chosen and hindered by ethical issues, poor reproducibility, toxicological aspects and high costs (Legin et al. 2004).

Thus, to overcome all ethical, regularity, reproducibility and statistical aspects of in-vivo testing methods to evaluate taste-masking, the use of in-vitro procedures for the taste assessment of orally administered drug formulations might be an encouraging alternative option. The electronic tongue is a promising tool that can be potentially used for in-vitro taste evaluation studies (Pein et al. 2014). The instrument is generally equipped with a sensor array and it works on the principles of electrochemical measurement including potentiometry, amperometry and voltammetry (Maniruzzaman and Douroumis 2015). In most of the instruments used today, electronic tongue sensors are made up of potentiometric membrane electrodes which follow the Nernst law, and their membrane potentials are correlated to at least one reference electrode (Woertz et al. 2011). Due to the interaction of test sample molecules and components of the electrode membrane, sensor responses are generated. Though it is an important tool for in-vitro taste evaluation studies in development of taste masked formulations, the results obtained are only a relative interpretation of taste. A correlation between in-vitro electronic tongue measurements and in-vivo human taste has been established to some extent in adults (Bastiaans et al. 2017; Kim et al. 2013).

2.2.2 In-Vitro Mucoadhesion Tests

Several mucoadhesive test setups have been explored in the literature based on modified Wilhelmy plate surface techniques, modified dual tensiometry, texture analysis, and rotating cylinder methods. Among these methods, the use of a texture analyzer is explored intensively in the recent literature due to the varied experimental setup, precision, and reproducibility of results (Patel et al. 2017). The adhesive strength at such a bonding interface can be measured by evaluating the force required to detach one entity from the other through the application of an external force. As such, the destruction of the adhesive bond usually occurs under the application of either a shearing, tensile, or peeling force. Texture profile analysis and atomic force microscopy (AFM) has been used for studying oral bioadhesive systems. Standardized buccal cells were added to polymer solutions and bioadhesion was evaluated using AFM by examining the changes in surface topography of the cells, indicating the presence of bound polymers. Such imaging methods are relatively simple, in that they make use of the contact mode, topographic operation, but are only able to provide qualitative and semi-quantitative information regarding cell coverage (Patel et al. 2012).

2.2.3 In-Vitro Dissolution Testing of Intraoral Dosage Forms

2.2.3.1 Dissolution of Sublingual Tablets

The official USP apparatus 2 (paddle) and method for ISDN SL tablets were used as a control for dissolution testing; the dissolution medium was 900 mL of distilled water at 37 ± 1°C. The paddle rotations were set at 50 rpm, and the samples were withdrawn as recommended at the 20-minute timepoint for the analysis of the API content. At each appropriate timepoint, the full vacuum was applied by opening the on/off switch causing the total volume of dissolution medium to be withdrawn instantly through a 0.45-µm filter membrane into the collection tube and terminating any further dissolution. To obtain the percent of drug released (DR%), the API content (mg) in the filtrate was compared with the mean content uniformity of 10 individual dosage forms of the SL tablets being tested (Dasankopa and Ningagowdar 2012; Narang and Sharma 2011; Verma and Garg 2002; Mia and Noriyasu 2017).

2.2.3.2 Dissolution of Chewing Gum

Unofficial Single-Module Chewing Apparatus

- The apparatus for determining the drug release from medicated chewing gum products consists of a conical Teflon™ base and a rotating ribbed Teflon™ plunger suspended in a dissolution vessel. A prototype gum product containing phenylpropanolamine hydrochloride was used to evaluate the product. A chew-out study was carried out in six human volunteers. Level A correlation was found between in-vitro dissolution studies and in-vivo chew-out studies.
- The other apparatus set up consists of two pistons, a reservoir, a regulator of the rate of the chewing chamber, and a thermostat. Ribs provide swirling of the dissolution medium. The machine provides the rotation speed of 20 rpm and cycle frequency of 30 cycles per minute.
- The other design of the apparatus considered the effect of occlusal surfaces, rotary, and shearing movements and the medium temperature on drug release. The apparatus has six chewing modules, as shown in Figure 2.1. Each module consists of a thermostatted test cell of glass, in which two vertically oriented pistons holding an upper and a lower chewing surface, respectively, are mounted. The cells are filled with an appropriate test medium, usually 25–50 ml of an aqueous medium, and the chewing gum is loaded onto the lower chewing surface. The chewing procedure consists of up and down strokes of the lower surface in combination with a twisting movement of the upper surface which provides mastication of the chewing gum and, at the same time, an adequate agitation of the test medium. Drug release profiles for a number of commercially available products: Nicorette® and Nicotinell® (active substance nicotine), Travvell® (dimenhydrinate), V6® (xylitol) and an experimental formulation containing meclizine were studied. The results suggested that the apparatus developed for in-vitro release of medicated chewing gums and delivered satisfactory results for a number of varied gum formulations (Morjaria et al. 2004; Ames et al. 1992; Christrup and Moeller 1986; Kvist et al. 1999; Cherukuri and Bikkina 1998; Chaudhary and Shahiwala 2012; Cherukuri et al. 1988).

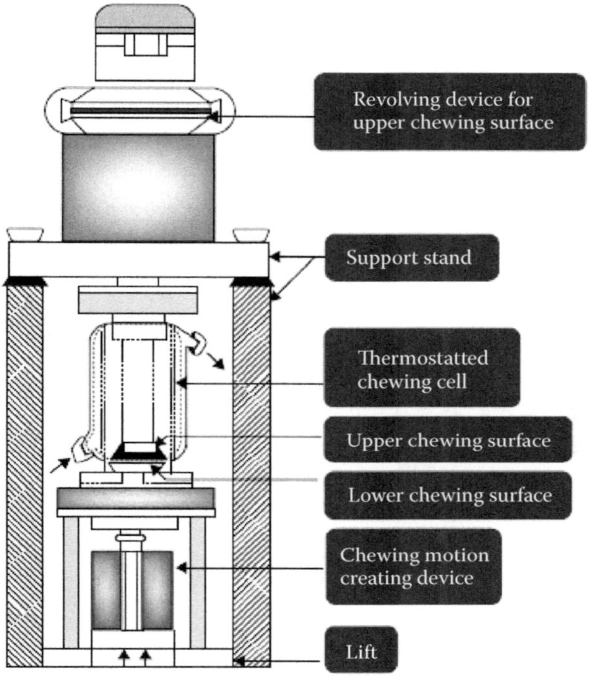

FIGURE 2.1
Schematic representation of an unofficial single-module chewing apparatus. (Adapted with permission from Taylor and Francis Publisher Limited; *Expert Opinion on Drug Delivery*, Chaudhary, S. A. and Shahiwala, A. F. 2010.)

The official modified dissolution apparatus for assessing drug release from MCG, as per *European Pharmacopoeia*, is depicted in Figure 2.2. In this apparatus, in addition to the pair of horizontal pistons ("teeth"), the chewing chamber is supplied with a vertical piston ("tongue") working alternate to the horizontal pistons, which ensures that the gum is always positioned in the correct place during the mastication process. The temperature of the chamber can be maintained at 37 ± 0.5°C and the chew rate can be varied. Other adjustable settings include the volume of the medium, the distance between the jaws and the twisting movement. The European Pharmacopoeia recommends 20 ml of unspecified buffer (with a pH close to 6) in a chewing chamber of 40 ml and a chew rate of 60 strokes a minute. This most recent device seems promising, competent and uncomplicated to operate.

Morjaria et al. (2004) studied the release of nicotine from conventional gums and from gums made using a directly compressible gum base using the European Pharmacopoeia apparatus. The compressible formulations exhibited a faster release rate compared to the conventional gums. The authors concluded that the apparatus can be useful as a routine quality control tool

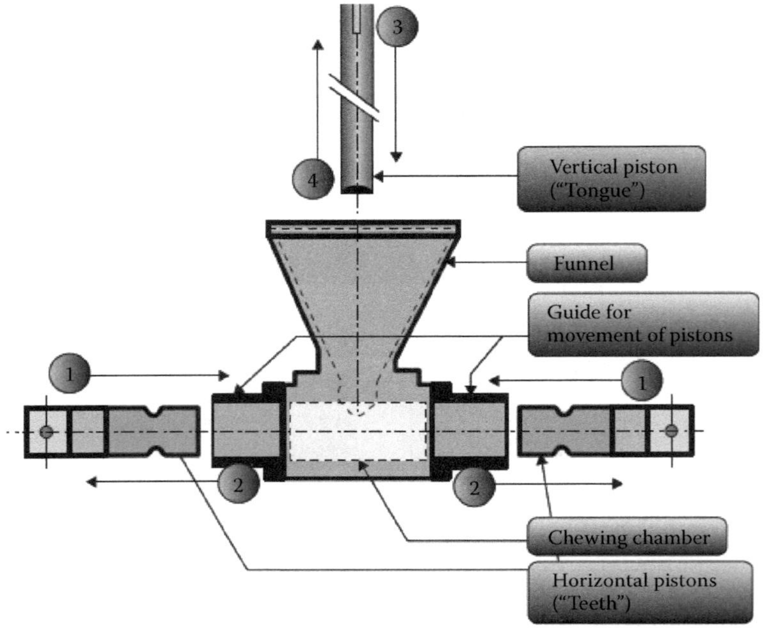

FIGURE 2.2
Schematic representation of modified dissolution apparatus as per the *European Pharmacopoeia*, where numbered arrows indicate sequence of motion. (Adapted with permission from Taylor and Francis Publisher Limited; *Expert Opinion on Drug Delivery*, Chaudhary, S. A. and Shahiwala, A. F. 2010.)

and in the development of newer gum-based formulations (Morjaria et al. 2004).

Shete et al. (2015) developed a medicated chewing gum of diphenhydramine hydrochloride using a natural gum base of prolamin. The drug release study in a phosphate buffer (pH 6.0) and simulated saliva was performed using the self-designed apparatus based on European Pharmacopoeia. The drug release of chewing gum was found to be dependent on the concentration of the gum base and calcium carbonate (texturizer). There was no significant deviation in the release of diphenhydramine hydrochloride when compared in simulated saliva and the phosphate buffer (pH 6.0). Thus, based on the findings of the authors, it can be concluded that phosphate buffer of pH 6.0 can be used as a routine dissolution medium for the study of drug release of medicated chewing gum (Shete et al. 2015).

Sander et al. (2013) prepared microparticles of nicotine bitartrate, which were incorporated into compressed medicated chewing gum. In-vitro release was determined using the method mentioned in the European Pharmacopoeia. The equipment was set at 60 chews per minute and the release medium was 20 mL of 20 mM phosphate buffer (pH 7.4) (Sander et al. 2013).

2.2.3.3 Dissolution of Buccal Patches

Karode and Prajapati carried out dissolution studies by using a USP type 2 rotating paddle dissolution test apparatus. A total of 100 mL of ethanol and simulated saliva solution (pH 6.2) mixture (20:80) was used as dissolution medium at (37 ± 1) °C, and stirred at 50 rpm. A 3-cm diameter buccal patch was fixed on the glass disc with the help of cyanoacrylate adhesive. The disc was put into the bottom of the dissolution vessel, so that the patch remained on the upper side of the disc. Samples (2 mL) were withdrawn at half-hour intervals and replaced with an equal volume of dissolution medium. The samples were filtered through 0.45-μm membrane filter and analyzed by suitable techniques (Gorle and Patil 2017; Achhra and Lalla 2015).

2.2.3.4 Dissolution of Buccal Films

Ammar et al. (2017) have performed the in-vitro dissolution of buccal film containing fluticasone propionate as a drug. USP XXIV six station dissolution apparatus type 1 with 900 ml pH 6.6 phosphate buffer (PB) (dissolution medium) was used to carry out dissolution. One film of each formulation was fixed to the central shaft using a cyanoacrylate adhesive. During the release study, they maintained the temperature and rotation speed of the apparatus at 37 ± 0.5°C and 50 rpm, respectively. After every hour, samples were withdrawn from each station, filtered, diluted suitably, and then analyzed by suitable methods (Ammar et al. 2017).

2.2.3.5 Dissolution of Lozenges

Raghvendra and Kumar performed the in-vitro dissolution studies of lozenges by using USP dissolution test apparatus type 2 (paddle type) at 100 rpm and 37 ± 0.5°C. They have utilized pH 6.8 buffer containing 2% sodium lauryl sulfate as a dissolution medium for studies. A lozenge was placed in each flask of the dissolution apparatus and samples of 5 ml were withdrawn at predetermined time intervals for 60 minutes. In order to maintain sink conditions, an equal volume of medium was replaced. The samples were analyzed by using a UV-visible spectrophotometer at suitable λ_{max} and the percentage of drug released was calculated. This experiment was done in triplicate and the average percentage release was calculated (Raghvendra and Kumar 2017).

2.2.3.6 Dissolution of Buccal Mucoadhesive Tablets

Ahmed et al. (2014) performed the studies using a dissolution test apparatus USP 24 paddle rotating at 100 rpm. The dissolution medium was 900 ml phosphate buffer (pH 6.8) and the temperature was set at 37°C. Samples of the solution were withdrawn at definite time intervals. The dissolution media was then replaced by fresh dissolution fluid to maintain a constant volume. The solution was passed through a filter, after which the concentration of drug was measured by an ultraviolet spectrophotometer or HPLC. The expression of results were in the form of a mean dissolution time (MDT) ± standard deviation (SD) (Ahmed et al. 2014; Brahmankar and Jaiswal 1995; Patel and Poddar 2009).

2.2.4 In-Vitro Drug Permeation Studies

Another important and widely used tool for the quantification of permeability of drug substances is an in-vitro model that emphasizes the use of Franz diffusion cells, thus reducing the cost of estimation and shortening the time to market. These models provided the appropriate quality standards that were maintained, similar to those models and techniques used in animal and clinical studies. These models are often used to determine the barrier characteristics of membranes as diffusion is the prime mechanism of passing of drug molecules through a membrane. The diffusion of drugs can be studied in variable environmental conditions such as temperature, pH, and osmolarity, which can be easily controlled variables to perform experiments. While choosing an in-vitro method, an appropriate animal model must be chosen on the basis of its similarity in structure and permeability to the human buccal mucosa in order for a closer interpretation of the permeability characteristics of drug molecules. The advantage of in-vitro diffusion cells is that the amount of drugs that have actually diffused across the membrane can be determined with respect to time, and thus the kinetics of tissue transport may be easily assessed. There are various diffusion cells that are used in the preclinical screening of compound permeability, including Franz-type diffusion cells, flow-through cells, and side-by-side diffusion cells. The use of animal buccal epithelium as a model for its human counterpart for the purpose of predicting transbuccal drug absorption kinetics is reported to be acceptable because of their similarity in the nature of keratinization, thickness, and lipid composition (Patel et al. 2012). Usually, buccal mucosa is removed immediately from the animals, further isolated from the underlying connective tissue and stored in ice-cold (4°C) buffers, usually a Krebs buffer, until it is used for the experiments. For longer storage, the buccal mucosa is stored below –20°C and frozen specimens should be equilibrated in phosphate buffer saline (PBS, pH 7.4) at room temperature to thaw completely before the start of experiments. For the permeation studies, buccal mucosa is mounted between the donor and receptor compartment of the diffusion cell or flow-through cell for the in-vitro permeation experiments. In the donor compartment, formulation with or without simulated saliva or human natural saliva is placed, while the receptor compartment is filled with PBS (Caro et al. 2017). Samples from the receptor compartment are collected from time to time and analyzed for the drug. The most important consideration in in-vitro experiments using animal mucosa is its viability and integrity. The use of human buccal tissue cultures, such as TR146 cell culture models, have also been proposed for buccal permeation studies (Sander et al. 2013; Nielsen et al. 2002).

2.2.5 In-Vivo Methods

Buccal absorption and perfusion studies usually provide an insight into the ability of a formulation to deliver a drug but these lack real in-vivo systemic evaluation and other physiological factors which may contribute to the overall performance of the dosage forms. Hence, it is often essential to perform an in-vivo pharmacokinetic study. Rabbits, dogs, and pigs are the animal models that have been used so far to perform such initial in-vivo studies prior to human assessment (Patel et al. 2012).

2.2.5.1 Residence Time

In-vivo residence time tests are usually performed in animal models or humans themselves. Patel and Poddar (2009) have used rabbits for measuring in-vivo residence time. The rabbits were anesthetized and the experiment was carried out with plain as well as medicated patches. The mucoadhesive patch was placed on the buccal mucosa between the cheek and gingiva in the region of the upper canine and gently pressed onto the mucosa for about 30 seconds. The time required for complete erosion of the patch or the time at which failure of adhesion occurred between the patch and mucosa was noted as the in-vivo residence time. The repetition of application of the mucoadhesive patches using the same animal was allowed after a 5-day rest period (Patel and Poddar 2009).

Nafee et al. (2004) used four healthy human volunteers aged between 25–50 years old. Plain bioadhesive tablets with optimized properties were selected for this in-vivo evaluation. The tablet and the inner upper lip were carefully moistened with saliva to prevent the sticking of the tablet to the lip. The volunteers were asked to monitor the ease with which the system was retained on the mucosa and note any tendency for detachment. The time necessary for complete erosion of the tablet was simultaneously monitored by carefully observing for residual polymer on the mucosa. The repeated application of the bioadhesive tablets was allowed after a 2-day period for the same volunteer (Nafee et al. 2004).

2.2.5.2 Buccal Absorption Test

The most common in-vivo method that has been widely used by researchers to access the permeability is absorption test (buccal absorption test) introduced by Beckett and Triggs (1967). A known volume of drug solution has been taken orally in the mouth by the subject, for a particular period of time while swirling the drug solution in mouth and expelling it in a container.

To make sure, subject will take a measured aliquot of distilled water, rinse, and expel the aliquot in the same drug solution container. These expelled solutions were analyzed for drug content. The difference between the initial and final drug concentration will give the permeability characteristics and amount of dug taken up by buccal mucosa (Tucker 1988).

Over a period of advancement, the test proposed by Beckett and Triggs has been modified and presented by many researchers with modifications in order to have accurate and well-defined permeability profiles. Tucker et al. (1988) have made an improvement in the absorption test by adding phenol red that accounts for the salivary dilution and accidental swallowing of the solution by the subject. This method proves to be advantageous over the traditional one, which requires a separate experiment for each time point, thus taking days for each kinetic profile. By this improvement, the kinetics of oral mucosal drug absorption from solutions can be studied in a single experiment within 15–20 min (Hoogstraate and Bodde 1993; Beckett and Triggs 1967).

These methods are widely used as a tool for determining the permeability characteristics but they have their own limitations. By this method, only the drug remaining in the oral cavity can be quantified; no accurate data can be predicted as drug loss may occur, like drugs that enter into the systemic circulation, are accidentally swallowed by the subjects and the metabolic properties of the drug itself. Also, the absorption of drug may occur from all the mucosal surface once the subject swirls the drug solution in the mouth, so no specific site permeation characteristics can be defined (Hoogstraate et al. 1996).

2.3 Models for Assessing Intestinal Permeability

At present, three groups of methods have been used widely by researchers for investigating the pathway/principal mechanism of absorptions, namely in-vitro, *in situ*, and in-vivo. To choose a correct model for the study also relies on the type of molecule under investigation and type of study protocol.

Permeability models play an important role and can rightfully assist in the suitable drug candidate selection for in-vivo clinical studies at early stages in drug discovery and development, along with regulatory applications. They not only serve as a tool in deciding the drug candidate, but also help researchers to prevent the loss of the drug that may incur in clinical trials (Donna 2010).

However, researchers can only rely on experimental permeability models by their ability to accurately predict a drug's in-vivo intestinal absorption. All experimental conditions shall be optimized, scrutinized and controlled for the physiological environment that drugs come across in the intestinal tract to obtain the satisfactory In-Vitro–In-Vivo Correlations (IVIVC). There is a need to establish a close correlation between experimental and in-vivo absorption with standardized methods for the quantification of the permeability of data, so that the model can serve as useful tool in drug discovery and drug development. The validation and standardization of these models assays helps pharmaceutical industries to reduce intra-laboratory variations in permeability protocols and thus can serve as qualitative and quantitative tools for assessing the behavior of a drug at the in-vivo intestinal barrier (Ungell 2002; Fagerholm et al. 1996; Lee et al. 2005; Borchardt 1996; Borchardt et al. 1996; Bock et al. 2004).

2.3.1 In-Vitro Methods

The most widely used apparatus for oral controlled release formulations are the paddle and the basket method; flow-through cell, the reciprocating cylinder, and others are also used. Regulatory guidance and compendial methods are widely available in case of controlled-release dosage forms, hence, not described here. Drug permeability models are for the testing of nanoparticulate/lipidic systems and are discussed further.

2.3.1.1 Flow-Through Diffusion Cells

Some researchers are using flow-through diffusion cells. The diffusion cell consist of an upper donor chamber and a lower receiver compartment. This model has an advantage that there is a continuous supply of fresh solvent which is directed to an automatic fraction collector.

Advantages:

1. The flow rate can be changed within a single run.
2. The sink conditions can be maintained due to the continuous flow of fresh medium when the system operates in an open-loop configuration. This feature is important for the study of poorly soluble drugs.
3. The study of samples with low drug loading is feasible when the system operates in the closed-loop configuration, as a small volume of medium can be used.
4. The release from dosage forms over extended periods can be studied, as this set-up eliminates the evaporation issue that can be observed with other apparatuses (Modi 2016).

The release of model drugs, i.e., vitamin B_{12}, from polymeric hydrogels that consist of N-vinyl-2-pyrrolidone and acrylic acid in simulated gastric fluid (pH 1.2) and intestinal fluid (pH 7.4), at 37°C, showed the hydrogels to be pH sensitive. An in-vitro release study by the "traditional dissolution test" (TDT) showed that percent drug released from the hydrogel was nearly

8.6 ± 2.1 and 83.2 ± 4.8 in the media of pH 1.2 and 6.8, respectively. However, for in-vivo, a new test model called flow-through diffusion cell (FTDC) was also used in the study. This study enabled researcher to mimic in-vivo GI conditions like acidic pH, high water content, and presence of a solid mass in the large intestine. It was observed that the two approaches yielded almost different release profiles (Bajpai et al. 2006).

2.3.2 *In Situ* Models

In situ perfusion models have been used to directly measure the drug permeability in the intact intestine by single-pass and recirculating methods (Schanker et al. 1958; Doluisio et al. 1969; Komiya et al. 1980; Eric et al. 2001). Parameters to standardize for *in situ* models are like animal species and age, fed/fast status of animal, anaesthesia regimen, intestinal region, perfusion buffer composition, osmolarity and pH, perfusion rate and, most importantly, drug analysis and P_{eff} calculation.

This model has been developed and is based on the stable, vascularly perfused preparations of the small intestine. Through the help of laparotomy, the abdominal cavity of an anesthetized animal is exposed, through which the drug solution is introduced. It can be either a closed-loop or an open-loop. For the single-pass (open-loop) perfusion method, a section of intestine in an anesthetized animal is cannulated proximally and distally, rinsed with suitable buffer solution and then perfused completely with a drug solution of choice. The amount of drug in the perfusate is measured at defined time points. The perfusion assays are normalized for inlet drug concentration, flow rate, and drug aqueous diffusion coefficient. Intestinal permeability (P_{eff}) is calculated from the difference between solute concentration entering and leaving the cannulated region through the intestine.

In situ methods have significant advantages over in-vivo models. For example, with *in situ* models there is a direct bypass of the stomach. That means that the acidic compounds are not likely to precipitate, which is likely to happen in the stomach due to pH, and thus dissolution rates do not confuse the intestinal drug concentrations and therefore plasma levels. Second, *in situ* instillation enables researchers to predominantly assess the formulation-independent breakdown in the stomach due to acidic conditions. Researchers can directly analyze the implication of gastrointestinal conditions on dosage forms. Although the animal has been anesthetized and surgically manipulated, the mesenteric blood flow is intact. However, one of the important considerations in this model is the choice of anesthetic as the animal has to go through anesthesia, which can have profound effects on intestinal drug absorption (Yuasa et al. 1993).

At present, *in situ* methods includes intestinal perfusion, intestinal loops, and intestinal vascular cannulation, which have been discussed below.

2.3.2.1 Intestinal Perfusion

Particularly in this method, animals like rats, rabbits, or mice are initially fasted for approximately 3 days and then subjected to anesthesia for surgical procedure. An incision will be made in the midline of abdomen to expose the small intestine. Small portions of the intestinal segment were sliced, held with a clamp, and washed to remove any particulate matter. Sliced segments were then ligated using suture to form a loop, however, precaution shall be taken to maintain the normal blood flow and movement to the looped segment.

An alternative method with slight modification is to make an incision at the intended location of intestine and then ligating that end to form a loop. To determine the particulate transport or to trace the system, delivery systems like nanoparticles (NPs) are injected into the loop and then the intestinal loop is returned to the body cavity of the animal for up to 2 hours to maintain the moist conditions while the animals are carefully maintained under anesthesia to resume normal digestive processes and movement. After a particular period of time, intestinal loop segment is removed from the body cavity of the animal and analyzed for drug content perfused into the intestine.

After the removal of intestinal segment, the tissue is flash frozen and the morphology of the intestine can be determined by microscopy/confocal. Tissue can be subjected to studies in order to evaluate the drug content and permeability characteristics. In order to measure the uptake of nonfluorescent peptides or drugs, fluorescein isothiocyanate (FITC) is often used as a tool to label the peptide or drug and further observed under fluorescent microscope which not only measures the fluorescence of the cryo-frozen tissue samples taken from the intestine but also gives information measuring the uptake or distribution of drugs in the intestine (Sahana et al. 2008; McClean et al. 1998; Cryan and O'Driscoll 2003; Desai et al. 1996).

Circular Perfusion Method

The circular perfusion method is an easy and simple method using an intestine sample and a tracer component that will help to determine the absorption constant. Likewise, the perfusion fluid containing a nanoparticle formulation were poured from one end of intestine with help of a peristaltic pump and were collected at the other end of intestine at a slow rate of 2.5 $mL \cdot min^{-1}$. The drug concentration and tracer concentration was measured at before and after perfusion thus calculating the absorption constant (Wei et al. 2016).

With the help of the above model, researchers plotted a drug-release kinetic curve and calculated the release kinetics, thus obtaining the absorption rate constant (ka) and the permeability coefficient (P), expressed as distance per unit time that has been used represent the velocity of drug movement across a heterogeneous medium, such as skin and intestinal epithelium (Chiou 1996).

Single-Pass Perfusion

Single pass perfusion, is very narrowly different from circular perfusion. As discussed in the earlier section, in single pass perfusion, the perfusate does not return to the original drug container but to another collector vessel. The perfusate is collected from the drainage tube. The amount of decrease in drug concentration over time helps to calculate pharmacokinetic parameters. The single-pass perfused rat intestinal model is an *in situ* perfusion method that can be used to determine the regional disposition of drugs. It is useful for selecting a development candidate from a series of active compounds and for studying the mechanisms of absorption and excretion. It is also useful for determining if a compound may be appropriate for a sustained-release control formulation. The Food and Drug Administration (FDA) recognized the model system as a useful model to classify a compound's absorption characteristics in the biopharmaceutics classification system. This model may be modified to determine the contribution of the intestines versus the liver in the disposition of a specific compound, both of which may be useful to determine the enteric and enterohepatic recycling of drugs (Holenarsipur et al. 2015).

2.3.2.2 Intestinal Loop Method

This method was first introduced by Punyashthiti in 1971. Rats were anesthetized, the abdomen was incised and the desired segment of intestine was washed, ligated, and then injected with specific concentration of NPs. After this process, the portion of intestine was placed back into abdominal cavity and the incisions were closed by clamps. Precautions were taken in order to keep the model moist by covering it with gauze pads which were presoaked in normal saline. After particular period of time 24 hours, the rats were euthanized and the intestines were removed and analyzed for permeability dynamics (Gamboa and Leong 2013; Punyashthiti and Finkelstein 1971).

This model has been used to evaluate potential of PEGylated solid lipid NPs as mucus-penetrating particles (MPP) for oral delivery across gastrointestinal mucus. Intestine was injected with specific concentration of NPs and fluorescent culture solutions and after a specific period of time, intestinal loops were observed under inverted two-photon confocal microscopy. This study revealed the potential of this model, as it has demonstrated the NPs can rapidly penetrate mucus secretions, whereas the solid lipid NPs were trapped by high viscoelastic mucus barriers (Yuan and Chen 2013).

2.3.2.3 Intestinal Vascular Cannulation

This method works on administration of drug loaded formulation to healthy rats which were fasted for 12 hours and further extraction of mesenteric artery or vein blood (Hamid et al. 2009). The collected fraction is thus analyzed to know the direct absorption of drugs from the lumen into the circulation. Mesenteric vein cannulation in combination with intestinal perfusion experiments cans improve insights into intestinal drug absorption mechanisms.

2.3.3 In-Vivo Methods

2.3.3.1 Use of X-Ray

This simple and traceable procedure involving the use of radioopaque markers, e.g., barium sulfate, was encapsulated in mucoadhesive dosage forms to determine the effects of mucoadhesive polymers on GI transit time. This study can be performed either on human volunteers or rabbits. X-ray photographs were taken at different time intervals (Hemant et al. 2011; Gopalkrishnan et al. 2010).

2.3.3.2 Use of Gamma Scintigraphy

In order to deliver drugs to a specific target area of the body, researchers need carrier systems that can be monitored easily in order to have actual insight on movement of a dosage form within body. Gamma scintigraphy is a neat method which enables researchers to gain actual insights of dosage forms within the body indicating the in-vivo behavior of the delivery system. This technique involves the use of radioactive tracers loaded into the delivery system so as to enable an optimum detection by a gamma ray camera. However, the choice of a convenient label enables what will actually determine the in-vivo targeting of the formulation administered. This method is non-invasive and providing reliable information on the transit time of dosage forms in different regions of the gastrointestinal (GI) tract and various other body organs. Gamma scintigraphy is widely used now for predicting the in-vivo behavior of delivery systems.

Radiolabeling is a process in where a suitable radionuclide such as technetium-99m or indium-111 is being incorporated into formulation or by addition of a non-radioactive isotope into the product, such as samarium-152 followed by neutron activation of the final product. These resulting insights can enable researchers to speed up the product development process at all developmental stages and help in supporting product license applications with success in early clinical trials (Soraya et al. 2003).

2.3.3.3 Use of Pharmacoscintigraphy

Pharmacoscintigraphic techniques provide a non-invasive method to monitor the in-vivo behavior of a different type of pharmaceutical dosage form. This technique combines gamma scintigraphy and pharmacokinetic information to predict the behavior of a dosage form inside the subject. The advantage of gamma scintigraphy with the usual pharmacokinetic methods enables researchers to correlate the information between the distribution of the drug inside the body with their absorption profile.

Naproxen tablets and pH-sensitive (enteric coated) radio telemetry capsules were developed and radiolabelled, which were administered to 6 healthy volunteers following breakfast. Results showed that five of the naproxen tablets disintegrated in the small intestine and one in the stomach which laid down close correlation between tablet disintegration and detection of naproxen in the blood concentration. It was concluded that gastric emptying is the main factor influencing the onset of drug release from enteric-coated tablets (Hardy et al. 1987).

2.3.3.4 In-Vivo Imaging Systems

There have been continual updates and new findings about the permeation of drugs in GIT. A new and innovative technique for in-vivo analysis after oral delivery is a form of optical imaging termed IVIS, which allows for in-vivo imaging without radiation on the body. It is an optical imaging technique that uses fluorescence and results in a high signal-to-noise ratio due to the low bioluminescence of mammalian tissue (Chen et al. 2010; Lee et al. 2012).

2.4 In-Vitro–In-Vivo Correlations

The in-vivo profile from in-vitro data can be predicted if drug exhibits dissolution rate limited absorption and absolute bioavailability of drug is known (Ghosh and Choudhary 2009; Cardot et al. 2007; Emami 2006; Nagpal et al. 2010). Satisfactory IVIVC can be developed in case of Biopharmaceutical Classification System class II drugs because these exhibit dissolution rate limited absorption. Controlled release formulations allow for the profiling of drug release over extended time periods in-vitro and in-vivo, which substantiates the value of the correlation. If the release from the dosage from is the rate-controlling step, the permeability of the drug must be sufficiently high enough to guarantee sink conditions in the gastrointestinal tract during the in-vivo dissolution process. In other words, the controlled release formulation with the potential for a successful IVIVC have to be class I or II drug products. For this purpose, a dissolution test with multiple time points is required to simulate bioavailability. The surfactants, pH, buffer capacity and food components can alter the solubility of drugs in GIT. The solubility and absorption of poorly water-soluble drugs is influenced by the physiological variation of the bile salt concentration and pH during fed and fasted states. Therefore, the composition of dissolution media, volume, and hydrodynamics of the contents in the lumen are the main factors that affect dissolution of

TABLE 2.1
Review of IVIVC Studies for Different Formulations

Formulation	Drug	In-Vitro Drug Release Method	Drug Release Media	In-Vitro Drug Release Period	In-Vivo Study Period	Level of IVIVC	References
Buccal mucoadhesive tablets	Buspirone hydrochloride	Modified USP 1	pH 6.0, 0.05 M PB	12 h	12 h	A	Kassem et al. 2014
Buccal films	Rivastigmine hydrogen tartarate	Franz diffusion cell	Phosphate buffer pH 7.4	24 h	24 h	A	Kapil et al. 2013
Buccal patches	Pravastatin sodium	USP apparatus 5	Phosphate buffer pH 6.8	10 h	10 h	A	Maurya et al. 2012
Chewing gums	Nicotine	Official and unofficial chewing apparatuses	Artificial saliva pH 6.2				Gajendran et al. 2012
Gastro retentive mucoadhesive tablets	Simvastatin	USP apparatus 1	0.1N HCl, 0.1N HCl/pH 6.8 buffer and pH 7.0 buffer containing 0.5% SDS	8 hr	8 hr	A	Bhalekar et al. 2016
Floating tablets	Lamivudine	USP apparatus 2	Simulated gastric fluid (SGF) pH 1.2	6 hr	6 hr	B multiple level C	Singh et al. 2012
Multiparticulate system	Metformin hydrochloride	USP apparatus 1	Simulated gastric fluid with pH 1.2 with 0.02% v/v Tween 80	12 hr	12 hr	Multiple level C	Pandit et al. 2013
Unfolding system	Levodopa	USP apparatus 2	Simulated gastric fluid without pepsin	24 hr	24 hr		Klausner et al. 2003
Osmotic system	Flurbiprofen	BP apparatus 2	Simulated gastric fluid followed by phosphate buffer as simulated intestinal fluid	30 hr	30 hr	Level A	Philip and Pathak 2008

Note: BP, British Pharmacopeia; HCl, hydrochloric acid; PB, phosphate buffer; SDS, sodium dodecyl sulfate; USP, United States Pharmacopeia.

drug in GIT. The dissolution rate with limited absorption can be accurately predicted only when these factors are adequately controlled and reproduced in-vitro. Compendial dissolution media are best suited for quality control purposes, but cannot be used for IVIVC in all cases, as the composition of these cannot differentiate between the fed and fasted state.

The complex and changing environment along the entire gastrointestinal tract has to be taken into consideration for evaluating the potential for IVIVC for controlled release products, due to their extended release time profile. The factors to be considered are:

- The impact of available fluid volume, presence of surfactants, and motility pattern on drug product dissolution and drug solubility
- The regional differences in intestinal permeability
- The regional differences in intestinal metabolism and secretion
- The transit time dependence of all of the above factors

Site-dependent differences in drug permeability for example will not only lead to absorption profiles that deviate from the in-vitro release characteristics, but will also exhibit a high degree of variability in the absorption data due to transit time variations among subjects. Obviously, permeability and metabolism characteristics are difficult to simulate accurately in an in-vitro setting, but they can be successfully addressed with an appropriate mathematical absorption model, if some regional permeability and metabolism data are available.

The different IVIVC studies performed for oral and intraoral controlled drug delivery systems (Table 2.1). Factors such as the dissolution medium, dissolution apparatuses used for in-vitro dissolution and time of sample collection are considered during IVIVC studies. Some of the IVIVC studies are shown in the table below for different formulations (Lipkaa and Amidon 1999; Kostewicz et al. 2014).

References

Achhra, C.V., and Lalla, J.K. 2015. Formulation Development and Evaluation of Sucrose-Free Lozenges of Curcumin. *International Journal of Pharmaceutical and Phytopharmacological Research* 5(1):46–55.

Ahmed, A., Goyal, N.K., and Sharma, P.K. 2014. Effervescent Floating Drug Delivery System: A Review. *Global Journal of Pharmacology* 8(4):478–485.

Ames, N.R., Walter, G.C., and Robert, W.C. 1992. Development and Evaluation of a Novel Dissolution Apparatus for Medicated Chewing Gum Products. *Springer link, Pharmaceutical Research* 9:255–259.

Ammar, H.O., Ghorab, M.M., Mahmoud, A.A., and Shahin, H.I. 2017. Design and in Vitro/in Vivo Evaluation of Ultra-Thin Mucoadhesive Buccal Film Containing Fluticasone Propionate. *AAPS PharmSciTech* 18(1):93–103.

Anand Vikas, Mahesh Kataria, Vipin Kukkar, Vandana Saharan, and Pratim Kumar Choudhury. 2007. The latest trends in the taste assessment of pharmaceuticals. *Drug Discovery Today* no. 12(5):257–265. doi: https://doi.org/10.1016/j.drudis.2007.01.010.

Bajpai, S.K., Dubey, S., and Saxena, S. 2006. Flow through Diffusion Cell Method: A Better Approach to Study Drug Release Behavior as Compared to Traditional Dissolution Test Method. *Journal of Macromolecular Science, Part A: Pure and Applied Chemistry* 43(4–5):627–636.

Bastiaans, D.E.T., Immohr, L.I., Zeinstra, G.G., Strik-Albers, R., Pein-Hackelbusch, M., van der Flier, M., de Haan, A.F.J., Boelens, J.J., Lankester, A.C., and Burger, D.M. 2017. In vivo and in vitro palatability testing of a new paediatric formulation of valaciclovir. doi: 10.1111/bcp.13396.

Beckett, A.H. and Triggs, E.J. 1967. Buccal Absorption of Basic Drugs and Its Application as an in Vivo Model of Passive Drug Transfer through Lipid Membranes. *The Journal of Pharmacy and Pharmacology* 19:31–41.

Bhalekar, M.R., Bargaje, R.V., Upadhaya, P.G., Madgulkar, A.R., and Kshirsagar, S.J. 2016. Formulation of Mucoadhesive Gastric Retentive Drug Delivery Using Thiolated Xyloglucan. *Carbohydrate Polymers* 136:537–542.

Bock, U., Thomas, F., and Eleonore, H. 2004. Validation of Cell Culture Models for the Intestine and the Blood Brain Barrier and Comparison of Drug Permeation. *Altex* 21(3):57–64.

Borchardt, R.T., Smith, P.L., and Wilson, G. 1996. General Principles in the Characterization and Use of Model Systems for Biopharmaceutical Studies. *Models for Assessing Drug Absorption and Metabolism* 21(3):1–11.

Brahmankar, D.M. and Jaiswal, S.B. 1995. *Biopharmaceutics and Pharmacokinetics – A Treatise*. Vallabh Prakashan.

Brown, D. 2003. Orally Disintegrating Tablets-Taste over Speed. *Drug Del Tech* 3(6):58–61.

Cardot, J.M., Beyssac, E., and Alric, M. 2007. *In Vitro-In Vivo* Correlation: Importance of Dissolution in IVIVC. http://www.dissolutiontech.com/DTresour/200702Articles/DT200702_A02.pdf.

Caro, V. De, Giandalia, G., Siragusa M.G., Campisi, G., and Giannola, L.I. 2017. Galantamine Delivery on Buccal Mucosa: Permeation Enhancement and Design of Matrix Tablets. *Journal of Bioequivalence & Bioavailability Open Access* 1(14):127–134. doi:10.4172/jbb.1000020.

Chaudhary, S.A. and Aliasgar, F.S. 2012. Directly Compressible Medicated Chewing Gum Formulation for Quick Relief from Common Cold. *International Journal of Pharmaceutical Investigation* 2(3):123–133.

Chen, C., Tsai, T.H., Huang, Z., and Fang, J. 2010. Effects of Lipophilic Emulsifiers on the Oral Administration of Lovastatin from Nanostructured Lipid Carriers: Physicochemical Characterization and Pharmacokinetics. *European Journal of Pharmaceutics and Biopharmaceutics* 74(3):474–482.

Cherukuri, S.R. and Bikkina, K. 1998. Tabletted Chewing Gum Composition and Method of Preparation. US4753805 A.

Cherukuri, S.R., Hriscisce, F., and Wei, Y.C. 1988. Reduced Calorie Chewing Gums and Method. US4765991 A.

Chiou, W.L. 1996. New Perspectives on the Theory of Permeability and Resistance in the Study of Drug Transport and Absorption. *Journal of Pharmacokinetics and Pharmacodynamics* 24(4):433–442.

Christrup, L.L. and Møeller, N. 1986. Chewing gum as a drug delivery system. *Arch Pharm Chem, Sci Ed.* 14:30–36.

Cryan, S.A. and O'driscoll, C.M. 2003. Mechanistic Studies on Nonviral Gene Delivery to the Intestine Using In Vitro Differentiated Cell Culture Models and an in Vivo Rat Intestinal Loop. *Pharmaceutical Research* 20(4):569–575.

Dasankoppa, S., Sanjeri, F., Ningangowdar, M., and Sholapur, H. 2012. Formulation and Evaluation of Controlled Porosity Osmotic Pump for Oral Delivery of Ketorolac. *Journal of Basic and Clinical Pharmacy* 4(1):2–9.

Desai, M.P., Labhasetwar, V., Amidon, G.L., and Levy, R.J. 1996. Gastrointestinal Uptake of Biodegradable Microparticles: Effect of Particle Size. *Pharmaceutical Research* 13(12):1838–1845.

Doluisio, J.T., Billups, N.F., Dittert, L.W., Sugita, E.T., and Swintosky, J.V. 1969. Drug Absorption I: An In Situ Rat Gut Technique Yielding Realistic Absorption Rates. *Journal of Pharmaceutical Sciences* 58(10):1196–200.

Donna, A.V. 2010. Application of Method Suitability for Drug Permeability Classification. *AAPS J* 12(4):670–678.

Emami, J. 2006. In Vitro-in Vivo Correlation: From Theory to Applications. *Journal of Pharmaceutical Sciences* 9(2):169–189.

Eric, L.F., Christophe, C., Per, A., David, B., Gérard, F., Pierre, G., Francois, G., and Monique, R. 2001. In *In Vitro Models of the Intestinal Barrier, The Report and Recommendations of The European Centre for the Validation of Alternative Methods (ECVAM) Workshop* 46., Paris, France, ATLA 26: Paris, France. pp. 649–668.

Fagerholm, U. Johansson, and M. Lennernäs, H. 1996. Comparison between permeability coefficients in rat and human jejunum. *Pharmaceutical Research* 13:1336–1342.

Gajendran, J., Kraemer, J., and Peter, L. 2012. In Vivo Predictive Release Methods for Medicated Chewing Gums. *Biopharmaceutics & Drug Disposition* 33(7):417–424.

Gamboa, J.M. and Leong, K.M. 2013. In Vitro and in Vivo Models for the Study of Oral Delivery of Nanoparticles. *Advanced Drug Delivery Reviews* 65(6):800–810.

Ghosh, A. and Choudhury, G.K. 2009. In Vitro-in Vivo Correlation (IVIVC): A Review. *Journal of Pharmacy Research* 2(8):1255–1260.

Gopalakrishnan, R., Sureshkumar, N.J., and Ganesh G.N. 2010. Mucoadhesive slow-release tablets of theophylline: Design and evaluation. *Asian Journal of Pharmaceutics* 64–68.

Gorle, A. and Patil, G. 2010. Design, Development and Evaluation of Fast Dissolving Film of Amlodipine Besylate. *International Journal of ChemTech Research* 10(4):334–344.

Hamid, K.A., Lin, Y., Gao, Y., Katsumi, H., Sakane, T., and Yamamoto, A. 2009. The Effect of Wellsolve, a Novel Solubilizing Agent, on the Intestinal Barrier Function and Intestinal Absorption of Griseofulvin in Rats. *Biological and Pharmaceutical Bulletin* 32(11):1898–1905.

Hardy, J.G., David F.E., Zaki, I., Clark, A.G., Tønnesen, H.H., and Gamst, O.N. 1987. Evaluation of an Enteric Coated Naproxen Tablet Using Gamma Scintigraphy and pH Monitoring. *International Journal of Pharmaceutics* 37(3):245–250.

Hemant, H.G., Mayur, A.C., and Sivakumar, T. 2011. Formulation and evaluation of sustained release bioadhesive tablets of ofloxacin using 3^2 factorial design. *International Journal of Pharmaceutical Investigation* 1(3):148–156.

Holenarsipur, V.K., Gaud, N., Sinha, J., Sivaprasad, S., Bhutani, P., Subramanian, M., and Singh, S.P. 2015. Absorption and Cleavage of Enalapril, a Carboxyl Ester Prodrug, in the Rat Intestine: In Vitro, In Situ Intestinal Perfusion and Portal Vein Cannulation Models. *Biopharmaceutics & Drug Disposition* 36(6):385–397.

Hoogstraate, A.J. and Boddé, H.E. 1993. Methods for Assessing the Buccal Mucosa as a Route of Drug Delivery. *Advanced Drug Delivery Reviews* 12(1–2):99–125.

Hoogstraate, A.J., Senel, S., Cullander, C., Verhoef, J., Junginger, H.E., and Bodde, H.E. 1996. Effects of Bile Salts on Transport Rates and Routes of Fitc-Labelled Compounds across Porcine Buccal Epithelium in Vitro. *Journal of Controlled Release* 40(3):211–221.

Kapil, R., Dhawan S., Beg S., and Singh, B. 2013. Buccoadhesive Films for Once-a-Day Administration of Rivastigmine: Systematic Formulation Development and Pharmacokinetic Evaluation. *Drug Development and Industrial Pharmacy* 39(3):466–480.

Karode, N.P., Prajapati, V.D., Solanki, H.K., and Jani, G.K. 2015. Sustained Release Injectable Formulations: Its Rationale, Recent Progress and Advancement. *World Journal of Pharmacy and Pharmaceutical Sciences* 4(4):702–722.

Kassem, M.A., El Meshad A.N., and Fares, A.R. 2014. Enhanced Bioavailability of Buspirone Hydrochloride Via Cup and Core Buccal Tablets: Formulation and in Vitro/in Vivo Evaluation. *International Journal of Pharmaceutics* 463(1):68–80.

Katsuragi, Y., Kashiwayanagi, M., and Kurihara, K. 1997. Specific inhibitor for bitter taste: Inhibition of frog taste nerve responses and human taste sensation to bitter stimuli. *Brain Res Protoc* no. 1(3):292–298.

Kim, J.I., Cho, S.M., Cui, J.H., Cao, Q.R., Oh, E., and Lee, B.J. 2013. In vitro and in vivo correlation of disintegration and bitter taste masking using orally disintegrating tablet containing ion exchange resin-drug complex. *International Journal of Pharmaceutics* no. 455 (1–2):31–39. doi: 10.1016/j.ijpharm.2013.07.072.

Klausner, E.A., Eyal, S., Eran, L., Friedman, M., and Hoffman, A. 2003. Novel Levodopa Gastroretentive Dosage Form: In-Vivo Evaluation in Dogs. *Journal of Controlled Release* 88(1):117–126.

Komiya, I., Park, J.Y., and Kamani, A. 1980. Quantitative mechanistic studies in simultaneous fluid flow and intestinal absorption using steroids as model solutes. *Int J Pharm* 4:249–262.

Kostewicz, E.S., Bertil, A., Marcus, B., Joachim, B., James, B., Sara, C., and Dickinson, P.A. 2014. In Vitro Models for the Prediction of in Vivo Performance of Oral Dosage Forms. *European Journal of Pharmaceutical Sciences* 57:342–366.

Kulkarni, U., Mahalingam, R., Pather, S.I., Li, X., and Jasti, B. 2009. Porcine Buccal Mucosa as an in Vitro Model: Relative Contribution of Epithelium and Connective Tissue as Permeability Barriers. *Journal of Pharmaceutical Sciences* 98(2):471–83.

Kvist, C., Andersson, S.B., and Fors, S. 1999. Apparatus for studying *in vitro* release from medicated chewing gums. *International Journal of Pharmaceutics* 189:57–65.

Lee, C., Jeong, H., Yun, K., Kim, D., Sohn, M., Lee, J., Jeong, J., and Lim, S. 2012. Optical Imaging to Trace near Infrared Fluorescent Zinc Oxide Nanoparticles Following Oral Exposure. *International Journal of Nanomedicine* 7:3203–3209.

Lee, K.J., Johnson, N., Castelo, J., Sinko, P.J., Grass, G., Holme, K., and Lee, Y.H. 2005. Effect of Experimental pH on the in Vitro Permeability in Intact Rabbit Intestines and Caco-2 Monolayer. *European Journal of Pharmaceutical Sciences* 25(2):193–200.

Legin, A., Rudnitskaya, A., Clapham, D., Seleznev, B., Lord, K., and Vlasov, Y. 2004. Electronic tongue for pharmaceutical analytics: Quantification of tastes and masking effects. *Analytical and Bioanalytical Chemistry* no. 380(1):36–45. doi: 10.1007/s00216-004-2738-3.

Lipka, E. and Amidon, G.L. 1999. Setting Bioequivalence Requirements for Drug Development Based on Preclinical Data: Optimizing Oral Drug delivery systems. *Journal of Controlled Release* 62(1):41–49.

Maniruzzaman, M. and Douroumis, D. 2015. An in-vitro-in-vivo taste assessment of bitter drug: Comparative electronic tongues study. *Journal of Pharmacy and Pharmacology* no. 67 (1):43–55. doi: 10.1111/jphp.12319.

Maurya, S.K., Bali, V., and Pathak, K. 2012. Bilayered Transmucosal Drug Delivery System of Pravastatin Sodium: Statistical Optimization, in Vitro, Ex Vivo, in Vivo and Stability Assessment. *Drug Delivery* 19(1): 45–57.

McClean, S., Prosser, E., Meehan, E., Malley, D., Clarke, N., Ramtoola Z., and Brayden, D. 1998. Binding and Uptake of Biodegradable Poly-Dl-Lactide Micro-and Nanoparticles in Intestinal Epithelia. *European Journal of Pharmaceutical Sciences* 6(2):153–163.

Mia, Y., Kamei, N., Muto, K., Kunisawa, J., Takayama, K., Peppas, N.A., and Morishita, M.K. 2017. Complexation Hydrogels as Potential Carriers in Oral Vaccine Delivery Systems. *European Journal of Pharmaceutics and Biopharmaceutics* 112:138–142.

Miyanaga, Y., Inoue, N., Ohnishi, A., Fujisawa, E., Yamaguchi, M., and Uchida, T. 2003. Quantitative prediction of the bitterness suppression of elemental diets by various flavors using a taste sensor. *Pharmaceutical Research* no. 20(12):1932–1938.

Modi, P.B. 2016. A Review Article: In Vitro Release Techniques for Topical Formulations. *Indo American Journal of Pharmaceutical Research* 6(5):5634–5640.

Morjaria, Y., Irwin, W.J., Barnett, P.X., Chan, R.S., and Conway, B. 2004. In Vitro Release of Nicotine from Chewing Gum Formulations. *Dissolution Technologies* 11(2):12–15.

Nafee, N.A., Ismail, F.A., Boraie, N.A., and Lobna, M.M. 2004. Mucoadhesive Delivery Systems. I. Evaluation of Mucoadhesive Polymers for Buccal Tablet Formulation. *Drug Development and Industrial Pharmacy* 30(9):985–993.

Nagpal, M., Rakha, P., Dhingra, G., and Gupta, S. 2010. Evaluation of Various Dissolution Media for Predicting the in Vivo Performance of BCS Class II Drug. *International Journal of Chemical Sciences* 8(2):1063–1070.

Narang, N. and Sharma, J. 2011. Sublingual Mucosa as a Route for Systemic Drug Delivery. *International Journal of Chemical Sciences* 3(2): 18–22.

Nielsen, Hanne Mørck, and Margrethe Rømer Rassing. 2002. Nicotine Permeability across the Buccal TR146 Cell Culture Model and Porcine Buccal Mucosa in Vitro: Effect of pH and Concentration. *European Journal of Pharmaceutical Sciences: Official Journal of the European Federation for Pharmaceutical Sciences* 16(3):151–157.

Pandit, V., Pai, R.S., Yadav, V., Devi, K., Surekha, B.B., Inamdar, M.N., and Sarasija, S. 2013. Pharmacokinetic and Pharmacodynamic Evaluation of Floating Microspheres of Metformin Hydrochloride. *Drug Development and Industrial Pharmacy* 39(1):117–127.

Patel, A., Gajera, N., Shah P.J., and Shah, S.A. 2014. Formulation Development and Evaluation of Sublingual Tablets of Enalapril Maleate Using Co-processed Superdisintegrants. *Research and Reviews: A Journal of Pharmaceutical Science* 5(3):24–29.

Patel, H., Panchal, D.R., Patel, U., Brahmbhatt, T., and Suthar, M. 2011. Matrix Type Drug Delivery System: A Review. *Journal of Pharmaceutical Sciences Research and Bioscientific Research* 1(3):143–151.
Patel, P.S., Parmar, A.M., Doshi, N.S., Patel, H.V., Patel, R.R., and Nayee, C. 2013. Buccal Drug Delivery System: A Review. *International Journal of Drug Development and Research* 5(3):35–48.
Patel, R.R. and Patel, J.K. 2010. Novel Technologies of Oral Controlled Release Drug Delivery System. *Systematic Reviews in Pharmacy* 1(2):128–132.
Patel, R., Gandhi, J.K., and Shah, P.J. 2017. Floating In-situ gelling systems for stomach specific oral Delivery of Valacyclovir. *International Research Journal of Pharmacy* 8(9):25–33.
Patel, R.S. and Poddar, S.S. 2009. Development and Characterization of Mucoadhesive Buccal Patches of Salbutamol Sulphate. *Current Drug Delivery* 6(1):140–144.
Patel, V.F., Fang, L., and Brown, M.B. 2012. Modeling the Oral Cavity: In Vitro and in Vivo Evaluations of Buccal Drug delivery systems. *Journal of Controlled Release* 161(3):746–756.
Pein, M., Preis, M., Eckert, C., and Kiene, F.E. 2014. Taste-masking assessment of solid oral dosage forms—A critical review. *International Journal of Pharmaceutics* no. 465(1–2):239–254. doi: 10.1016/j.ijpharm.2014.01.036.
Philip, A.K. and Pathak, K. 2008. Wet Process-Induced Phase-Transited Drug Delivery System: A Means for Achieving Osmotic, Controlled, and Level a IVIVC for Poorly Water-Soluble Drug. *Drug Development and Industrial Pharmacy* 34(7): 735–743.
Punyashthiti, K. and Finkelstein, R.A. 1971. Enteropathogencity of Escherichia coli. *Infect. Immun.* 4(4):473–478.
Raghavendra, H.L. and Kumar, G.P. 2017. Development and Evaluation of Polymer-bound Glibenclamide Oral Thin Film. *Journal of Bioequivalence & Bioavailability* 9(1):2–8.
Sahana, D.K., Mittal, G., Bhardwaj, V., and Kumar, M.N. 2008. PLGA Nanoparticles for Oral Delivery of Hydrophobic Drugs: Influence of Organic Solvent on Nanoparticle Formation and Release Behavior in Vitro and in Vivo Using Estradiol as a Model Drug. *Journal of Pharmaceutical Sciences* 97(4): 1530–1542.
Sander, C., Nielsen, H.S., Sogaard, S.R., Stoving, C., Yang, M., Jacobsen, J., and Rantanen, J. 2013. Process Development for Spray Drying of Sticky Pharmaceuticals; Case Study of Bioadhesive Nicotine Microparticles for Compressed Medicated Chewing Gum. *International Journal of Pharmaceutics* 45(2):434–437.
Sayani, A.P. and Chien, Y.W. 1996. Systemic Delivery of Peptides and Proteins across Absorptive Mucosae. *Critical Reviews in Therapeutic Drug Carrier Systems* 13(1–2):85–184.
Schanker, L.S., Tocco, D.J., Brodie, B.B., and Hogben, A. 1958. Absorption of Drugs from the Rat Small Intestine. *Journal of Pharmacology and Experimental Therapeutics* 123(1):81–88.
Shah, P., Gadetheriya, Y., and Shah, S.A. 2014. Formulation and Evaluation of Buccal Mucoadhesive Bilayered Tablets of Enalapril Maleate. *Trends in Drug Delivery* 1(1):1–15.
Shete, R.B., Muniswamy, V.J., Pandit, A.P., and Khandelwal, K.R. 2015. Formulation of Eco-Friendly Medicated Chewing Gum to Prevent Motion Sickness. *AAPS PharmSciTech* 16(5):1041–1050.

Singh, B., Garg, B., Chaturvedi, S., Arora, S., Mandsaurwale, R., Kapil, R., and Singh, B. 2012. Formulation Development of Gastroretentive Tablets of Lamivudine Using the Floating-Bioadhesive Potential of Optimized Polymer Blends. *Journal of Pharmacy and Pharmacology* 64(5):654–669.

Sohi, H., Sultana, Y., and Khar, R.K. 2004a. Taste masking technologies in oral pharmaceuticals: Recent developments and approaches. *Drug Development and Industrial Pharmacy* no. 30 (5):429–448. doi: 10.1081/ddc-120037477.

Sohi, Harmik, Yasmin Sultana, and Roop K. Khar. 2004b. Taste Masking Technologies in Oral Pharmaceuticals: Recent Developments and Approaches. *Drug Development and Industrial Pharmacy* no. 30(5):429–448. doi: 10.1081/DDC-120037477.

Soraya, S., Davood B., and Mohammad, E. 2003. Gamma Scintigraphy in the Evaluation of Drug delivery systems. *Iranian Journal of Nuclear Medicine* 11(2):21–33.

Tucker, I. 1988. A Method to Study the Kinetics of Oral Mucosal Drug Absorption from Solutions. *Journal of Pharmacy and Pharmacology* 40(10):679–683.

Ungell, A. 2002. Transport Studies Using Intestinal Tissue Ex-Vivo. In *Cell Culture Models of Biological Barriers: In Vitro Test Systems for Drug Absorption and Delivery*. Taylor & Francis, New York. pp. 166–184.

Verma, R.K., Krishna, D.M., and Garg, S. 2002. Formulation Aspects in the Development of Osmotically Controlled Oral Drug delivery systems. *Journal of Controlled Release* 79(1):7–27.

Wei, L, Hao, N., and Caiyun, Z. 2016. Developments in Methods for Measuring the Intestinal Absorption of Nanoparticle-Bound Drugs. *International Journal of Molecular Sciences* 17:1171.

Woertz, K., Tissen, C., Kleinebudde, P., and Breitkreutz, J. 2011. Taste sensing systems (electronic tongues) for pharmaceutical applications. *International Journal of Pharmaceutics* no. 417 (1–2):256–271. doi: 10.1016/j.ijpharm.2010.11.028.

Yuan, H. and Chen, C.Y. 2013. Improved transport and absorption through gastrointestinal tract by PEGylated solid lipid nanoparticles. *Molecular Pharmaceutics* 10:1865–1873.

Yuasa, H., Matsuda, K., and Watanabe, J. 1993. Influence of Anesthetic Regimens on Intestinal Absorption in Rats. *Pharmaceutical Research* 10(6):884–888.

Zhang, H., Zhang, J., and Streisand, J.B. 2002. Oral Mucosal Drug Delivery. *Clinical Pharmacokinetics* 41(9):661–680.

3

Transdermal Drug Delivery Systems

Anuj Garg and Sanjay Singh

CONTENTS

3.1 Introduction

The skin offers a large and accessible surface for drug delivery. Typically dosage forms applied to skin can be utilized to achieve either through local action or systemic action of drugs. The topical drug delivery systems deliver drug molecules to particular locations within the skin for local action whereas transdermal drug delivery systems (TDDS) deliver the drug molecules through the skin direct into the blood circulation for their systemic effect. Figure 3.1 represents the difference between topical and transdermal drug delivery.

TDDS produces the stable drug concentration in plasma over extended time periods and may reduce the dosing frequency. In 1979, US FDA approved the first transdermal patch containing scopolamine (for three days) indicated for prevention of travel sickness. The application of "transdermal drug delivery systems" requires simple adhesion of a "patch" or "hydrogel". The patient compliance is excellent, as people are quite comfortable with the application of patches on the skin instead of swallowing medicine for systemic application. An additional advantage possessed by the transdermal route over the

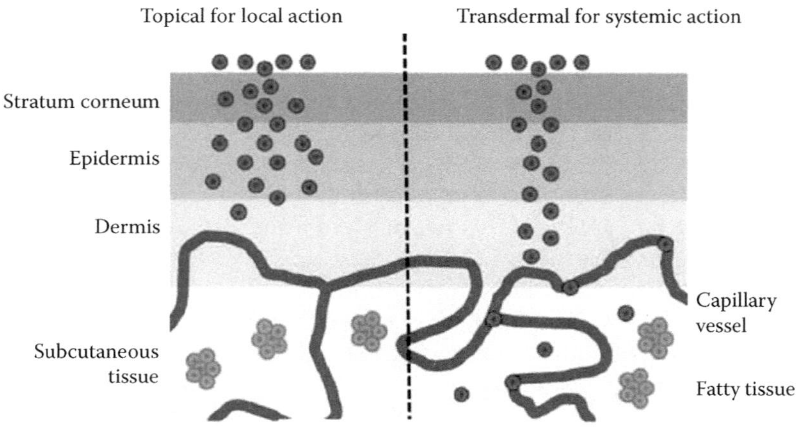

FIGURE 3.1
Representation of difference between topical and transdermal drug delivery.

TABLE 3.1

Advantages and Challenges of Transdermal Drug Delivery

Advantages

a. Avoid first pass metabolism, reduce gastrointestinal drug absorption variability due to presence of food, different pH conditions of GIT, and enzymatic activity.
b. Provide controlled release and prolonged therapy with single application, lead to reduce fluctuation of drug concentration peaks and troughs in plasma, and hence improve therapeutic benefit-risk ratio, as well as minimize dose-dependent systemic toxicity.
c. Improved patient convenience and acceptance, offers non-invasive route for the potent drugs like hormone and avoids the inconvenience of parenteral therapy, gives the option of painless and self-administration.
d. Ease of drug therapy termination in case of any adverse reactions or contraindication with other drugs.
e. Provide an alternative route for drugs which are degraded in pH of stomach and intestines.

Challenges

a. Only applicable for relatively potent drugs as high dose of drug is difficult to administer via the transdermal route.
b. Drugs with high molecular weight (more than 500 Dalton) are difficult to administer using the transdermal route.
c. Log P (octanol/water) value between 1 and 3 is optimal for the permeant to successfully cross the main barrier, i.e., stratum corneum and its underlying aqueous layers for systemic action.
d. Some drug molecules or excipients in patches could cause erythema, itching, local edema, and irritation on the skin.
e. Permeability of drug molecules across the intact and diseased human skin is varied and exhibits intrasubject and intersubject variability in the biological response.

oral route of administration is their ease of removal either at the end of an application period, or in the case that continued delivery is required to be stopped due to undesirable effects. The advantages and challenges of TDDS are summarized further in Table 3.1.

3.1.1 Skin Structure: Anatomy, Physiology and Barrier Functions

The human skin comprises of three layers: epidermis, dermis, and subcutaneous. The epidermis layer is stratified and vascular in nature, while the dermis layer is made of connective tissues with blood capillaries underlying subcutaneous fatty tissues (Figure 3.2). The thickness of the epidermis varies from 0.8 mm on palms and soles down to 0.06 mm on the eyelids. It consists of an outer stratum corneum (nonviable) and viable epidermis. The thickness of dermis is 3–5 mm and is a highly vascularized layer of the skin. It is composed of connective tissues and fibrous proteins. It is adjacent to the epidermis and provides the skin with mechanical support and underlying subcutaneous tissues, with sebaceous and sweat glands running throughout (Menon 2002).

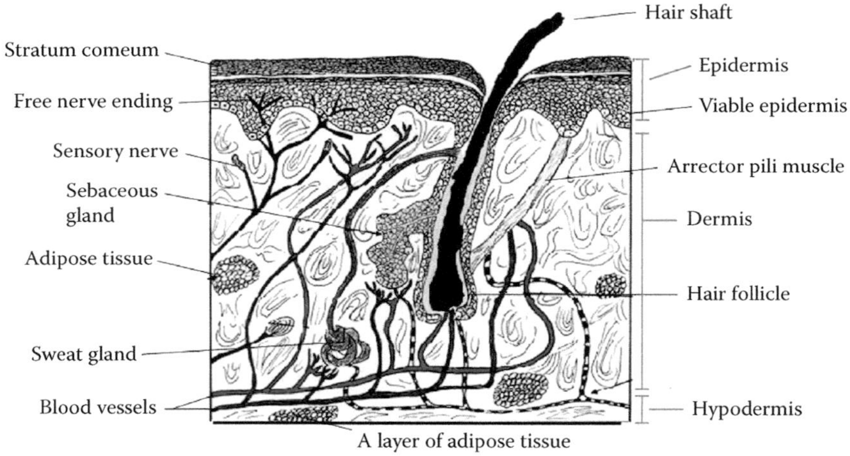

FIGURE 3.2
Schematic diagram of skin structure.

The stratum corneum (also known as the horny layer) is the outermost layer of skin and considered to be the main barrier of penetration for the drug. The structure of stratum corneum can be well explained by a "brick and mortar" model. In this model, the keratinized cells (corneocytes) represent the "bricks" embedded in intercellular lipid which represent the "mortar" (Elias 1983). It contains 10 to 25 layers of dead, keratinized cells called "corneocytes." The stratum corneum (SC) provides its utmost barrier function against hydrophilic compounds, whereas the viable epidermis is most resistant to highly lipophilic compounds. It is approximately 10 microns thick when dry but swells several times when fully hydrated (Scheuplein and Morgan 1967; Anderson and Cassidy 1973).

3.1.2 Principles and Different Routes of Penetration Across the Skin

The main principle of percutaneous absorption is passive diffusion of the drug molecules across the skin (Flyn et al. 1974; Scheuplein and Blank 1971). The diffusion is best described by the Fick's law of diffusion in the following equation:

$$dQ/dt = -D.\Delta Cv.A/h \tag{3.1}$$

where dQ/dt is the amount of drug diffused per unit time (i.e., drug flux), D is the diffusion coefficient, h the diffusional path length (thickness of SC), ΔCv is the concentration gradient of permeant, and A is the surface area of treated skin.

There are two major routes for the permeants to penetrate normal intact skin, the transepidermal and transfollicular route.

3.1.2.1 Transepidermal Route

The transepidermal route include two pathways such as intracellular (transcellular) and intercellular for the transport of drug molecules across the skin (Figure 3.3).

The intracellular (transcellular) route involves the transport of the drug molecule across the corneocytes. Initially, drug molecules partition into a keratin-filled corneocyte, an aqueous environment, and afterwards diffuse through the corneocyte before distributing in the lipoidal region present in between the cells. To continue this transport, the drug molecule must then diffuse through the lipoidal region and again partition into and diffuse through the aqueous keratin in underlying corneocytes and then their intercellular lipids. In the intercellular route, drug molecules are transported across the skin via spaces between the corneocytes. Drug molecules partition into the lipid bilayers between the corneocytes and diffuse within the continuous lipid domain of underlying corneocytes (Hueber et al. 1992).

3.1.2.2 Transfollicular Routes

Transfollicular routes are also referred as the "shunt route" as molecules can pass across the stratum corneum barrier by using the appendages like hair follicles (Figure 3.3). Hair follicles are the largest appendages on the skin. Generally, the skin has 50 to 100 hair follicles per cm^2. In addition to hair follicles, sweat glands (eccrine glands) and sebaceous glands also offer the shunt route to transport the drug molecules across the skin. These shunt

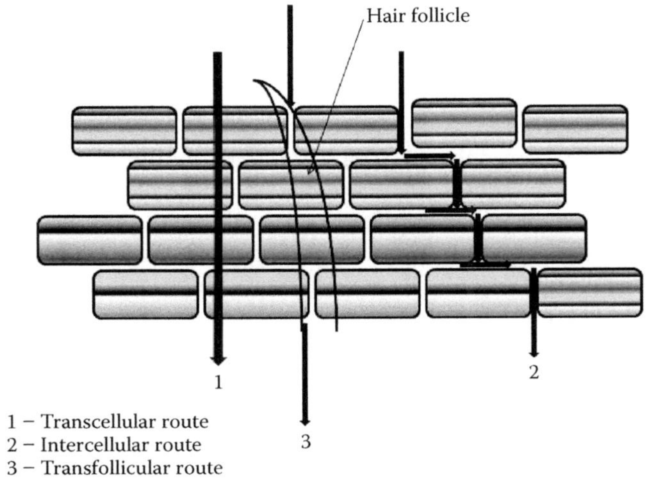

FIGURE 3.3
Illustration of different route of percutaneous absorption.

routes can be utilized for enhancement of transdermal drug delivery via iontophoresis, but the fractional area occupied by the appendages are relatively very small (Otberg et al. 2004).

3.2 Factors Affecting Permeation Across the Skin

3.2.1 Physiochemical Factors of Permeant

The major physiochemical factors which can influence the permeation of drug molecules across the skin include molecular weight, partition coefficient or log P and pKa of drug molecules. Transcellular, as well as the intercellular, routes of penetration involve the partitioning of drug molecules into the lipid bilayer and then diffusion from the lipid bilayer between the corneocytes. Therefore, the molecular weight and partition coefficient of drug molecules affect their permeation across the skin. Thus, drug molecules with a low molecular weight (less than 500 daltons) and lipophilic nature are usually considered to be better candidates for transdermal delivery than those with a high molecular weight and hydrophilic compounds. The structure of the skin provides barrier to external chemicals and thus prevents the entry of large molecules, such as larger peptides, proteins, and antigens. In addition to its molecular weight, the molecular structure (in particular hydrogen-bonding potential) is also an important factor in diffusion and can control the extent of binding to skin constituents and hence affect their bioavailability.

3.2.2 Formulation Factors

The drug molecule should be in the dissolved form in the formulation to transport across the stratum corneum. Generally, the low aqueous solubility of the drug molecule is the rate-limiting step in transdermal delivery and poor drug release from a formulation can occasionally limit drug transport. Additionally, an important factor is the effective dose of the drug, which must be relatively low to allow the application of aptly sized patches/formulations. The stability of the drug in the transdermal patch is also essential. The above factors restrict the variety of compounds that can be delivered via the transdermal route for therapeutic benefits.

3.2.3 Physiological Factors

The different physiological or pathological conditions of the skin can influence the permeation of drug molecules across the skin. The percutaneous absorption in human and animals in-vivo shows individual differences due to multiple factors such as skin thickness, blood flow, lipid content, number of hair follicles,

TABLE 3.2

Different Factors Affecting the Permeation of Drug Molecules Across the Skin

Factors	Description
Physiochemical factors	• Molecular weight • Permeability coefficient • Molecular structure • pKa
Formulation factors	• Release characteristic • Concentration of drug in formulation • Composition of formulation • pH
Physiological factors	• Skin condition (hydration, temperature) • Skin age • Regional skin site • Blood flow rate • Skin metabolism • Injury to skin • Species difference

etc. Racial differences in skin exist and the literature indicates that the skin of peoples from different races may have different responses to various chemical stimuli. Williams et al. (1991) suggested that racial differences have an effect on the absorption of nitroglycerin upon transdermal application. The skin of racial groups with higher amount of melanin showed lower drug permeation than skin of racial groups having less amount of melanin. It was found that there is an increased resistance to tape stripping in skin having high content of melanin than that of low melanin content (Weigan and Gaylor 1974). Different factors affecting permeation of drug molecules across the skin are summarized in Table 3.2.

3.3 Methods to Enhance Skin Permeability

The poor permeability of drug molecules across the skin is the major limitation associated with a transdermal drug delivery system. This limits the drug candidates for transdermal delivery. Over the last 30 years, several studies have been carried out to overcome the permeability barriers of the skin and increase the opportunity to develop TDDS for drugs with poor permeability. The different techniques that have emerged over the years are broadly classified into passive or active methods.

3.3.1 Passive Methods

In passive methods dosage forms have been developed and/or tailored to augment the diffusion of drug and/or enhance their permeability across the skin. These methods include the use of penetration enhancers (Williams and Barry 2004), supersaturated systems (Pellet et al. 2003), prodrugs or metabolic approaches (van Heerden et al. 2010; Tsai et al. 1996), liposomes, and other vesicles or nanocarriers (Honeywell and Bouwstra 2005; Trotta 2004; Escobar-Chávez et al. 2012; Shakeel and Ramadan 2010). However, these methods cannot lead to significant changes in the barrier properties of the skin, and thus, the number of drugs which can penetrate are limited. The passive methods do not involve the application of external forces on the skin to manipulate its barrier properties, while active methods use external forces to overcome the barriers. These are useful for low molecular weight drug molecules with poor aqueous solubility and permeability and also improve the dose control, patient acceptance and compliance in comparison to semisolid formulations. Penetration enhancers can increase the permeability of the skin. The classification of different penetration enhancers with their examples are summarized in Table 3.3 (Dragicevic et al. 2015; Escobar-Chávez et al. 2012).

3.3.2 Active Methods

Passive methods are incapable of enhancing the permeation of drug molecules with a high molecular weight, such as therapeutics peptides and proteins. The advent of recombinant biotechnology technique in last decade has led to the generation of therapeutically active macromolecules like peptides, proteins, and nucleic acids which possess a large molecular weight (>500 Da). Moreover, these are generally polar and hydrophilic in nature and show poor permeability across the skin. Further, the oral delivery of such macromolecules is also difficult, as they have a tendency to extensively degrade by enzymes in the gastrointestinal tract. Therefore, the successful administration of these macromolecules requires an alternative route and suitable drug delivery system. The active methods showed potential to reduce the barrier nature of SC. These methods include the use of external energy, such as electric currents or ultrasonic energy, to act as a driving force to enhance the permeation of drug molecules. The latest expansion in these technologies has occurred as a consequence of advances in different areas in the engineering and material sciences, which lead to the development of a new generation of small, delivery-modulated devices for the transdermal delivery of drugs to produce their desired clinical response.

3.3.2.1 Iontophoresis

Iontophoresis utilizes a low-level electric current between a medicated solution and the patient's skin to enhance the permeation of a topically applied therapeutic agent (Banga 1998; Singh et al. 2008). The enhancement of drug

TABLE 3.3

Summary of Different Type of Passive Methods to Increase the Permeation of Drug Molecules Across the Skin

Type	Subclass and Examples	Comments
Penetration enhancer	***Fatty acids/Fatty acid ester,*** e.g., Oleic acid, lauric acid, isopropyl palmitate	Promote the permeation of drugs by increasing their partitioning into intercellular lipid domain of SC.
	Surfactants like sodium lauryl sulphate (SLS), Tween 80, Brij 36T, cetrimide	Surfactants interact strongly with both lipids and keratin and alter the permeability of the skin. In general, anionic surfactants are more effective penetration enhancers than cationic and nonionic surfactants.
	Dimethylsulphoxide and related compounds (Decylmethyl sulphoxide)	One of the earliest and most widely studied penetration enhancers; may denature the intercellular structural proteins of the stratum corneum. Also promotes lipid fluidity by disruption of the ordered structure of the lipid chains.
	Fatty alcohol, e.g., Olyl alcohol, lauryl alcohol	Promote the partitioning of permeants into the intercellular lipids like fatty acids and increase the penetration across the skin.
	Azone (laurocapram) and related compounds (N-Dodecyl-2- pyrrolidone, 1-Geranylazacyclo-heptan-2-one	It can interact with the lipids present in stratum corneum and results in increase in the degree of fluidity of the hydrophobic regions of the intercellular lamellar structure, explaining its ability to decrease the diffusional resistance of the skin.
	Pyrrolidone and related compounds, e.g., 2-Pyrrolidone, N-methyl-2-pyrrolidone, Poly	May affect both structure of lipids and proteins barriers in the stratum corneum.
	Terpenes, e.g., α-terpineol, Carvol, α-pinene, β-carene, D-limonene, Carvone, Menthol, 1,8- cineole	Possible reversible disruption of intercellular lipids barriers in the stratum corneum and results in an increase in the permeation.
	Solvent and related compounds like ethanol, propylene glycol and polyethylene glycol 400 (PEG-400)	May extract the lipid component of horny layer and increase the permeation.

(*Continued*)

TABLE 3.3 (CONTINUED)

Summary of Different Type of Passive Methods to Increase the Permeation of Drug Molecules Across the Skin

Type	Subclass and Examples	Comments
Prodrugs and derivatives	Lipophilic derivatives of water soluble drugs like steroid derivatives	A prodrug with an optimal partition coefficient that enhances the permeation across the skin barrier.
Novel formulations (Vesicular and nanoparticulate-based)	Liposomes	Liposomes are bilayered vesicles and are frequently made of phospholipids and cholesterol. They contain both a hydrophilic and lipophilic core and can serve as carriers for both polar as well as nonpolar drugs. The liposomes were accumulated within the cells of uppermost stratum corneum and can be interacted with lipid components of skin to release their loaded drugs.
	Transferosomes	Transferosomes are modified liposomes containing surfactant molecules, such as sodium cholate. The surfactant molecules act as edge activators and confer ultradeformability to them. This ultradeforrmable nature permits them to squeeze through pores of the stratum corneum, which are less than one-tenth of their diameter.
	Ethosomes	Ethosomes are liposomes composed mainly of phospholipids, alcohol in relatively high concentrations, glycols (sometimes) and water. These can lead to increase the penetration of drug molecules deeper in the skin and the systemic circulation. The alcohol fluidizes the ethosomal lipids as well as stratum corneum intercellular lipids, thus allowing the soft flexible ethosome to penetrate the stratum corneum and enhance penetration.
	Solid lipid nanoparticles or nanostructured lipid carriers (NLCs)	Upon application of solid lipid nanoparticles or NLCs form layers on the skin. The occlusion due to layering causes hydration of stratum corneum and loosens the highly packed corneocytes. Thus, hydration increases the partitioning into the stratum corneum and release of their content. The SLN or NLC also interact with the intercellular lipids and may disrupt the barrier properties of layer.

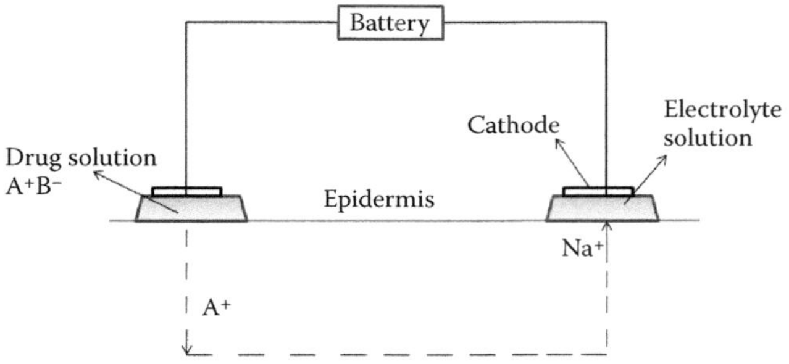

FIGURE 3.4
Schematic diagram of iontophoretic system for transdermal drug delivery.

permeation across the skin as a result of this method can be ascribed to either one or an amalgamation of the following mechanisms:

a. Electrorepulsion (for charged solutes)
b. Electro-osmosis (for uncharged solutes)
c. Electropertubation (for both charged and uncharged)

Figure 3.4 shows a simple schematic diagram of the iontophoresis experiment.

Transdermal iontophoresis would be especially valuable in the delivery of small peptides, like hydrophilic molecules, in humans. However, it is evident that this technique may not be useful for the systemic delivery of larger peptides (with a molecular weight more than 8000 dalton). Factors such as current intensity, type of electrode, pH, the competitive ion effect, and the nature of the permeant can affect the design of transdermal iontophoretic delivery system (Banga et al. 1999). Extensive literature is available on the various kinds of permeants using this technique (Kalia et al. 2004; Kanikkannan 2002; Ita 2016).

3.3.2.2 Electroporation

Electroporation is a process, in which brief, intense electric charges (voltage more than 100 V pulses for milliseconds) create small pores in the phospholipid bilayer of cell membranes that assists in the transdermal delivery of drugs. Electroporation appears to disrupt the lipid bilayers in the stratum corneum, and the channels it creates are said to promote the passage of hydrophilic drugs through the skin. Skin hydration increases, skin resistance decreases, and cutaneous blood flow increases (ephemerally).

It has been anticipated that the produced transitory pores are accountable to increase in skin permeability (Weaver et al. 1999). Electrical pulse parameters like waveform, rate, and number can affect the delivery via this technique (Banga et al. 1999). This technique has been found to be valuable for improving the skin permeability of hydrophilic molecules with a molecular weight greater than 7kDA (Denet et al. 2004). The simultaneous use of iontophoresis and electroporation was found to be much more successful than either technique used alone in the transdermal delivery of molecules (Chang et al. 2000; Badkar and Banga 2002).

3.3.2.3 Sonophoresis and Phonophoresis

Sonophoresis is an active form of transdermal delivery, which enhances the transport of drugs through cell membranes employing ultrasonic energy. This method is capable to enhance the penetration of several low molecular weight drugs as well as high molecular weight therapeutic proteins (Smith et al. 2003). Ultrasonic sound waves cause acoustic cavitation, the resultant effects of which microscopically disrupt the lipid bilayers of the stratum corneum and thereby influencing the flux of drugs (Polat et al. 2011; Mitragotri 2017).

3.3.2.4 Magnetophoresis

In this technique, the magnetic field is used as an external driving force to increase the permeability of drugs across the skin. The magnetic field induces the alterations in the structure of skin and leads to enhancement in the permeability of drugs that are particularly diamagnetic in nature. A previous in-vitro study showed that the linear relationship in between permeability flux across the skin and the strength of the magnetic field. Murthy et al. (2010) showed an enhancement in the lidocaine flux across the skin with an increase in the strength of the magnetic field (Murthy et al. 2010).

3.3.2.5 Thermophoresis

The surface temperature of skin is typically maintained at 32°C in humans. The consequence of high temperature (nonphysiological) on percutaneous absorption was reported in the literature which demonstrated the enhancement in flux with increase in the temperature of skin (Hull 2002). The increased flux across the skin could be attributed to the enhanced diffusion of drug in the vehicle and also in the skin at a higher temperature. Moreover, hyperthermia also induces the vasodilation of blood vessels present in the dermis and subcutaneous tissues as a homeostatic mechanism.

3.3.2.6 Skin Abrasion

Disruption or removal of the top layer of skin is known as "skin abrasion" that leads to increase in the flux of drug molecules across the skin. The physiochemical properties of drug molecules cannot restrict the delivery potential of this technique. In the literature, it has been shown that this technique is able to enhance and control the delivery of a hydrophilic permeant like ascorbic acid and different biopharmaceuticals, such as therapeutic peptides (Mikszta et al. 2001; 2003). Different methods of abrasion include using a stream of aluminium oxide crystals, motor-driven fraises, and pretreating the skin before delivery with a microabrader device to abrade the skin at a site of interest for enhancing permeability (Lee et al. 2003).

3.4 Experimental Methods to Assess Skin Permeability

Experiments should be designed and performed to assess the (i) permeation flux of drug molecules through the skin; (ii) rate the limiting step in the permeation flux across the skin layers; and (iii) assess the major penetration route across the skin.

3.4.1 In-Vitro Methods

In-vitro methods are the most commonly used technique to screen and assess the flux of drug molecules through the skin using diffusion cells. In Section 3.4.1, the different parameters to assess the transport of drug molecules, various diffusion cells, and membrane-separating donor and receptor compartments are described.

3.4.1.1 Different Parameters for Assessing the Transport or Permeation Across the Skin

3.4.1.1.1 Steady State Flux

Steady state flux can be defined as the number of atoms or mass per unit area and per unit time (e.g., atoms or kg/m^2.second). Equation 3.2 describe the steady state flux along direction x,

$$J = -D\, dC/dx \tag{3.2}$$

where dC/dx represents the concentration gradient and D is the diffusion constant. The negative sign showed that the diffusion process is down the concentration gradient.

3.4.1.1.2 Permeability Coefficient

The strength of the percutaneous absorption of drug molecules is determined by the total amount of drug penetrated across the skin, the percentage of absorption of applied dose, the flux and the permeability coefficient (Kp).

It represents the ratio of steady-state flux and the concentration of the test compound (Kp = Jss/C) where Jss is the steady-state flux and C the concentration of the test compound in the donor phase. Generally, it is accepted that the permeability coefficient is more reliable than the flux to describe the extent of a dermal absorption potential of chemical substances. Kp has the advantage of being fairly constant over a range of concentrations and can be calculated for other concentrations than the ones used in the experiment. Moreover, it can also be used for the direct comparison of various species or vehicle effects on their dermal absorption or the barrier integrity of human skin.

The flux can be calculated from the slope of the linear portion of the plot of cumulative amount permeated per square centimeter of skin (on y axis) against time (on x axis). The permeability coefficient Kp can be calculated at the peak absorption rate according to Fick's first diffusion law as:

$$Kp = J/C \tag{3.3}$$

where Kp is the permeability coefficient (cm/h), J is the flux ($mg/cm^2/h$) and C the concentration of permeant in the donor solution (mg/cm^3).

3.4.1.1.3 Lag Time

The time required to obtain steady state conditions under infinite dose conditions is known as lag time. Lag time is a function of the drug loading the SC and dermis, diffusivity, and thickness of the skin.

3.4.1.2 Diffusion Cells

The most common technique to assess the permeation parameters of the drug molecules for transdermal drug delivery is the employing of any of the following diffusion cells. Diffusion cells are designed to simulate in-vivo or clinical conditions by using agitated receptor solutions to correspond to blood and unstirred donor phases to represent the formulation.

3.4.1.2.1 Franz Diffusion Cell

A Franz diffusion cell is a static and vertical type diffusion cell. It consists of two compartments: the donor compartment and receptor compartment separated by a membrane, as shown in Figure 3.5A. The donor compartment may be closed or open to the ambient temperature. The receptor compartment is stirred to mimic the in-vivo conditions, and is sampled at predetermined time intervals for the analysis of the drug permeated across the membrane used in the study.

3.4.1.2.2 Jacketed Franz Diffusion Cell

The Franz diffusion cell is jacketed and used to maintain the temperature of receptor compartment at physiological temperature by the circulation of

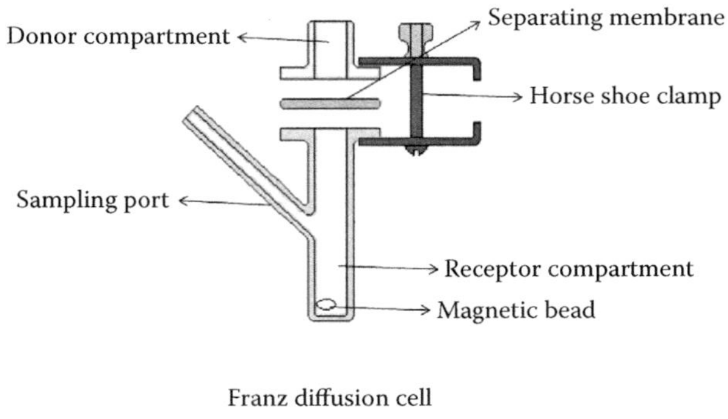

FIGURE 3.5A
Schematic representations of Franz diffusion cell.

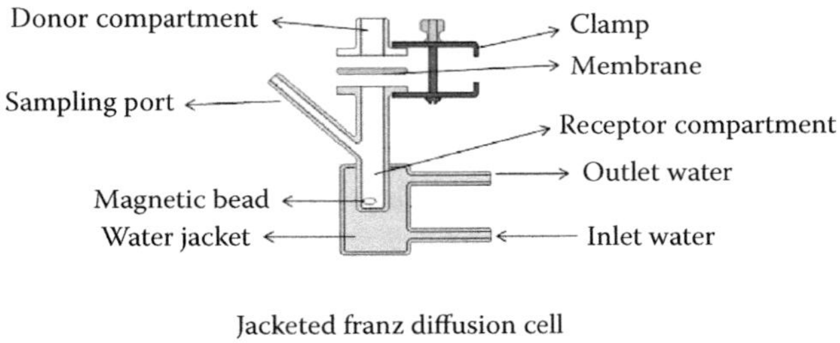

FIGURE 3.5B
Schematic representations of jacketed Franz diffusion cell.

hot water at a corresponding temperature (Figure 3.5B). The jacketed Franz diffusion cell mimics the in-vivo conditions more than that of a non-jacketed Franz diffusion cell. The temperature can significantly influence the permeation parameters, like flux and the permeability coefficient.

3.4.1.2.3 Side-by-Side Diffusion Cell

Next to the Franz diffusion cell, this is most commonly used for permeation study. In this cell, both the receptor and donor compartment have the capacity of approximately 3.5 ml and is constantly rotated by matched set of star head magnets at 600 rpm. The membrane area is about 0.64 cm^2. The temperature is controlled by the circulation of hot water at body temperature through the water jacket surrounding the both compartment. The side-by-side cells are used mainly for the measurement of permeation from one stirred solution across the membrane and into another stirred solution. The use of a side-by-side cell is more common during iontophoresis, electroporation, and sonophoresis.

3.4.1.2.4 Flow-Through Cells

This system consists of a reservoir containing the dissolution/release medium, a pump that forces the medium upwards through the vertically positioned flow cell, and a water bath to control the temperature in the cell. It can be used to study in-vitro release from topical formulations.

3.4.1.3 Different Type of Membrane

3.4.1.3.1 Artificial Membrane

Artificial and reconstructed skin models are valuable tools in definite circumstances, motivated by the need to find well-suited and reproducible alternatives to other existing skin models, like animal and human. These artificial skin models include simple homogeneous polymer materials, such as silicone membranes, e.g., poly (methoxysilane) and phospholipid vesicle-based membranes. These artificial membranes are designed to mimic the stratum corneum (Flaten et al. 2015). The artificial membranes are mainly useful to study the primary controlling mechanisms of passive transport through the membrane, as it lacks the complexity of human or animal skin (Zhang et al. 2013; Oliveira et al. 2012). However, artificial membranes do not replicate the multitude of in-vivo skin properties (Dabrowska et al. 2016). The permeation flux of the hydrocortisone was evaluated using the artificial membrane, made of different polymers like cellulose, cellulose acetate, and polysulfone of similar pore size and thickness, and was found to be steady irrespective of the membranes used in evaluation (Shah et al. 1989).

3.4.1.3.2 Animal Skin

A broad range of animal skin models can be used as a substitute to human skin to evaluate the percutaneous absorption of drug molecules through the skin, for example. Animal skin models include pig, mouse, rat, guinea pig, and snake. The characteristics of different animal skin models are summarized in Table 3.4.

3.4.1.3.3 Human Cadaver/Human Equivalent Skin (In-Vitro Human Skin Model)

It is best to use human/human cadaver skin to obtain the penetration or permeation data. However, current scenario regulatory restrictions are rising on the use of animals in experimentation. In addition, the availability of excised human cadaver skin is also limited. Moreover, advent of tissue engineering, the application of an artificial in-vitro human skin models are emerged as valuable tools in the evaluation of dermal absorption and mimics both healthy and diseased skin conditions. Furthermore, these models permit the assimilation of particular disease distinctiveness in a controlled and rather reproducible manner. These models have been developed for a large range

TABLE 3.4

Characteristics of Different Animal Skin Models

S. No.	Animal Skin Model	Characteristics
1.	Porcine Ear Skin Model	• Histologically similar to human skin. • Thickness of SC comparable to human skin. • The average hair follicle density in porcine skin is $20/cm^2$ while human forehead skin has $14–32/cm^2$. • Easy to acquire • Lipids in porcine stratum corneum are arranged primarily in a hexagonal lattice, whereas denser orthorhombic lattice arrangement are shown by lipids in human SC.
2.	Rodent Skin Model	• This is the most widely used skin model because of their ease of availability and low cost. • Rat skin is the most frequently used among rodent models because of its structural similarity to human skin. • Rat skin usually more permeable than human skin for a variety of permeants with diverse physicochemical properties.
3.	Shed Snake Skin Model	• Can be obtained without killing the animal. • Little leaching of chemicals to confuse UV assays. • It possesses thin, flat squamate cells encircled by intercellular phospholipids which is similar to human skin. • However, dissimilar as it lacks hair follicles.

of skin diseases, such as psoriasis, atopic dermatitis, fungal infections, skin cancer, photodamaged skin, and wounds. A key challenge in the utilization of these models is to evaluate whether they are appropriate and predictable in the in-vivo situation. For further detailed study about the different skin models for testing of transdermal drug delivery was reviewed by Abd et al. (2016).

3.4.1.4 Integrity Testing of Membrane

The visual observation for physical damage is the foremost step in evaluating the integrity of skin preparation. Other methods to check the integrity of the skin barrier include, physical methods, like transepidermal water loss (TEWL), or transcutaneous electrical resistance (TER) (Davies et al. 2004). It has been suggested that before (and in some cases after) the experiment, the integrity of skin barrier should be checked by these methods. The integrity of skin samples may also be checked using the tritium method, where the permeation of triturated water through the skin is determined and compared with standard values (Bartosova and Bajgar 2012).

3.4.2 Ex Vivo Methods

3.4.2.1 Tape Stripping Methods

Tape stripping is a simple and efficient *ex vivo* method for the assessment of the permeation of drug molecules across the skin layers with respect to their depth after a topically applied formulation (Lademann et al. 2009). This method is useful to establish the importance of the stratum corneum in skin barrier functions. The cell layers of the stratum corneum are sequentially removed from the skin area, where the formulation is applied and permeated using adhesive tape or films. The tape strips hold corneocytes and the equivalent amount of the penetrated drug molecule, which can be determined by a suitable analytical technique. Different formulations can strongly persuade the depth of stratum corneum removed with every tape strip. Hence, it is critical for the comparison of the penetration of different formulations that the quantity of drug molecules detected on the single tape strip is not correlated to the tape strip number as a relative measure of the penetration depths, but to their coordinated definite position in the stratum corneum. This method can be utilized to determine the dermal pharmacokinetics of formulations applied upon topical administration. In addition, this method can be used to obtain information about the homogeneity and the distribution of formulations on the skin and in the stratum corneum.

However, the tape stripping method is an invasive technique, only assesses the depth of penetration of drug in to the skin layers and can only determine a single concentration time point per administration site. Therefore, it is difficult to determine the elimination of drugs from the skin layer by using this method.

3.4.2.2 Confocal Scanning Laser Microscopic Techniques

The confocal scanning laser microscopic (CSLM) technique has been utilized to observe selectively stained structures in the skin *ex vivo* and hence, can be utilized as a tool to illustrate and establish the penetration route for different kind of drug molecules across the skin (Dayan 2005). The most important advantage of this technique over optical microscopy is that CSLM facilitates the observation of dynamic processes in real time and sequential imaging of the same site and this is not possible with the light microscopy of skin samples. Histological sections for optical microscopy showed their structure only at one point of time. The extensive preparative protocol increases the potential risk for artefacts and disruption of their native structure and environment. This technique in association with fluorescent ultra-deformable lipid vesicles allowed for visualization of penetration pathway in intact skin (Dayan 2005).

3.4.3 In-Vivo Methods

Generally, the in-vivo methods used animals and these models are important for studying the anatomy, physiology, and biochemistry of the skin, for screening topical agents, for detecting possible hazards and for preliminary biopharmaceutical investigations. Animal skin models differ from humans in the thickness and nature of the SC, the density of hair follicles and sweat glands, the nature of the pelt, the papillary blood supply, and biochemical aspects. Hence, data obtained with animals cannot entirely surrogate for human studies, and regulatory authorities will typically demand additional data from human studies before granting a product license.

3.4.3.1 Microdialysis

Microdialysis is technique which continuously collects the sample fluid from the target tissues/organs. The probe is hollow tube like and semipermeable in nature, as shown in Figure 3.6.

These probes are inserted into the dermis and can be used to monitor the time course of drugs in the dermis layer. The microdialysis tube is slowly perfused typically with 0.1–5.0 microliter/minute with physiological solution, which equilibrates with the extra cellular fluid of the surrounding tissues. The exchange of substance occurs due to the concentration gradient as per Fick's second law of diffusion. Drug molecules diffuses from the extracellular fluid into the dialysate buffer through the pores in the semipermeable membrane of the probe, excluding large molecules, particularly proteins. The resulting drug solution is collected and analyzed. A very sensitive and robust analytical method is required to determine the drug concentration in the samples particularly for highly protein-bound drugs, as the free drug concentrations in the samples are reduced due to protein binding. Protein-bound drugs are not able to cross the membrane of microdialysis tube.

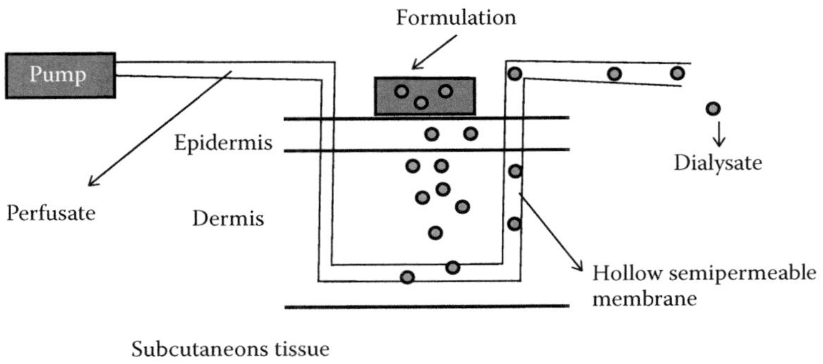

FIGURE 3.6
Illustration of microdialysis techniques.

3.4.3.2 Plasma Concentration Time Profile after Transdermal Delivery (Pharmacokinetic)

This is a direct method to determine the amount of drug reached to the systematic circulation after penetration through the skin. This method determines the pharmacokinetic factors inherent in the absorption, distribution and elimination process upon transdermal drug delivery. Due to the advancement in analytical techniques, it can be possible to detect the drug concentration at nanogram level in the blood or plasma. Hence, this is a very important in-vivo tool to determine the efficacy of the drug administered by the transdermal route.

3.4.3.3 Pharmacodynamic Activity Relationship

In this method, pharmacodynamic responses like a change in blood pressure (for topical application of antihypertensive agents), reduction in pain threshold and production of convulsion, etc. correlate with their absorption across the skin after transdermal application. Dose-response relationships can be established by using this tool and can be useful for determining the efficacy and safety of TDDS.

3.4.4 In Silico Methods (Software-Based Prediction of Permeation for Transdermal Delivery)

In silico methods are emerging as a promising tool to evaluate efficacy and safety of compounds after their dermal or transdermal application. The concentration of compounds in the skin can be estimated by using this technique upon completion of modeling for permeation process. Recently, researchers have developed *in silico* methods based on a three-layered diffusion model for estimation of skin concentration of dermally metabolized chemicals (Hatanaka et al. 2017). The three layers consist of stratum corneum, viable epidermis and dermis. The model was developed based on the Fick's second law of diffusion with the Michaelis-Menten equation and plasma clearance in the viable epidermis and dermis, respectively.

SKIN CAD® is one of such registered software developed by Biocom System Inc. to predict the skin permeation amount and blood concentration by the input of various model parameters. These input parameters include skin structure (thickness of each layer); release characteristics from formulation, diffusion, partition coefficient in skin; and body pharmacokinetic parameters.

This software can help to predict clinical performance using model parameters obtained from in-vitro release or skin permeation study. *In silico* studies provide useful data to design transdermal drug delivery systems.

3.5 Skin Sensitization and Irritation Studies

Skin sensitization and irritation can be classified into the different categories, as described in Table 3.5.

3.5.1 Evaluation of Skin Sensitization and Irritation

A sensitization test is done to evaluate the allergic potential of chemicals and to assess the potential sensitization properties. Animals are treated with one or several doses of chemicals (the induction phase) by intradermal or cutaneous application. The skin irritation studies are also performed to evaluate the compound for irritation in animals which correlates well with the degree of skin response in humans. Different tests for skin sensitization and irritation are summarized in Table 3.6.

TABLE 3.5

Different Categories of Skin Sensitization and Irritation

Type of Skin Sensitization	Description
Allergic contact dermatitis	Allergic contact dermatitis or dermatitis venenta, is the result of an interaction between the complicated pathophysiological mechanisms of type IV cell-mediated immunity and environmental allergens.
Light-induced cutaneous toxicity	The wavelength of light found in the UV-B spectrum is generally considered the primary source of toxic changes in the skin. The specific wavelength responsible for a particular biological response is termed as "action spectrum" for that effect. In some cases, xenobiotics play a role in these effects, while in others, the interaction of light with the normal compound of the skin is responsible in either case; the adverse reaction of the skin to light (UV or visible) is termed as photosensitization.
Cutaneous carcinogenesis	The skin is the most common site of cancer in humans. Both benign and malignant tumors may be derived from viable keratinocytes and melanocytes of the epidermis, and rarely from skin appendages, blood vessels, peripheral nerves, and the lymphoid tissue of the dermis.
Acne-like eruptions	These reactions are initiated by the proliferation of the epithelium of sebaceous glands and the formation of keratin cysts, resulting in the development of a pustule filled with fatty compounds and other products of sebaceous origin.

TABLE 3.6

Different Tests for Skin Sensitization and Irritation

Name of the Technique	Description
Draize test (intradermal technique)	First test to describe standardized irritation and sensitization tests. The Draize test is the simplest and most predictive. The demerits of this test includes high incidence of negatives with weak sensitizers. In addition, the test recommends a constant induction concentration of 0.1% w/v injected intradermally without regard to use pattern or exposure potential of the chemicals.
Freund's complete adjuvant test (intradermal technique)	The test substance is mixed with Freund's complete adjuvant (FCA) which is a mixture of heat-killed mycobacterium tuberculosis, paraffin oil and mannide monooleate prior to the intradermal injection for induction. The use of FCA increases the immunological response and aids in the detection of weak sensitizers. It is considered as sensitive as the optimization test and is of low cost to perform.
Open epicutaneous test	Klecak et al. (1977) found this test to be a sensitive, highly predictive test. The open epicutaneous test includes an induction phase of repeated applications of an undiluted test substance (which may be a formulation of a final product for consumer exposure) over several weeks. The challenge phase is separated into initial and rechallange phases.
Buckler's test	The Buckler's test was designed to reproduce a human patch test in animals and therefore allows the deviation of conditions to optimize the detection of moderately strong sensitizers prior to testing in humans. This test utilizes the induction and challenge phases of occluded epidermal doses of the test substances that may be allergenic chemicals or final product formulations.
Single application irritation testing	The test described by Draize et al. (1944) or the slight modification of this test is the most widely used for predicting the potential skin irritation of chemicals and chemical mixtures. In this test, the hair is clipped from the back of a rabbit and four distinct areas for the application of test substances are identified. Two of four areas were abraded by making four epidermal incisions in the appropriate areas. All four areas were covered by gauze that is held in place with adhesive tape and the test substance is applied to the appropriate area under the gauze. The entire trunk of the rabbit is wrapped in impervious cloth or plastic to hold the patches in place and decrease the evaporation of volatile test substances. The rabbits remained wrapped for 24 hrs after treatment and are evaluated for irritation at the time of unwrapping and 24 and 48 hrs after being unwrapped.
Repetitive application irritation testing	The repetitive application over at least 7–14 days appears to be better able to predict the irritability of test substances than the single application. Additionally, the repetitive application test to assess irritability can be combined with an assessment for systemic toxicity by a cutaneous route.

3.6 Emerging Technologies for Transdermal Drug Delivery

3.6.1 Microneedle-Based Transdermal Drug Delivery

Microneedles have been intended to go through the epidermis up to a depth of 70–200 mm. Microneedles are thin and short (Figure 3.7) and do not reach to the dermis layer with its nerves; hence they are almost pain free. In comparison to other transdermal delivery methods, they are more competent to increase the permeation of drug across the skin and could be attributed to overcoming of the stratum corneum barrier. Ita (2016) reviewed the potential and challenges of microneedles in transdermal drug delivery systems (Ita 2016).

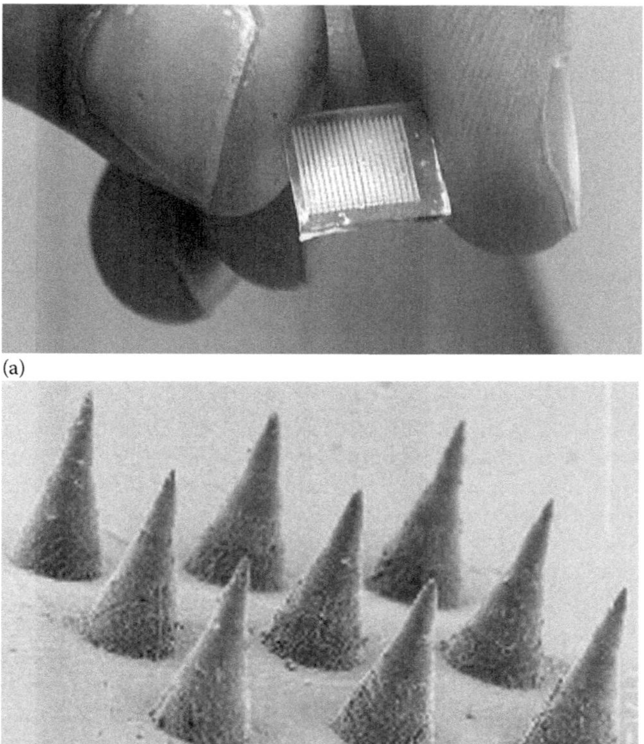

(a)

(b)

FIGURE 3.7
(a) Microneedle for transdermal drug delivery (Source: www.britishgeriatrisociety.com). (b) Microscopic view of microneedles.

3.6.2 Feedback-Controlled Transdermal Drug Delivery

In feedback-controlled transdermal drug delivery, the release of the drug molecule from the transdermal system is facilitated by an agent that triggers their release, such as biochemical in the body, which is also regulated by its concentration through some of the following feedback mechanisms:

a. Bioerosion-regulated drug delivery system
b. Bioresponsive drug delivery system
c. Self-regulated drug delivery system

Recently, a graphene-based wearable electronic patch has been developed for diabetes control. In this system, glucose was monitored in sweat by enzyme-based glucose sensors non-invasively. The glucose concentration in sweat can be well correlated with that of blood glucose level. The glucose concentration provides feedback-controlled drug delivery via heating responsive coated microneedles containing antidiabetic agents like metformin, insulin, and coated with heat-responsive polymers. Hyperglycemia (high concentration of glucose) induces the thermal actuation by embedded microheaters and lead to melt the coating layer which activates the microneedles to release the antidiabetic agents (Lee et al. 2017).

References

Abd, E., Yousef, S.A., Pastore, M.N. et al. 2016. Skin models for the testing of transdermal drugs. *Clin. Pharmacol. Adv. Appl.* 8: 163–176.

Anderson, R.L. and Cassidy, J.M. 1973. Variation in physical dimensions and chemical composition of human stratum corneum. *J. Invest. Dermatol.* 61: 30–32.

Badkar, A.V. and Banga, A.K. 2002. Electrically enhanced transdermal delivery of a macromolecule. *J. Pharm. Pharmacol.* 54: 907–912.

Banga, A.K. 1998. *Electrically Assisted Transdermal and Topical Drug Delivery*. London: Taylor and Francis, pp. 13–28.

Banga, A.K., Bose, S., and Ghosh, T.K. 1999. Iontophoresis and electroporation: Comparisons and contrasts. *Int. J. Pharm.* 179: 1–19.

Bartosova, L. and Bajgar, J. 2012. Transdermal Drug Delivery *In Vitro* Using Diffusion Cells. *Current Medicinal Chemistry*. 19: 4671–4677.

Chang, S.L., Hofmann, G.A., and Zhang, L. et al. 2000. The effect of electroporation on iontophoretic transdermal delivery of calcium regulating hormones. *J. Control. Rel.* 66: 127–133.

Dabrowska, A.K., Rotaru, G.M., Derler, S. et al. 2016. Materials used to simulate physical properties of human skin. *Skin Res Technol.* 22(1): 3–14.

Davies, D.J., Ward, R.J., and Heylings, J.R. 2004. Multi-species assessment of electrical resistance as a skin integrity marker for *in vitro* percutaneous absorption studies. *Toxicol. In Vitro.* 18(3): 351–358.

Dayan, N. 2005. Pathways for skin penetration. *Cosmetic Toiletries*. 120(6): 67–76.
Denet, A.R., Vanbever, R., and Préat, V. 2004. Skin electroporation for topical and transdermal delivery. *Adv. Drug. Del. Rev.* 56: 659–674.
Dragicevic, N., Atkinson, J.P., and Maibach, H.I. 2015. Chemical Penetration Enhancers: Classification and Mode of Action. In: Dragicevic N., Maibach H. (eds.) *Percutaneous Penetration Enhancers Chemical Methods in Penetration Enhancement*. Springer, Berlin, Heidelberg.
Draize, J.H., Woodard, G., and Calvery, H.O. 1944. Methods for the study of irritation and toxicity of substances applied to the skin and mucous membranes, *J. Pharmacol. Exp. Therap.* 82: 377–390.
Elias, P.M. 1983. Epidermal lipids, barrier function and desquamation. *J. Invest. Dermatol.* 80: 44–49.
Escobar-Chávez, J.J., Díaz-Torres, R., Rodríguez-Cruz, I.M. et al. 2012. Nanocarriers for transdermal drug delivery. *Res. Report Trans. Drug Del. (Dovepress)* 1: 3–17.
Flaten, G.E., Palac Z., and Engesland, A. et al. 2015. In vitro skin models as a tool in optimization of drug formulation. *Eur J Pharm Sci.* 75: 10–24.
Flynn, G., Yalkowsky, S.H., and Roseman, T.J. 1974. Mass transport phenomena and models. *J. Pharm. Sci.* 63: 479–510.
Hatanaka, T., Yamamoto, S., Kamei, M., Kadhum, W.R., Todo, H. et al. 2017. *In Silico* Estimation of Skin Concentration of Dermally Metabolized Chemicals. *Int. J. Pharm. Sci. Dev. Res.* 3(1): 007–016.
Honeywell-Nguyen, P.L. and Bouwstra, J.A. 2005. Vesicles as a tool for transdermal and dermal delivery. *Drug Del. Tech.* 1: 67–74.
Hueber, F., Wepierre, J., and Schaefer, H. 1992. Role of Transepidermal and Transfollicular Routes in Percutaneous Absorption of Hydrocortisone and Testosterone: In vivo Study in the Hairless Rat. *Skin Pharmacol.* 5: 99–107.
Hull, W. 2002. Heat-enhanced transdermal drug delivery: A survey paper. *J. Appl. Res.* 2: 1–9.
Ita, K. 2016. Transdermal iontophoretic drug delivery; advances and challenges. *J. Drug Target* 24(5): 386–391.
Kalia, Y.N., Naik, A., Garrison, J., and Guy, R.H. 2004. Iontophoretic drug delivery. *Adv. Drug Del. Rev.* 56: 619–658.
Kanikkannan, N. 2002. Iontophoresis based transdermal delivery systems. *Biodrugs* 16: 339–347.
Kleck, G., Geleick, H., and Frey, J.R. 1977. Screening of fragrance materials for allergenicity in the guinea pig. I. Comparison of four testing methods. *J. Soc. Cosmet. Chem.* 28: 53–64.
Lademann, J., Jacobi, U., Surber, C. et al. 2009. The tape stripping procedure – evaluation of some critical parameters. *Eur. J. Pharm. Biopharm.* 72: 317–323.
Lee, H., Song, C., Seok, Y. et al. 2017. Wearable/disposable sweat-based glucose monitoring device with multistage transdermal drug delivery module. *Science Advances* 3(3): e160131.
Lee, W.R., Shen, S.C., Wang, K.H. et al. 2003. Lasers and microdermabrasion enhance and control topical delivery of vitamin C. *J. Invest. Dermatol.* 121: 1118–1125.
Menon, G.K. 2002. New insights into skin structure: Scratching the surface. *Adv. Drug Deliv. Rev.* 54: S3–17.
Mikszta, J.A., Britingham, J.M., and Alarcon, J. 2001. Applicator having abraded surface coated with substance to be applied. Patent (serial number *WO 01/89622 A1*).

Mikszta, J.A., Britingham, J.M., and Alarcon, J. et al. 2003. Topical delivery of vaccines. Patent (serial number *U.S. 6, 595, 947 B1*).

Mitragotri S. 2017. Sonophoresis: Ultrasound-Mediated Transdermal Drug Delivery. In: Dragicevic N., I. Maibach H. (eds.) *Percutaneous Penetration Enhancers Physical Methods in Penetration Enhancement*. Springer, Berlin, Heidelberg.

Murthy, S.N., Sammeta, S.M., and Bowers, C. 2010. Magnetophoresis for enhancing transdermal drug delivery: Mechanistic studies and patch design. *J. Control Release*. 148(2): 197–203.

Oliveira, G., Hadgraft, J., and Lane, M.E. 2012. The role of vehicle interactions on permeation of an active through model membranes and human skin. *Int. J. Cosmet Sci*. 34(6): 536–545.

Otberg, N., Richter, H., and Schaefer, H. 2004. Variations of hair follicle size and distribution in different body sites *J. Invest. Dermatol*. 122: 14–19.

Pellet, M., Raghavan, S.L., Hadgraft, J., and Davis, A.F. 2003. The application of supersaturated systems to percutaneous drug delivery. In: *Transdermal Drug Delivery*, eds. Guy, R. H., and Hadgraft, J., 305–326. New York: Marcel Dekker.

Polat, B.E., Hart, D., Langer, R., and Blankschtein, D. 2011. Ultrasound-mediated transdermal drug delivery: Mechanisms, scope, and emerging trends. *J. Control. Release* 152: 330–348.

Scheuplein, R.J. and Morgan, L.J. 1967. "Bound water" in a keratin membranes measured by microbalance technique. *Nature* 214(29): 456–458.

Scheuplein, R.J. and Blank, I.H. 1971. Permeability of the skin. *Physiol Rev*. 51: 702–747.

Shah, V.P., Elkins, J.S., Lam, S.Y. et al. 1989. Determination of *in vitro* drug release from hydrocortisone creams. *Int. J. Pharm*. 53: 53–39.

Shakeel, F. and Ramadan, W. 2010. Transdermal delivery of anticancer drug caffeine from water-in-oil nanoemulsions. *Colloids Surf B Biointerfaces* 75: 356–362.

Singh, S., Mahendra, K., Srivastava, A.K. et al. 2008. Effect of Permeation Enhancers on the Iontophoretic Transport of Timolol Maleate through Rat Skin. *Acta Pharmaceutica* 49(3): 235–252.

Smith, N.B., Lee, S., Maione, E. et al. 2003. Ultrasound mediated transdermal transport of insulin in vitro through human skin using novel transducer designs. *Ultrasound. Med. Biol*. 29: 311–317.

Trotta M. 2004. Deformable liposomes for dermal administration of methotrexate. *Int. J. Pharm*. 270(1–2): 119–125.

Tsai, J.C., Guy, R.H., Thornfeldt, C.R. et al. 1996. Metabolic approaches to enhance transdermal drug delivery. 1. Effect of lipid synthesis inhibitors. *J. Pharm. Sci*. 85: 643–648.

Van Heerden, J., Breytenbach, J.C., N'Da, D.D., du Plessis, J., Breytenbach, W.J., and du Preez, J.L. 2010. Synthesis and in vitro transdermal penetration of methoxypoly(ethylene glycol) carbonate and carbamate derivatives of lamivudine (3TC). *Med. Chem*. 6: 91–99.

Vanbever, R., Langers, G., Montmayeur, S., and Préat, V. 1998. Transdermal delivery of fentanyl: Rapid onset of analgesia using skin electroporation. *J. Control. Rel*. 50: 225–235.

Weaver, J.C., Vaughan, T.E., and Chizmadzhev, Y.A. 1999. Theory of electrical creation of aqueous pathways across skin transport barriers. *Adv. Drug. Del. Rev*. 35: 21–39.

Weigan, D.A. and Gaylor, J.R. 1974. Irritant reaction in Negro and Caucasian skin. *South. Med. J.* 67: 548–551.

Williams, A.C. and Barry, B.W. 2004. Penetration enhancers. *Adv. Drug Del. Rev.* 56: 603–18.

Williams, R.L., Thakker, K.M., and John, V. et al. 1991. Nitroglycerin Absorption from Transdermal Systems: Formulation Effects and Metabolite Concentrations. *Pharm. Res.* 8: 744. https://doi.org/10.1023/A:1015802101272.

Zhang, J., Sun, M., Fan, A., Wang, Z., and Zhao, Y. 2013. The effect of solute-membrane interaction on solute permeation under supersaturated conditions. *Int. J. Pharm.* 441(1–2): 389–394.

Zhang, L., Nolan, E., Kreitschitz, S., and Rabussay, D.P. 2002. Enhanced delivery of naked DNA to the skin by non-invasive in vivo electroporation. *Biochim. Biophys. Acta* 1572: 1–9.

4

Nasal and Pulmonary Drug Delivery Systems

Pranav Ponkshe, Ruchi Amit Thakkar, Tarul Mulay, Rohit Joshi, Ankit Javia, Jitendra Amrutiya, and Mahavir Chougule

CONTENTS

4.1 Introduction

The delivery of drugs through the pulmonary route has been explored over a decade due to advantages such as a large absorptive surface area, thin mucosal lung lining, extensive vascularization, low proteolytic activity, and highly perfused lung tissue, which helps us achieve local and systemic delivery (Labiris and Dolovich 2003). Inhalation therapy for pulmonary delivery is gaining increasing attention for various lung diseases such as lung cancer, chronic obstructive pulmonary disorder (COPD), asthma, chronic bronchitis, etc. (Muralidharan, Hayes et al. 2015). New technological developments in delivery devices have led to the efficient delivery of large doses of the drug to the lungs with deposition at the desired site. This advancement has made the pulmonary drug delivery system more efficacious with respect to its dosing regimen while making it patient compliant (Rau 2005).

The nasal route has received considerable attention for local and systemic delivery due to its convenience, patient compliance, and reliability not only for local drugs but also for systemic activity (Chien and Chang 1987; Harris 1993). The high density of blood vessels in the nasal cavity offers excellent drug absorptivity and the rapid onset of action (Illum 2002; Graff and Pollack 2002; Rhidian and Greatorex 2015). In recent years, a wide range of nasal products have been developed and/or are in development, mainly aimed for the rapid onset of action for various diseases like migraines, pain, memory diseases, etc. Nowadays, vaccines are also being developed and delivered via the nasal route in the treatment of respiratory tract infections like influenza (Illum 2003; Türker, Onur et al. 2004; Vyas, Shahiwala et al. 2005). Due to this wide range of applicability and potential advantages, this route has emerged as the novel route for drug delivery.

The ability of nanoparticulate systems to display enhanced cellular uptake, prolonged drug release, and the capability to overcome limitations of active pharmaceutical ingredients has awakened the interest of scientists (Wilczewska, Nienirowicz et al. 2012). Additionally, the proficiency of nanoparticles to target and deliver protein and genetic material to specific cells has made it the locus of attraction (Bivas-Benita, Romeijn et al. 2004). Nanoparticulate systems, and methods for the preparation of various drug delivery systems such as spray drying, solvent preparation, etc. are described in Chapter 4 to bolster its credibility. Along with the aforementioned part, Chapter 4 also includes few examples pertaining to in-vitro, in-vivo, and *ex vivo* models used to determine the efficacy of the nanoparticles. Chapter 4 touches upon certain aspects such as lung and nasal clearance mechanisms, alveolar macrophage clearance, and mucocilliary clearance; followed by formulation strategies and devices used for drug delivery.

4.2 Anatomy and Physiology of the Respiratory Tract

The respiratory system and circulatory system work together simultaneously to deliver oxygen and remove carbon dioxide from cells, which is exhaled through the lungs. Respiration is also known as the transport of oxygen from the outer environment to blood and body tissue while the movement of carbon dioxide is in the opposite direction. Each minute, 1 pint of air is inhaled by healthy lungs for about 12–15 times. The total amount of blood present in the human body is passed through the lungs every single minute (Rang, Dale et al. 2007). The gas exchange and supply of oxygen to all the cells are the predominant functions of the lungs. The respiratory tract is bifurcated into two main parts; the first part is the upper respiratory tract which consists of the nose, nasal cavity, and pharynx; while the other part which also known as lower respiratory tract, which consists of the larynx, trachea, bronchi, and lungs (Figure 4.1). The right lung consists of three lobes while the left lung consists of two lobes. The intramural part of the lung is composed of smaller air passages and bronchi, blood vessels, alveoli, and lymph tissue. The division of bronchi takes place into primary and secondary bronchi and bronchioles and eventually into alveoli (which amounts to over 300 million). There is a conducting zone in the respiratory tract that consists of nose, pharynx, larynx, trachea, bronchi, bronchioles, and terminal bronchioles. The general function of this zone is to filter, warm, and moisten the air and conduct it into the lungs. The lungs have a total surface area of 70 m^2. A vast network of 280 billion pulmonary capillaries is formed in alveolus which provides a total surface area is around 50–75 m^2 for the blood–gas barrier.

The lung consists of around 40 cell types. The alveolar epithelium, endothelium, and interstitial cell layer are present at the interface where the alveolar

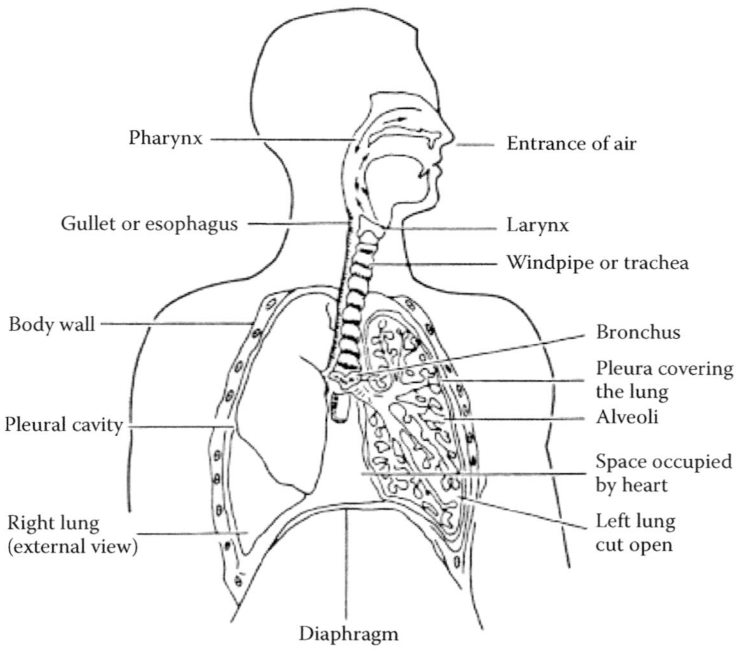

FIGURE 4.1
Major respiratory structure and its anatomy. The respiratory system starts at the nasal cavity and culminates in the diaphragm. The functional unit of the lung is alveoli.

gas exchange primarily transpires. Pneumonocytes (alveolar epithelial cells, type I and type II) together form the analveolar wall. The gas exchange is the most important function of type I cells, while the production and secretion of lung surfactant is the function of type II cells. There is 0.5-μm distance between the air within the alveoli and capillaries, which leads to the gas exchange via diffusion due to a partial pressure gradient at the interface. Along with mucus that is coated onto the alveoli, phospholipids and surface protein form analveolar fluid. Pulmonary surfactants lower the surface tension and are vital for operating the proper gas exchange. Pulmonary surfactants are a mixture of 90% lipids and 10% protein secreted by type II alveolar cells, whose primary role is to decrease the surface tension formed at the air/liquid interface within the alveoli of the lung. The additional functions of these surfactants include elevating the pulmonary compliance, facilitating and recruiting collapsed airways, and preventing the collapse of the lung at the end of expiration. The surfactants present in lungs are dipalmitoylphosphatidylcholine (DPPC), phosphatidylglycerol (PG), lysophosphatidic acid (PA), phosphatidylethanolamine (PE), phosphatidylinositol (PI), and cholesterol. Apart from surfactants, two groups of proteins are present in pulmonary surfactants. Surfactant proteins A, B, C, and D are the designated proteins, of which surfactant protein B and surfactant protein C are small hydrophobic proteins, while surfactant protein A and surfactant protein D are large hydrophilic proteins (Patton 1996).

Surfactant protein A was the first protein lung surfactant to be discovered. It possesses a short, N-terminal, proline rich collagen like domain, a neck region, and carbohydrate recognition domain. Functions of this protein include tubular myelin formation, the enhancement of surfactant protein B surface activity and regulation of uptake and secretion of surfactants by type II alveolar cells.

Surfactant protein B and surfactant protein C are hydrophobic, and show more interaction with lipids. They provide the promotion of lipid adsorption to the air/liquid interface, the formation of tubular myelin, respreading of films from collapse phase, reuptake of surfactants by type II cells, stabilization of monolayer lipid films, membrane fusion, and lysis. A deficiency of protein B may lead to lethal consequences.

During breathing, there is continuous exposure to materials of myriad sizes and sources, such as bacteria (0.2–200 μm), tobacco smoke (0.01–1 μm), and pollen (20–90 μm). These particles tend to deposit along the respiratory tract from conducting airways down to the lower region of the respiratory tract. Particles are cleared rapidly by cilia in the upper airway, within the mucous layer that lines to the throat via the epithelia in the upper airways, while getting swallowed and metabolized. A barrier to absorption is posed by the pulmonary epithelium which is thick (50–60 μm) in the trachea. The epithelium diminishes to a thickness of 0.2 μm in the alveoli as we move towards the lower airways. In the lower airways, a gas exchange occurs. There is a vast surface area of alveoli approximately 43–102 m^2 (in an adult) which provides access to the entire systemic circulation along with highly vascularized expanse (Gehr, Bachofen et al. 1978). Alveolar macrophages also are known as cells of the immune system that protect alveoli by scavenging foreign materials along the surface of lungs. However, there is a possibility of particles that are too small or too large to escape phagocytosis. Dendritic cells are also present throughout the airways where they evaluate pathogens and foreign substances.

The lungs provide an interesting approach for non-invasive drug delivery while proving to be beneficial for both systemic and local administration. Pulmonary drug delivery is advantageous against other methods such as injections, or oral delivery, as the lungs have a large surface area for absorption, limited proteolytic activity, high solute permeability, and the delivery systems are non-invasive. Targeting the alveolar region of the lung can provide systemic drug delivery as the drugs get absorbed via the thin layer of epithelial cells and into the systemic circulation (Labiris and Dolovich 2003). This provides an avoidance of the first-pass metabolism, rapid onset of action, and the delivery of biotherapeutics (i.e., proteins and peptides) which is difficult to deliver orally (owing to poor intestinal membrane permeability and enzymatic degradation). Along with aforementioned cells, alveoli are also comprised of fibroblasts, nerves, as well as lymph vessels, which makes it an ideal location for delivery of drugs, which require access to the pulmonary system as well as the lymphatic system (Patton 1996; Rang, Dale et al. 2007).

4.3 Anatomy and Physiology of the Nasal Mucosa

The nasal cavity (from the nose to thepharynx) is entirely lined with nasal mucosa and forms the physical barrier of the body's immune system. It provides mechanical protection against pathogens and harmful substances. The human nasal cavity is divided by the septum into two symmetrical parts. Each of the parts made up of three regions: (a) the vestibule region (inside the nostrils, area of 0.6 cm^2) (b); olfactory region, situated at top of the nasal mucosa (covers only 10% of the total nasal mucosal area of 150 cm^2); and (c) respiratory region, made up of 3 nasal turbinates (superior, middle, and inferior) (Illum 2003; Illum 2012) (Figure 4.2).

About a third of the initial nasal cavity is lined by stratified squamous epithelium, which does not contain ciliated cells. The posterior part consists of pseudostratified columnar ciliated epithelium and mucus-producing goblet cells over a basement membrane (Rang, Dale et al. 2007). About 20% of total cells in the lower turbinate portion is ciliated on the apical surface with fine projections (100 per cell) and of 60–70% of the mucosa is lined with non-ciliated cells which may be responsible for fluid transport and metabolic activity (Baroody 2007).

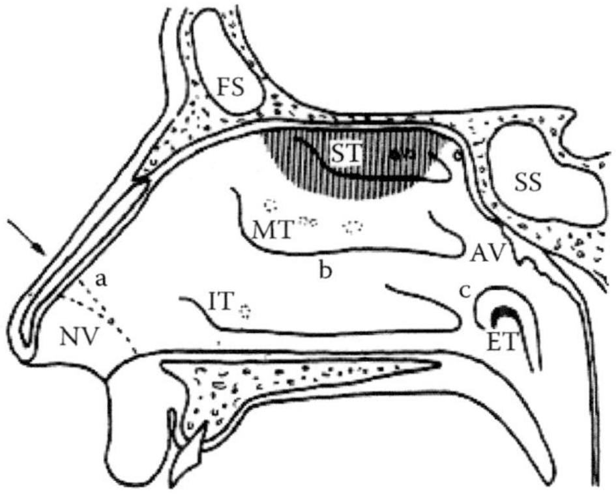

FIGURE 4.2
Lateral wall of the nasal cavity, and cross-sections through (a) the internal ostium, (b) the middle of the nasal cavity, and (c) the choanae. The hatched area in the upper figure is the olfactory region; NV, nasal vestibule; IT, inferior turbinate and orifice of the nasolacrimal duct; MT, middle turbinate and orifices of frontal sinus, anterior ethmoidal sinuses and maxillary sinus; ST, superior turbinate and orifices of posterior ethmoidal sinuses; FS, frontal sinus; SS, sphenoidal sinus; AV, adenoid vegetations; ET, orifice of eustachian tube. (Reprinted from Mygind N, Dahl R. Anatomy, physiology and function of the nasal cavities in health and disease. *Advanced Drug Delivery Reviews* 1998; 29(1): 3–12. Copyright 1998, with permission from Elsevier.)

4.4 Challenges and Desired Target Site for Drug Deposition

4.4.1 Challenges and Desired Target Site for Drug Deposition from Nasal Delivery

The drug transport across the nasal mucosa is assumed to be dependent on the transport mechanisms, i.e., paracellular and transcellular diffusion and carrier-mediated transcytosis (Illum 2003). However, factors that may affect the drug absorption/permeation across the nasal cavity include anatomy and physiologic conditions (mucociliary clearance, enzymatic activity, characteristic of drug molecules, and formulation considerations), which are described in the following section.

4.4.1.1 Mucociliary Clearance

Nasal mucociliary clearance (MCC), a function of the upper respiratory tract (UTI) prevents the access of noxious stimuli, i.e., bacteria, viruses, allergens, etc. to the lungs by means of eventual discharge of these substances in the gastrointestinal tract, via the nasopharynx. It may work as the harmonized interface between the metachronal rhythm of underlying cilia and the mucous layer (Jadhav, Gambhire et al. 2007). Considering the interindividual variability in the MCC rate, it has been projected at 5 mm/min (Schipper, Verhoef et al. 1991). However, several factors like temperature (above or below 23°C), cigarette smoking, air pollutants (sulfurdioxide) and pathological conditions, i.e., asthma, rhinitis, sinusitis, common cold, allergy, etc. may affect the ciliary beat frequency (CBF) and mucus production resulting in an increase in the MCC rate. In a nutshell, both the environmental and pathological conditions ultimately alter the performance of the nasal formulations, which should be taken into consideration during the product development stage.

Generally, the quantity of drug that permeates through the nasal mucosa depends upon the contact time in which the drug interacts with the absorbing tissues (Türker, Onur et al. 2004). Formulations are usually administered through the human respiratory epithelium, and can be cleared from the nasal cavity with a half-life of 15–20 min (Illum 2002; Djupesland 2013). Therefore, the efforts have been made to increase the residence time of the formulations in the nasal cavity, which may ultimately enhance the bioavailability of drugs when administered nasally. This includes the use of mucoadhesive polymers like chitosan, carbomers, hydroxypropyl methylcellulose (HPMC), hydroxypropyl cellulose (HPC), etc., which have been used widely.

4.4.1.2 Enzymatic Activity

Although drugs administered to the nasal cavity are assumed to bypass the hepatic first-pass metabolism, the bioavailability of peptide/protein drugs can be limited owing to the presence of the metabolic enzymes such as carboxylesterase, aldehyde dehydrogenase, epoxy hydroxylase, UDP-glucuronyltransferase, cytochrome P450, etc. in the nasal cavity or in passage across the layer of epithelial cells (Hussain and Aungst 1994; Chung and Donovan 1996). The exopeptidases (mono-aminopeptidases and di-aminopeptidases) degrade the peptide at its N and C terminals while the endopeptidases (cystine, serine, aspartic proteinases, etc.) attacks the interior peptide bonds thus resulting in a low bioavailability of peptides. Moreover, cytochrome P450 can catalyze the NADPH-dependent mono-oxygenation of xenobiotics or endogenous substrates (Sarkar 1992). Peptides such as estradiol (Brittebo 1985), progesterone (Brittebo 1982), testosterone (Lupo, Lodi et al. 1986), octreotide (Kissel, Drewe et al. 1992), leuprolide (Adjei, Sundberg et al. 1992), phenacetin (Brittebo 1987), etc. have been shown to degrade via this mechanism when given via the nasal cavity. Also, also the high molecular weight of these peptides may affect their bioavailability. Enzyme inhibitors such as boroleucin, betastatin, borovaline, amastatin, puromycin, peptidase/protease inhibitors (nafamostat mesylate), and absorption enhancers (dimethyl β-cyclodextrin, bile salt, glycocholate, etc.) are some of the few approaches to overcome the degradation of peptides/proteins in the nasal cavity (Costantino, Illum et al. 2007).

4.4.1.3 Characteristic of the Drug Molecules

The physicochemical characteristics of the drug molecules such as molecular weight (MW), hydrophobicity/lipophilicity, partition coefficient, solubility, particle size, pKa, etc. (Behl, Pimplaskar et al. 1998) may affect the absorption and transport of drug molecules across the nasal mucosa. Moreover, the selection and/or development of a suitable nasal drug delivery system also depends upon characteristics of the drug molecules.

Several reports have suggested that molecular weight has an inverse relationship with the absorption of the drug molecules, i.e., drug molecules with molecular weight <300 Da in solution get absorbed faster and more efficiently across the nasal cavity compared to a significant reduction observed in the absorption of lipophilic drugs greater than of 1000 Da MW (Bahadur and Pathak 2012). Nevertheless, few drug molecules with a MW of >1000 D, such as salmon acetate, cyanocobalamin, desmopressin acetate, etc., are approved by the FDA in the form of nasal solutions and sprays that also show low bioavailability (10%). However, the use of absorption enhancers, surfactants, cyclodextrins, tight-junction modulators, etc. may result in better bioavailability than the drugs with a MW of >1000 Da. (Kumar, Pathak et al. 2009). Hydrophobicity/lipophilicity also has an important consideration in the transport of drug molecules via the nasal cavity. Generally, increasing the lipophilicity of drug molecules may often lead to increasing in permeation, e.g., propranolol, progesterone, testosterone, naloxone, pentazocine, etc. (Illum 2003; Costantino, Illum et al. 2007). Lipophilic drugs have a tendency of crossing biological membranes and generally permeate via the transcellular route, whereas hydrophilic drugs, such as metoprolol, permeate through the nasal mucosa via the paracellular route (Illum 2000). Moreover, a linear relationship was found between partition coefficient of the absorption rate of drug molecules, e.g., sulfanilic acid, sulfamethizole, sulfisoxazole, and sulfisomide, with an increasing order of partition coefficient of (0.012, 0.25, 0.26, 0.892, respectively) across the nasal mucosa (Behl, Pimplaskar et al. 1998). The use of cosolvents (the salt form of drugs), prodrugs, and complexation with cyclodextrins are the approaches to overcome the limited drug absorption across the nasal mucosa.

4.4.2 Challenges and Desired Target Site for Drug Deposition for Pulmonary Delivery

Although the lungs can be the ideal site for drug delivery, there are numerous challenges and obstacles which hamper the successful translation. The major challenges are mentioned in the following sections.

4.4.2.1 Mucociliary Clearance

Mucociliary clearance is the main clearance mechanism and a primary innate defense against the harmful effects of allergens, pollutants, and pathogens deposited in the conducting airways.

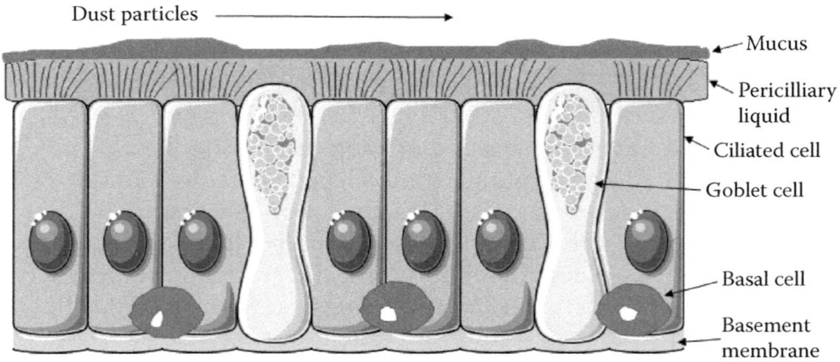

FIGURE 4.3
Cells involved in mucociliary clearance. The bronchial wall epithelium provides support. Mucus is secreted by a goblet and the mucous cells, together with cilia, clear the dust particles.

The mucociliary clearance involves three compartments viz. the cilia, protective mucus layer, and airway surface liquid (ASL) layer (Figure 4.3). All three components work together to expel or remove the inhaled particles from the lungs. Mucociliary clearance works by providing a mechanical and chemical barrier. It traps the particles in the ASL and removes them by ciliary action. ASL also prevents adherence and circulation of these particulates in the airway epithelium by providing a biological barrier (Mall 2008). Antioxidative properties of the mucus create a chemical barrier for the particulates (Wanner, Salathé et al. 1996). It is experienced by those particles, which are deposited in the tracheobroncheolar tree region. In mucus clearance, the second layer (the periciliary layer) provides liquid for clearance from the lungs towards the mouth. Particles more than 6 µm are generally expelled out by coughing or swallowing due to the mucociliary clearance. In the case of lung disease, this mucociliary balance is disturbed, and it hastens the clearance of inhaled particles, as well as reduces their residential time and effectiveness of the drugs (El-Sherbiny, El-Baz et al. 2015).

The interaction between mucus and cilia is an important factor in deciding the rate of mucociliary clearance. The three layers of mucus, namely the upper gel layer floating on middle aqueous periciliary layer, are separated by a layer of surfactants, which alter the pharmacokinetic properties of the drug. Every day, almost 20–40 mL of mucus is secreted and transported to the pharynx region, which removes the particles ranging in between size range of 4.0–12.5 µm. The average mucociliary clearance rate for an adult is 7–15 mins, which increases in case of rhinitis and allergies (Gizurarson 2015).

4.4.2.2 Macrophage Uptake and Alveolar Clearance

The alveolar macrophage clearance mechanism is seen in the areas of deep lung, wherein each alveolus is cleared of foreign particles by at least

12–14 alveolar macrophages. Alveolar macrophages are the primary clearance mechanism for insoluble particles deposited in the alveolar region (El-Sherbiny, El-Baz et al. 2015). They clear the deposited particles in the range of 1.5–3.0 μm, either by translocating to the mucociliary clearance towards the pharynx region or engulfed into the lymph nodes. The steps involved in size-dependent macrophage clearance pathways are the recognition of particles by macrophages, and the attachment, internalization, and movement of macrophages along with particles.

Macrophages sometimes transport across the epithelium with the engulfed particles and are ultimately deposited into the thoracic lymph nodes. This mechanism can alter the half-life or residence time of drug which can influence the pharmacokinetics and the efficacy of the particles.

4.5 Approaches to Overcome Challenges in Drug Delivery

4.5.1 Approaches to Overcome Challenges in Nasal Delivery

The nasal route is a very promising route to deliver drugs locally and to the brain. However nasal drug delivery is associated with several challenges and obstacles which need to be overcome for successful nasal delivery. To circumvent the challenges discussed above, several approaches and techniques have been sought in the research area. Very extensive research work has been carried out by various research groups in these areas and various approaches have been sought like permeation enhancers and natural and synthetic mucoadhesive polymers to make mucoadhesive-based drug retention in the nasal cavity for good absorption.

4.5.1.1 Absorption Promoters/Absorption Modulators

The nasal cavity is widely used as the local delivery of the drugs for the treatment of mostly upper respiratory diseases like nasal decongestion, allergic conditions, nasal infections, etc. Nevertheless, the nasal cavity is broadly exploited for systemic drug delivery for the treatment of migraines, pain, and certain other conditions as well, and several marketed products are also available. Drug molecules in these marketed products are adequately lyophilic, so they are easily available in systemic circulation at its therapeutic level, even though they don't contain absorption promoters or modulators. However, more hydrophilic molecules and small drug molecules have low membrane permeability, so they're in need of promoters or modulators to transport the adequate amount of the drug molecules across the nasal membrane. Studies have reviewed the commercially developed absorption promoters and modulators including cyclopentadecalactone, alkyl saccharides, chitosan polyglycol mono, and diesters of 12-hydroxystearate (70%), polyethylene glycol (30%), etc. (Illum 2012).

4.5.1.2 Nanotechnology-Based Carriers

In the recent years, nanotechnology-based drug delivery has gained popularity to overcome challenges like limited absorption, toxicity, and certain physiochemical properties to make drug delivery systems the most therapeutically efficacious. The nasal delivery of drugs, when associated with promising nanotechnology-based carriers, widen the scope of the systemic drug delivery and drug delivery to the brain. The systemic delivery of therapeutics to the brain to treat several diseases like migraines, schizophrenia, depression, etc. are limited by the delivery across the blood–brain barrier (BBB). The development of nanotechnology-based carriers' nasal delivery systems circumvent the barrier through the olfactory region, penetrate the brain, and opens the horizon for the understanding and possibilities of future therapies for disorders of the central nervous system (CNS). The following sections will describe the different nanocarriers, e.g., liposomes, polymeric/solid lipid nanoparticles, microemulsion, nanoemulsion, microspheres, etc.

4.5.1.2.1 Liposomes

For many years, liposomes have been demonstrated as promising potential carriers for drug delivery owing to their salient features of biocompatibility, vesicular size, loading of hydrophobic and hydrophilic drug molecules, and safety. In a view to nasal delivery, liposomes have also been extensively studied by several research groups for topical delivery, targeted delivery, and systemic delivery to the brain. Corace et al. (2014) have developed tacrine-loaded multifunctional intranasal liposomes for the treatment of Alzheimer's disease. Their results show enhanced permeation through the nasal mucosa and neuroprotective activity with the addition of α-tocopherol (Corace, Angeloni et al. 2014). Liposomes are also known to demonstrate the sustained release of the drug by entrapping the therapeutics in the bilayer or in the aqueous compartment.

In the recent times, for the delivery of the biomacromolecules like proteins, peptides, and vaccine as well, intranasal delivery has been demonstrated as promising by overcoming these challenges. Novel concepts have been evaluated to overcome the challenge of the low encapsulation efficiency of liposomes. A combination of liposomes and biodegradable polymer polyacrylic acid based in situ gelling system has been reported for mucosal immunization by intranasal delivery of the hepatitis B protein antigen. This work has been demonstrated that the protein antigens are encapsulated in the liposomes successfully and liposome *in situ* gelling systems induced a considerably high immune response to protein antigens confirmed by the immunization study in-vivo (Tiwari, Goyal et al. 2009).

In addition, poor bioavailability of the lipophilic drugs and avoidance of the first-pass metabolism and permeation challenge through the blood–brain barrier can be overcome by the combining the drug encapsulation into the liposomes and delivery via the intranasal route to address two challenges mentioned later. Upadhyay et al. have demonstrated the loading of the quetiapine fumarate, used to treat schizophrenic conditions, into the lipid bilayer vesicles and delivered intranasally to the brain to eliminate the problem of first-pass metabolism and the blood–brain barrier (Upadhyay, Trivedi et al. 2017).

4.5.1.2.2 Nanoparticles

Over the decades, nanoparticles have been explored for local drug delivery to the nose and nasal-to-brain delivery for various brain disorder treatments. The smaller size and tunable properties of the nanoparticles offer extra advantages, as drug delivery carriers overcome the nasal mucosa and permeation issues to deliver the drug at the site.

The utilization of mucoadhesive polymers like chitosan, cellulose derivatives, and carbopol in the nanoparticles are also promising carriers for intranasal drug delivery. Several research groups have extensively exploited chitosan in mucoadhesive nanoparticle formulation. Recently, *ex vivo* and in-vivo studies of tapentadol HCl mucoadhesive chitosan nanoparticles for pain management have demonstrated very promising results for brain delivery via the nasal route. Higher permeation through nasal mucosa is due to the ability of chitosan to open the tight epithelial cell junctions which results in higher drug biodistribution in the brain through olfactory region delivery (Javia and Thakkar 2017). Recently, several techniques have been investigated to improve the uptake of antigens in the mucosa and improve the immunogenicity of the vaccines. To deliver vaccines, polymeric nanocarriers have attracted the attention of researchers. Hepatitis B surface antigen–encapsulated PLGA and chitosan and glycol chitosan–coated PLGA nanoparticles for vaccine delivery has been developed. Immunogenicity studies performed in-vivo demonstrated that glycol chitosan–coated PLGA nanoparticles provoke elevated mucosal and systemic immune response and could be promising candidates as carriers for nasal vaccine delivery (Pawar, Mangal et al. 2013). Further, PLGA nanoparticles have been capably delivering the drugs such as olanzapine (Seju, Kumar et al. 2011), diazepam (Sharma, Sharma et al. 2015), and other many therapeutic agents to the brain via the intranasal route, which are efficient enough to treat CNS disorders.

4.5.1.3 Prodrug Approach and Structural Modifications

Generally, the prodrug approach is intended to modify the drug molecules for the favorable physicochemical and biological characteristics for better absorption. Various linkages on the drug moiety or modulation in the lipophilicity or hydrophilicity of the drug moiety can be utilized to target certain receptors or transporters. The altered chemistry and affinity of the prodrugs may lead to improved absorption. Limited solubility of the drug molecule can be challenging for nasal delivery; therefore, to improve the aqueous solubility, forming an aqueous soluble derivative may be the one of the approaches for enhancing drug diffusion via nasal mucosa delivery, utilizing the concentration gradient of the diffusion process. Several papers have been published involving such water-soluble prodrug approaches which include. Phosphocholine-linked prodrugs, succinate, and polyethylene glycol polymer conjugate, phosphate alkali metal salt of the drug moiety (Pezron, Mitra et al. 2002). Furthermore, transporter-specific and tissue-specific targeting prodrug approaches are the recent trend in the nasal drug delivery. Nasal absorption of relatively small, impermeable drug molecules can be enhanced by amino acid prodrug techniques. Different acyclovir prodrugs using L-aspartate beta-ester, L-lysyl, and L-phenylalanyl esters and *in situ* perfusion studies showed that L-aspartate beta-ester of acyclovir significantly enhanced the nasal absorption compared to other ester prodrugs, and suggested that this nasal uptake route involve an active transport (Yang, Gao et al. 2001). The derivatization of prodrugs has the great potential to enhance drug stability and target the transporters and membrane enzymes. Future strategies of prodrugs in nasal delivery may be targeted to achieve improved nasal-to-brain delivery exploring the transporter and enzyme-specific prodrug approaches (Pezron, Mitra et al. 2002).

4.5.2 Approaches to Overcome Challenges in Pulmonary Drug Delivery

Pulmonary drug delivery is quite complicated and challenging. Most of the solutions to overcome challenges in pulmonary drug delivery have been to focus on developing technology which will enable the efficient deposition of drug particles in the deep lung. Discussed below are some of the challenges and the potential ways to overcome obstacles.

4.5.2.1 Aerodynamic Particle Size

Aerodynamic particle size is one of the most critical attributes in determining the deposition mechanism of the particle and also in the deposition and distribution of the drug dose (Dolovich 1993). Large aerosol particles deposit a higher drug-per-unit surface compared to fine aerosol particles. Large aerosol particles also deposit drugs on the central airways, whereas fine particles deposit drugs on the peripheral airways (Ruffin, Dolovich et al. 1978). The aerodynamic diameter can be explained as the diameter of a sphere which has a unit density and which, in comparison to a nonspherical particle of random density, deposits with the same settling velocity. A small aerodynamic diameter can be obtained by reducing the particle size as well as the density, which can be achieved by spray drying or spray freeze drying (Edwards, Langer et al. 2003; Platz, Patton et al. 2003; Shekunov, Chattopadhyay et al. 2004; Shekunov, Chattopadhyay et al. 2007).

For pulmonary drug delivery in the peripheral lung area, the particle should have a diameter of 1–5 μm. It is also found that >50% of the 3-μm diameter particles are deposited in the alveolar region (Labiris and Dolovich 2003).

4.5.2.2 Limiting the Lung Clearance Process

Various innate clearance mechanisms of the lungs, like alveolar clearance or macrophage uptake, are obstacles encountered in the pulmonary drug delivery, which leads to the rapid removal of the drug from the lung. This results in increasing the frequency and level of dosing which, in turn, leads to an increase in unwanted side effects. The challenge is to increase the residence time of the drug particles in the lung. Some of the potential approaches to this problem include:

a. Modifying the physicochemical properties of drug that are chemically modified so that the dissolution process can be slowed. (For example, amphotericin.) Amphotericin can be formulated in polymeric micelles with block polymers or liposomes, which will enable its sustained-release pulmonary delivery (Lambros, Bourne et al. 1997; Kataoka, Harada et al. 2001).
b. Increasing the residence time of the formulation by the introduction of positively charged molecules. (For example, tobramycin.) (Patton and Byron 2007)
c. Synthesizing sustained-release formulation by encapsulating the active ingredients (Bailey and Berkland 2009).

4.6 Particle Deposition Mechanism and Factors Affecting Nasal and Pulmonary Drug Delivery

Suspensions comprised of liquids or solids in the air (gas) are known as aerosols. Aerosols contain a particulate portion denoted as particulate matter, or PM. These are chemically heterogeneous discrete solid particles or liquid droplets. The size of particulate matter in an aerosol can vary from 0.001 to greater than 100 microns (or micrometer 10^{-6} m) in diameter. Based on size, the three different types of particles are categorized for pulmonary administration (Labiris and Dolovich 2003):

- Coarse particles: size larger than 2 microns in diameter
- Fine particles: size in the range of 0.1 to 0.2 microns in diameter
- Ultrafine particles: size less than 0.1 micron

The respiratory deposition of particles inhaled varies in a different region of the respiratory system, which is very complex and depends on many factors. Mouth or nose breathing, breathing rate, respiration volume, lung volume, a constantly changing hydrodynamic flow field as an outcome of bifurcation in the airway, and the health of the individual are some of the factors that have an impact on respiratory deposition.

Depending on the airflow, particle size, and location in the respiratory system, particle deposition transpires via the following mechanisms (Cheng 2014).

4.6.1 Impaction

Suspended particles are inclined to travel along their original path due to inertia. Changes in the airflow due to a bifurcation in the airways may cause them to collide on an airway surface. The aerodynamic diameter particularly plays an important role in this mechanism, since the stopping distance is low for very small particles. Impaction for larger particles that are close to the airway wall mostly occurs near the first airway bifurcation. The deposition of particles led by an impaction mechanism tends to be deposited more on the bronchial surface. A majority of particle deposition on a mass basis is carried out by impaction.

Aerosol particles at a high velocity pass through the oropharynx and upper respiratory passages. Particles collide to the respiratory wall and tend to deposit at the oropharynx region owing to the centrifugal force, which is generally observed in a metered dose inhaler (MDI) and dry powder inhaler (DPI) with a particle size greater than 5 µm. In the case of dry powder inhaler, if the inhalation force is meager, the dry powder deposits on the upper airways due to the inertial force and mass of the particles, so the inspiratory effort of the patients is paramount.

4.6.2 Sedimentation

The settling out of particles in small airways of alveoli and bronchioles, where airway dimensions are small and air flow is low is known as sedimentation. Gravitational force plays an important role in this mechanism. Terminal settling velocity of particles is the dependent factor for the rate of sedimentation, for particles with larger aerodynamic diameters; this mechanism plays an important role in particle deposition. Particles that are hygroscopic tend to swell or grow in size as they pass through warm, humid air passages thereby increasing the likelihood of deposition by sedimentation. Particles with a size range between 1–5 µm are deposited in secondary bronchi and bronchioles.

4.6.3 Interception

An airway surface is exposed to particles due to its physical shape or size interception occurs. The deposition of particles through interception tends to avoid deviation from their air streamlines. When air streamlines are close to airway walls, there is a probability of interception. Fibers provide the most notable exposure to airway surface by interception mechanism; also due to its smaller aerodynamic diameter, it is easier for it to reach the smallest airway.

4.6.4 Diffusion

For particles less than 0.5 microns in diameter, the primary mechanism of deposition is diffusion. This mechanism is regulated by geometric rather than aerodynamic size. In diffusion, Brownian motion accounts for the transfer of particles from a region of high concentration to a region of low concentration. Diffusional deposition transpires when the particles have reached the tip of the nasopharynx and alveolar region where airflow is low.

4.7 Formulation Considerations

4.7.1 Nasal Formulation Considerations

Factors related to formulations that can affect the drug absorption via the nasal cavity need to be considered, such as the concentration and dose of the drug, the volume of administration, osmolarity, pH, and viscosity of the formulation, excipients (preservatives, permeation/absorption enhancers), etc. Also, absorption of the drug is dependant on the dosage form (spray, powder, drops), administration devices, and techniques (Ugwoke, Agu et al. 2005). A mucociliary clearance, lower residence time (15–20 min), allergic conditions, etc. are the major factors that may lead to low bioavailability/lesser absorption or permeation of drugs across the nasal mucosa. This can be overcome by adjusting pH (4.5–6.5), osmolarity (ideally of 285–310 mOsmol/l), use of permeation/absorption enhancers, and mucoadhesive agents.

4.7.2 Novel Pulmonary Drug Delivery Formulation Considerations

As mentioned earlier, the important properties of drugs (such as solubility, morphology, particle size, site of action, and stability) influence the formulation selection process. Nanoparticles have gained immense attention for pulmonary delivery. They have inherent advantages over conventional systems such as pressurized Metered Dose Inhalers (pMDI) and nebulizers, due to the advantages provided such as targeted deposition, bioadhesion, sustained release, and reduced dosing frequency for improving convenience to the patient. The increase in particle surface area due to decreased particle size is responsible for the increase in dissolution rate (Müller, Jacobs et al. 2001).

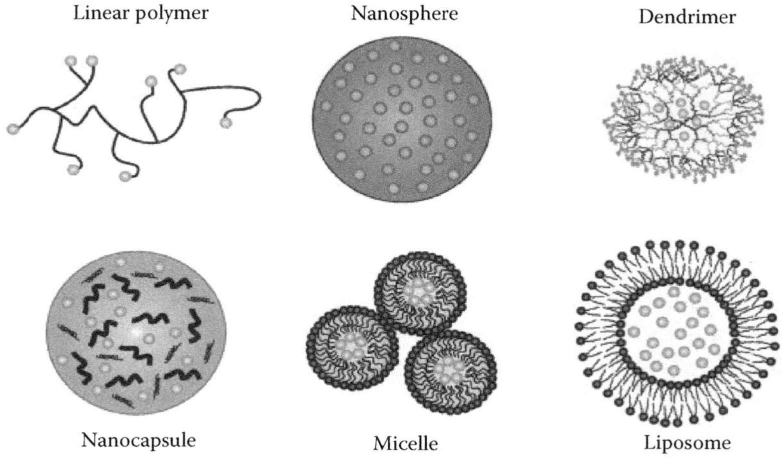

FIGURE 4.4
Types of nanocarriers used for Pulmonary delivery.

Nanoparticulate formulations show increased dissolution and saturation solubility, making it highly effective in terms of mass transfer to the surrounding environment (Müller, Jacobs et al. 2001). The efficacy of nanoparticulate systems is dependent on the excipients used. Currently, various excipients used to formulate different nanoparticulate systems are in different stages of regulatory approval, and the selection of delivery system and excipients would play a major role in decreasing the bench to clinic transition of developed systems. A few examples of different types of nanocarriers for pulmonary delivery are described shortly (Figure 4.4). In-depth reviews and methods are available elsewhere (Bailey and Berkland 2009; Jawahar and Reddy 2012).

4.7.2.1 Micelles

Micelles are the dispersed aggregates of surfactant molecules in their solution. Due to their characteristic of self-assembling into core and shell structure, they serve as a promising drug delivery system. Micelles have a sustained release effect, and they can circulate for a longer time in the system because of the hydrophilic shell, which decreases macrophage uptake. The size of 20–100 nm facilitates ready uptake by cells. Micelles in asthma and chronic obstructive pulmonary disease are found to cross the mucus and epithelial layer to reach the inflamed site. For example, poly-(ethylene oxide)-block-distearoyl phosphatidyl-ethanolamine (mPEG-DSPE) polymer based Beclomethasone dipropionate micelles have been reported with a higher drug solubility and a better Mass Median Aerodynamic Diameter (MMAD) (Gaber et al. 2006).

4.7.2.2 Liposomes

Liposomes are some of the most explored and considered drug delivery systems for the delivery of a wide range of therapeutic cargo in the form of controlled release. In addition to being biodegradable and biocompatible in nature, liposomes now have the advantage of being produced in the form of powders. This has proved to be of profound advantage in terms of delivery of the drug, as it avoids any leakage from the nebulizer while administering. Liposomes are amenable for all types of delivery devices, such as pMDI, soft mist inhalers, and dry powder inhalers for delivering does in small amounts, as well as medical nebulizers. These single- or bilayer-structured delivery systems are preferred because of the flexibility of their composition, size, materials used for preparation, and charge on liposomes. Cationic liposomes are generally used for gene delivery as they form compatible interaction with anionic oligonucleotides to hold them inside their structure. The rate and extent of pulmonary uptake of liposomes are a function of their composition; significantly faster rates occur when liposomes contain phosphatidylglycerol (Taylor and Newton 1992). A complete review of liposomes available for pulmonary delivery is beyond the scope of Chapter 4, but is available elsewhere (Elhissi 2017).

The stealth effect due to PEGylation of liposomes has proven to give a longer circulating half-life. The increased, as well as prolonged availability of the drug, has been established by comparing the intratracheal and oral administration of drugs. Taking the advantages and the amenability of liposomes in consideration, combined with the fact that liposomes are actively marketed for the intravenous route, it is expected that clinical trials of liposomal amikacin (Arikace® developed by Insmed) and Pulmaquine® (liposomal ciprofloxacin) will be successful (Elhissi 2017).

4.7.2.3 Micro and Nanoparticulate Polymeric Systems

Multiple natural and synthetic polymers have been employed for the formulation of the nanoparticulate delivery system. Microparticles have a size range between 1–100 μm and nanoparticles are within the size range of 1–1000 nm. Both these delivery systems are generally prepared by using natural or synthetic polymers such as chitosan, poly(lactic-co-glycolic acid), poly(lactic) acid, etc. Compared to natural polymers, synthetic polymers can sustain the drug release over a period of days to several weeks. Microparticles are generally used for delivery of the drug into deep lung. Insulin–PLGA microparticles show a comparatively longer residence time than free insulin. Rifampicin microparticles prepared by solvent evaporation and spray drying techniques are reported for tuberculosis. As the physicochemical properties of polymers are tunable, polymeric nanoparticles have also been utilized for gene, peptide, and vaccine delivery extensively. Chitosan, PEI (polyethylene amine), and PLGA are few of the widely used polymers for delivery of peptides like insulin, gene-like siRNA, and plasmids and vaccines against diphtheria toxoids.

The use of natural polymers such as chitin, chitosan, and cellulose makes the microspheres biodegradable and/or biocompatible. The method of preparation is an important parameter to achieve desired characteristics of microspheres. The delivery of drugs for the treatment of various diseases, with microspheres as drug delivery systems, have been explored. Some of the treatments include the delivery of antitubercular drugs; delivery of tetradine (an antisilicotic alkaloid) on the surface of albumin microspheres via dry powder inhalers for pulmonary delivery; delivery of peptides and proteins; and local delivery of antibiotics for local effects in tuberculosis.

Microspheres can be prepared in one step by spray drying method by mixing drug along with a polysaccharide solution. Terbutaline sulfate, a hydrophilic drug encapsulated inside the hydrophobic core of microspheres showed sustained release of approximately 32% after administration via a dry powder inhaler. Cisplatin-loaded microspheres for the treatment of lung cancer were prepared by emulsification ionotropic gelation and a heat crosslinking method. The microspheres are also suitable delivery systems for delivery of combination therapy such as in lung cancer treatment. Synergistic effect of doxorubicin and paclitaxel is obtained by delivering them in single-formulation consisting of PLGA microspheres (Feng, Tian et al. 2014). Inorganic materials like calcium carbonate–based composite microparticles have been investigated for delivery of multiple peptides. The spray-dried composite microparticles made of calcium carbonate/hyaluronic acid displayed four times higher bioavailability compared to a peptide solution after pulmonary administration in rats (Tewes, Gobbo et al. 2016).

Microparticles with low mass densities and higher diameters have been extensively investigated for their superior deep lung delivery capabilities. Testosterone microparticles were capable of maintaining the blood level of testosterone up to 24 hours compared to subcutaneous injection, and 53% higher bioavailability as compared to nonporous microparticles with a lower diameter (Edwards, Hanes et al. 1997). Porous microparticles have been investigated for the local delivery of doxorubicin resulting in a sustained delivery over 4 days and a decreased macrophage-mediated clearance as compared to nonporous small particles (Yang, Bajaj et al. 2009).

4.7.2.4 Supermagnetic Nanoparticles

Magnetic nanoparticles have many advantages, including thermal therapy, guidance to the target site, and a relatively nontoxic nature. Mesoporous silica nanoparticles were used targeting MRP1 and Bcl-2 by delivering doxorubicin/cisplatin along with 2 siRNA molecules. The nanoparticles were surface decorated with luteinizing hormone-releasing hormone (LHRH) peptide for targeted therapy (Chen, Zhang et al. 2009). Also, mesoporous nanoparticles conjugated with epidermal growth factor receptor (EGFR) ligands were used in the treatment of non-small cell lung cancer. Magnetic nanoparticles are targeted to the desired site of action by the effect of external magnetic field. As the external magnetic field intensity can be controlled, the increase in the temperature of nanoparticles kills the tumor cells in case of cancer. The most commonly used super magnetic nanoparticles are prepared by using superparamagnetic iron oxide and iron oxide with poly(lactic co-glycolic-acid). The concerns of pulmonary toxicity of the polymer-coated nanoparticles have been addressed recently, but uncoated iron oxide nanoparticles have been reported to cross the lung epithelium and cause serious inflammation to the lungs and liver (Bahadar, Maqbool et al. 2016).

4.7.2.5 Nanosuspensions

Nanosuspensions are the dispersion colloidal systems in which the drug is dispersed in the absence of polymeric matrix. The dispersion is stabilized by addition of surfactant in it. Nanosuspensions make the drug available rapidly after administration. For pulmonary delivery, nanosuspensions are suitable delivery systems, as they have many advantages over solution formulations. Nanosuspensions can be prepared without the use of organic solvents, hence lessen the possible toxic side effects as well as does not interfere in-vivo. In addition to this, nanosuspensions do not have vehicle-dependent solubility, which allows the use of a wider range of doses. The nanosuspension of azole derivative itraconazole was prepared to treat respiratory tract fungal infection. The effects were improved compared to oral or systemic administration (Wlaz, Knaga et al. 2015). A nanosuspension of budesonide administered has a more efficient distribution and deposition of drug in the lungs due to its micronized form (Jacobs and Müller 2002). Ketotifen and ibuprofen are some of the drugs which are administered as nanosuspensions (Patel and Agrawal 2011).

4.8 Manufacturing Methods for the Formulation of Nasal and Pulmonary Drug Delivery Systems

There is an increasing demand for inhalation therapy with appropriate dose delivery for chronic diseases like asthma or chronic obstructive pulmonary disease (COPD). Consequently, the quest for technologies that alleviate the efficient pulmonary delivery of these inhalable particles is also on rise. These technologies affect the particle size, shape, and morphology, thereby affecting the underlying site for deposition in the lungs. Therefore, it is essential to look at these methods meticulously and select the appropriate one for manufacturing particulate matter for efficient pulmonary drug delivery. Some of these important technologies which are presently in use are briefly explained below.

4.8.1 Spray Drying

Since the past few decades, this advanced manufacturing technology is being used extensively in the pharmaceutical, food, and biochemical industries. This method is a single step process of converting a liquid solution into the dried powder (Figure 4.5a). It is gaining popularity for pulmonary drug delivery due to the (i) ease of large-scale production, (ii) producing efficient inhalable colloidal particles in the solid state (Duddu, Bäckström et al. 2002; Mosén, Sisk et al. 2005), (iii) superior control of particle formation (Gilani et al. 2005), and (iv) ability to form particles of a uniform morphology. An SD cycle usually consists of the following steps (a) atomization of feed solution at room temperature by a nozzle which sprays dry air; (b) mixing of the atomized spray with dry air; (c) formation of dried droplets resulting from contact with dry gas at high temperature; and (d) removal of the dried solid product from the air (Chow, Tong et al. 2007). The SD process can also be utilized in various modes like open cycle, closed cycle, semi-closed cycle, with or without aseptic control. The moisture content of the drying gas affects the MMAD and the median geometric diameters. Despite its versatility, the significance of spray-dried nanoparticles for inhalation therapy has been undervalued. Proliposomes of Isoniazid (Rojanarat, Changsan et al. 2011) have been prepared by spray drying.

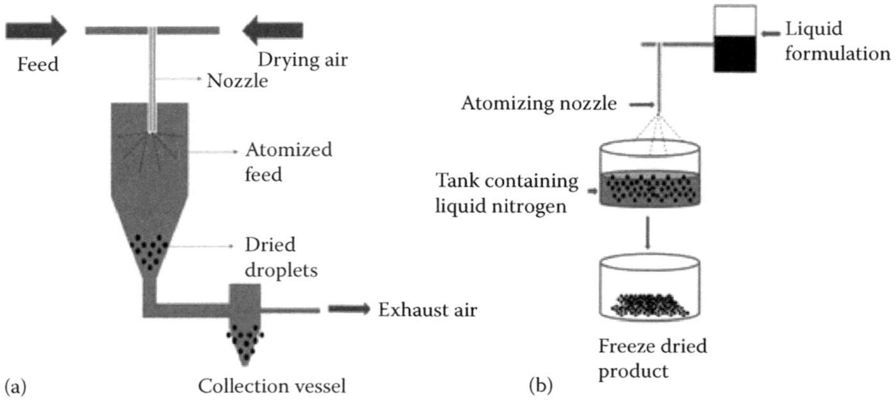

FIGURE 4.5
(a) Spray drying process is preferred for thermostable compounds and (b) spray-freeze-drying process are preferred in case of thermolabile compounds.

4.8.2 Spray-Freeze-Drying (SFD)

This technique has been used in the pharmaceutical industry since 1990. SFD is a combination of the spray drying and freezes drying processes. Spray drying involves atomization of liquid drug solution by spraying it in a chamber which contains a cryogenic liquid (Figure 4.5b) (Rogers, Johnston et al. 2001). SFD is capable of engineering particles in the desired target for respiration, i.e., below 5 µm and can also produce freeze-dried particles in the nano range (Yu, Rogers et al. 2002; Leach, Simpson et al. 2005; Williams III, Johnston et al. 2005; Shekunov, Chattopadhyay et al. 2006). The most critical parameter which governs the particle size in the process is the ratio of cryogenic liquid to the volume of liquid feed (Costantino, Firouzabadian et al. 2000). Insulin-loaded liposomes prepared by SFD were successful in controlling the blood glucose level and a relatively high bioavailability of around 38% after intratracheal administration (Bi, Shao et al. 2008). An in-depth review of spray-freeze-drying is available elsewhere (Wanning, Süverkrüp et al. 2015).

Some of the advantages of this technique for pulmonary drug delivery are listed below:

- Like SD, SFD too is capable of formulating thermolabile substances (like proteins and peptides) and biologicals (like monoclonal antibodies and vaccines).
- Another advantage of SFD is it results in particles which are light and porous at subambient temperature (>6 µm) which have better aerosol performance due to their high fine particle fraction compared to SD, which produces dense particles (~3 µm) (Maa, Nguyen et al. 1999; Maa and Prestrelski 2000; Maa and Nguyen 2001).
- SFD can also be used to prepare respirable dry powder from liposomal solutions for pulmonary drug delivery.
- SFD is reported to be superior to SD in terms of maintaining the polymeric and lipid nanocarriers particle size following reconstitution (Ali and Lamprecht 2014).

4.8.3 Supercritical Fluid (SCF) Technology

SCF technology is a single step process which leads to the straight production of respirable particles utilizing supercritical fluids (Figure 4.6a). This method also avoids the use of organic solvent. The use of SCF in this process leads to the not only elimination of drying and harvesting stages but also utilizes a minimum amount of solvent because SCF is in the gaseous state under ambient conditions (Palakodaty, York et al. 1998). The use of SCF also makes this method an environment-friendly, commercially viable, recyclable process. The most commonly used SCF in the pharmaceutical industry is supercritical

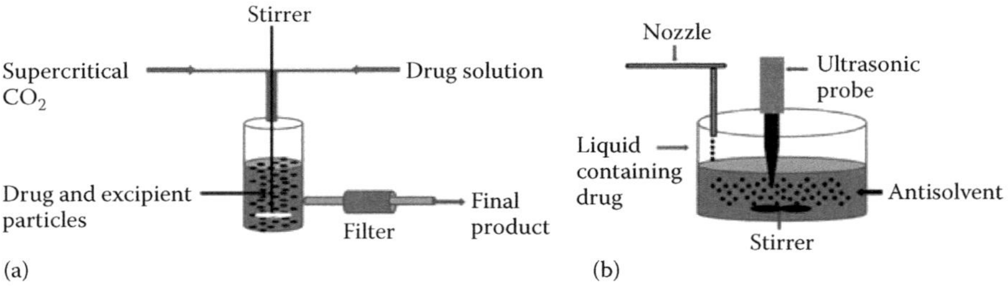

FIGURE 4.6
(a) Supercritical fluid technology; (b) sonocrystallization.

CO_2 because it is safe and inert. The application of this method in pulmonary drug delivery is it can be used to produce particulate systems containing proteins and peptides (Rehman, Shekunov et al. 2004; Chattopadhyay, Shekunov et al. 2007). The added advantage of these respirable particles is better purity and control of crystalline forms (Chow, Tong et al. 2007). In spite of being an attractive option, there is a paucity of literature available on the usage of SCF for generating nanoparticles for pulmonary delivery.

4.8.4 Solvent Precipitation Method

Based on this principle there are two techniques which are explained below:

4.8.4.1 Sonocrystallization

By using antisolvents, inhalable particles can be produced due to rapid precipitation from aqueous solutions (Figure 4.6b). However, the complexity in generating uniform inhalable particulate matter arises due to nucleation and crystal growth. Therefore, in order to avoid this problem, ultrasonic radiation has been applied to the precipitation process (Lancaster, Singh et al. 2000) Several other drugs like salbutamol sulfate, budesonide, and formoterol fumarate can also be prepared via this method for inhalation (Chan and Chew 2003). The critical parameters governing this process include the rate at which the two solutions move towards each other and the concentration of drug solution. The ratio of the drug solution to antisolvent also affect the precipitation process. It is very crucial for the appropriate functioning of this process that the two liquids approach each other with the same velocity (Patil and Sarasija 2012; Paliwal, Babu et al. 2014).

4.8.5 Emulsification/Solvent Evaporation

In this method, the first step is the formation of a single or double emulsion. The second step in this method is solvent evaporation, which is critical for particle morphology. In a single emulsion/evaporation method, the hydrophobic drug and lipids are dissolved in the organic solvent. This organic solvent which is immiscible in water is then dissolved in an aqueous phase to prepare an oil in water emulsion. Pressure is applied to evaporate the organic phase; the lipid also precipitates in the aqueous phase resulting in solid lipid nanoparticles. The double emulsion, on the other hand, generates particulate matter, which is highly porous and has the adequate shape and geometric diameter to be utilized for inhalation. This double emulsion method has the advantage that it is economical and simple.

4.8.6 Particle Replication in a Nonwetting Template (PRINT)

Particle replication in a nonwetting template (PRINT) is a technique developed by Dr. Joseph DeSimone et al. for fabrication of smart particles whose size, shape, and surface functionality and be modified and achieved as desired. Shape-specific monodisperse nanoparticles can be obtained using PRINT (Kelly and DeSimone 2008).

This top-down method is based on a photolithography molding technique is used to produce particles in the range of several nanometers to micrometers for various applications.

PRINT consists of the following process (Xu, Wong et al. 2013):

a. A silicon master template with cavities is generated by photolithography.
b. A liquid perfluoropolyether (PFPE) is poured on the template, which spreads easily on the silicon template due to its positive spreading coefficient giving rise to an elastomeric PRINT mold with nano and macro features.
c. A preparticle liquid is then filled into the cavities using the roll to roll technique in which a polyethylene terephthalate PET countersheet with high surface energy was passed through a roller, thereby applying pressure on the mold.
d. The liquid in the cavities is then converted into a solid by various processes, such as solvent evaporation and photocuring.
e. After solidification, the particles are removed from the cavities of PET by bringing them into contact with an adhesive layer.

The particles are then freed from the surface by dissolving the adhesive layer. PRINT particles can be made from materials which include hydrogels, such as crosslinked polyethylene glycol (PEG), and poly(silyl ether) thermoplastic polymers, such as polylactic acid (PLLA) and polylactic-co-glycolic acid (PLGA) (Rolland, Maynor et al. 2005; Enlow, Luft et al. 2011). An example of the application of PRINT technology is the formulation of butyrylcholinesterase (BuChE) using PRINT to evaluate the interactions with biologic lung fluid. The MMAD was found to be 2.77 µm which means it can be used for deposition in the deep airway via a DPI. Therefore, particles produced from

PRINT can be easily given by DPI, thereby providing a suitable treatment option (Rahhal, Fromen et al. 2016).

Even though the nanoparticulate toxicity evaluation is in its infancy, it is proving to be an excellent tool to identify and comprehend the set of parameters for a successful nanoparticulate formulation such as lethal dose, identification of a therapeutic window of the drug-loaded nanoparticles, and their immunological effects. With some modifications, many models have been established and investigated for the evaluation of nanoparticle toxicity post inhalational delivery. Notable efforts have been made to develop models utilizing air–liquid interface (ALI) with an in-depth review available elsewhere. This is a brief description of the various in-vitro, *ex vivo*, and in-vivo methods.

4.9 In-Vitro, In-Vivo and *Ex Vivo* Delivery: Considerations and Approaches

4.9.1 Nasal Drug Delivery

Difficulties in obtaining human nasal mucosal tissue along with ethical considerations may lead to use the alternative in-vitro characterization techniques, i.e., in-vitro excised tissue and cell line models to study the absorption and transport of the drugs through the nasal cavity. Although it is difficult to obtained intact excised nasal tissue, in-vitro cell culture techniques are more reliable and prominent, and are being actively developed with new advancements. Several studies were reported for drug transport, absorption, and metabolism across the nasal cavity using cultured nasal cells models, due to their inherent biological properties like tight junctions, cilia, mucin secretions, etc.

4.9.1.1 Cell Culture Models

RPMI 2650, is the only human nasal epithelial cell line that has been used prominently to evaluate the in-vitro permeation and metabolism studies for nasal drug delivery over the years (Peter 1998; Wengst and Reichl 2010). Numerous advantages have been associated with this in-vitro technique like rapid and ease of drug permeability evaluation, clarification of drug transport, and mechanisms involving the limited use of animals (Schmidt, Peter et al. 1998). Nowadays, primary culture cells have been widely used in transport and metabolism studies of drug molecules including peptide molecules across the nasal mucosa (Kissel and Werner 1998; Hoang, Uchenna et al. 2002). Several in-vitro studies have been carried out of primary cultures of human nasal mucosal cells on Transwell® inserts, especially for peptides, to evaluate transport/permeation studies (Werner and Kissel 1995; Agu, Jorissen et al. 2001). Moreover, a variety of animals, i.e., rats, rabbits, bovines, etc. (Steele and Arnold 1985; Audus, Bartel et al. 1990) were also assessed by this technique and provided useful information in regards with the transport/permeation and metabolism studies.

4.9.1.2 **Ex Vivo** *Permeation Models for Nasal Drug Delivery*

On the road of preclinical development of any drug molecules/formulations for nasal delivery, factors like metabolic stability, need for permeation/absorption enhancers, and compatibility of formulations need to be considered. Excised tissue models of different animals such as rabbits, pigs, sheep, etc. have played an important role in the investigation of permeability/absorption of drug molecules. These are summarized in the following sections.

4.9.1.2.1 Rabbit Tissue Model

Several studies have been reported by this model to evaluate permeation of drug molecules across the nasal mucosa (Kubo, Hosoya et al. 1994). Transport of macromolecules like insulin (Bechgaard, Gizurarson et al. 1992), apomorphine (Ugwoke, Agu et al. 2000), calcitonin, sucrose, and corticotrophin across the nasal epithelium were investigated using rabbit nasal mucosa in the Ussing chamber at a temperature of 37°C (Cremaschi, Rossetti et al. 1991).

Nakamura et al. performed *ex vivo* nasal permeation of FITC-dextran (FD-4) (MW 4400 Da) in rabbit nasal mucosa by means of using permeation enhancers such as β-sitosterol, b-D-glucoside (Sit-G), and β-sitosterol (Sit) (Nakamura, Maitani et al. 2002). The authors found an increased permeation of FD-4 with Sit-G when compared with Sit. Moreover, not only with FD-4 but Sit-G enhanced 2-fold nasal absorption with the insulin and verapamil, when given as a powder form via the nasal route (Ando, Maitani et al. 1998; Yamamoto, Maitani et al. 1998; Maitani, Nakamura et al. 2000).

4.9.1.2.2 Porcine Tissue Model

It is important to have such validated *ex vivo* tissue models that may have morphological as well as physicochemical similarities when compared to the human nasal mucosa. Wadell et al. developed and examined the permeation studies of ^{14}C-mannitol and D-(2-3H) glucose across the porcine nasal mucosa in an Ussing chamber (Wadell, Björk et al. 1999). The authors also evaluated the permeability coefficient, electrophysiological assessment, histological studies, and biochemical assay on the porcine nasal mucosa. Moreover, they found that the cavity porcine mucosa was the most suitable, and it also remained viable up to 8 hours after removal.

In contrast, no significant differences were observed in the bioelectrical characteristics as well as apparent permeability (P_{app}) when the release/permeation of dihydroalprenolol, testosterone, or hydrocortisone in Carbopol-934P gel formulations by using a horizontal Ussing chamber was examined (Östh, Paulsson et al. 2002). However, the porcine nasal mucosa model further showed the usefulness in the intranasal delivery of bevacizumab (Samson, de la Calera et al. 2012), thymoquinone (Alam, Khan et al. 2012), neuropeptide (Kumar, Pandey et al. 2013), venlafaxine (Haque, Md et al. 2012), etc.

4.9.1.2.3 Sheep Nasal Mucosa

In recent decades, several studies of permeation have been reported by using sheep nasal mucosa (ovine tissue). The transepithelial electrical properties and fluxes of insulin, mannitol, and propranolol in the presence/absence of permeation enhancers by using ovine tissue in Ussing chambers have been characterized (Wheatley, Dent et al. 1988). They conclude that the ovine tissue maintained viability up to 8 hours and showed normal electrophysiology and histology. The bovine tissue model has been widely used to evaluate the permeation characteristic of various drugs. This includes the intranasal delivery of carbamazepine (Gavini, Hegge et al. 2006), metoclopramide (Gavini, Rassu et al. 2005), dimenhydrinate (Belgamwar, Chauk et al. 2009), tacrine (Luppi, Bigucci et al. 2011), loratadine (Singh, Kumar et al. 2013), saquinavir (Mahajan, Mahajan et al. 2014), etc.

4.9.1.3 In-Vivo Animal Models for Nasal Drug Delivery

Isolated organ perfusion models or whole animal models are the only options available to assess the absorption/permeation as well as the pharmacodynamic and pharmacokinetic studies for nasal drug delivery. Animals like rats (Hussain, Hirai et al. 1981; Javia and Thakkar 2017), rabbits (Chen, Eiting et al. 2006; Russo, Sacchetti et al. 2006), dogs (Henry, Ruano et al. 1998), monkeys (Kelly, Asgharian et al. 2005), etc. have been used to evaluate nasal absorption studies. Even though the rat animal models have been widely used preliminary studies for nasal drug delivery due to their small size, ease of maintenance, and overall low cost; the small size of rats also limits blood withdrawal samples. However, researchers have widely used rat animal models due to increasing pressure and hesitance to use primate models from animal ethics committees. It should be obvious to know that the physiological and anatomical differences in the nasal cavity of human and alternate animal models such as rabbits, dogs, monkeys, etc. and should be considered in formulation development stages.

4.9.2 Pulmonary Drug Delivery

4.9.2.1 In-Vitro Pulmonary Epithelial Cell Models

The first and foremost step for testing any formulation is in-vitro cell culture models before pursuing *ex vivo* and in-vivo testing. Cell models offer a multitude of advantages, such as easy handling, continuous cell lines, and an abundant number of cells to carry out multiple experiments concurrently; also, they diminish the necessity of using live animals for experiments. A complete understanding of the cell nanoparticle interaction is a prerequisite prior to intensive in-vivo studies. Conventional in-vitro models cannot provide a correct picture for nanoparticle toxicity testing, as physiologically, the lung cells are supplied nutrition only from the basal side, and the basal side faces the environment. Cell culture models which can provide such a milieu should be preferred, as nanoparticle deposition in immersed cultures is completely different from lung deposition (Schön, Ctistis et al. 2017). Due to these factors, in-vivo pulmonary toxicity used to be preferred. This provides a strong rationale for carrying out various studies using isolated murine and human lung epithelial cell lines, along with the fact that epithelial cells are the first point of contact for inhaled nanoparticles. Cell models are generally procured from human or murine tissues that have been accepted and validated over the last two decades. Efforts have been focused on the development of a standard cell line, which can predict transporter mechanisms across the pulmonary epithelium, on the similar lines of Caco-2 cell lines, which is standardized to study transporter mechanism studies in the gastrointestinal tract.

4.9.2.1.1 Alveolar Epithelial Cells

Human lung AT-II cells are represented by A549 (adenocarcinoma) cell line that resembles it in terms of multilamellar cytoplasmic inclusion bodies. The use of this cell line is abundant in various biopharmaceutical tests. hAEpC, a primary human alveolar epithelial cell line with AT-I phenotype, these are isolated from lung tissue. hAEpC cell lines are a suitable model for AT-I required tests, but require a lot of time for procurement, and isolation is expensive. There is a necessity for a cell model of type I cells that develops a tight junction, drug transporters (as per drug delivery requirements), and grows for several passages, as A-549 do not form tight cellular junctions (Castell, Donato et al. 2005; Rothen-Rutishauser, Blank et al. 2008). The major pulmonary CYP enzyme is expressed by A-549, however, at a significantly lower level when compared to human lungs. Highly predictable value for the absorption of conventional substances through the lungs were provided by 16HBE14 cells cocultured with Calu-3 cells (Sakagami 2006). Drug transporters molecules such as P-GP (P-glycoprotein) and organic cation transporters were expressed by NCI-H441 cell lines along with significant transepithelial electrical resistance. The origin of this cell line is human lung adenocarcinoma with characteristic features of AT-II cells (de Souza Carvalho, Daum et al. 2014).

4.9.2.1.2 Macrophages

The internalization and digestion of particles and microorganisms are the primary mechanisms of macrophages. Mouse leukemic monocyte-derived continuous macrophage cells line (RAW 264.7), murine alveolar macrophages (MH-2), and mouse reticulum sarcoma-derived continuous macrophage cell lines (J774.A1) are some of the common cell lines preferred for carrying out cell internalization assays or any other pulmonary-related assay. Monocyte-derived macrophages, which are obtained from human blood as monocytes and are differentiated by the usage of granulocyte-macrophage-colony-stimulating factor (GM-CSF) are used in pulmonary cell culture while carefully evaluating physiological aspects as compared to in-vivo process (de Souza Carvalho, Daum et al. 2014).

Rat alveolar macrophage cells (NR8383) were reported to be used for in-vitro uptake studies. The cells were treated with microspheres containing ofloxacin; the study was to check the uptake of microspheres containing ofloxacin via phagocytosis. The conclusion of the experiment was that the enhanced uptake of ofloxacin-loaded microspheres was observed compared to free ofloxacin, which is beneficial against *M. tuberculosis* present in alveolar macrophages of a tuberculosis patient (Park, Jin et al. 2013).

Layered cocultures with epithelial cells in the middle, macrophages on top and dendritic cells on the other side have been used to study interaction with polystyrene particles (Rothen-Rutishauser, Kiama et al. 2005). Another coculture consisting of a mixture of A549 + THP-1 (monocytes) + HMC-1 (mast cells) in the ratio of 10:2:1 in the basal compartment, and an insert containing EAhy926 endothelial cells was reported to provide inflammation and dysfunction consistent with local and systemic effects (Alfaro-Moreno, Nawrot et al. 2008). A similar coculture model, utilizing A549 epithelial cells, HMC-1 mast cells, THP-1 monocytes on the apical side, and EAhy926 endothelial cells on the basal side of transwell, was prepared to mimic ALI and used for aerosol exposure (Klein, Serchi et al. 2013).

Membrane selection plays an important role in nanoparticle permeation studies using transwell, as membranes are known to hinder the penetration of nanoparticles, which may result in a faulty inference. Even in the absence of cells, the passage of 400 nm polystyrene particles was considerably hindered through the membrane.

4.9.2.1.3 ALI-Based Exposure Systems

Respiratory cells based on physiological exposure systems utilize transwell cultures wherein, the cells are cultured on a semipermeable insert, which is placed into a culture well which provides ALI. This system was adapted from Caco-2 models (Hidalgo, Raub et al. 1989). Call-3 sub-bronchial gland cell lines (human) have been used to study drug permeation by considering the transepithelial electrical resistance (TEER) values that are necessary for a tight monolayer. Additionally, the transport and efflux of small molecules using this cell line has been reported too. The expression of cystic fibrosis transmembrane conductance regulator (CFTR) protein, was provided by the tight monolayer present in cells, vital for studying cystic fibrosis. Call-3 cells can be obtained via transwell plates containing two compartments; an apical compartment being empty to denote air, and the basolateral compartment containing cell culture medium, this provides us an ALI culture. The pulmonary epithelium of this cultivation type is in contact with air, thereby correlating with the physiological conditions (Foster, Avery et al. 2000). Various other options are available for modeling epithelium. Immortalized cells are preferred over primary cells, as the latter has a limited lifespan and donor variability. Immortalized cells such as A-549 have been used. Other cell lines, like the 16HBE140 (human bronchial epithelium cell line), have been used for studying drug transport mechanisms and are used for carrying out different experiments, such as seed density, culture condition, TEER, which is necessary for optimal drug transport. Mannitol was used for studying the barrier properties of cell model. Higher permeability was observed in hydrophilic molecules present in 16HBE140 cell line as compared to alveolar epithelial cells, and permeation of lipophilic molecules showed a sigmoidal relationship between permeability and lipophilicity.

ALI has traditionally been used for assessment of environmental nanoparticles on the lungs. Multiple setups are commercially available to study particle deposition with an attached aerosol generator, such as CULTEX® (Cultex® Laboratories GmbH, Hannover, Germany), MucilAir®, and Vitrocell® (Figure 4.7).

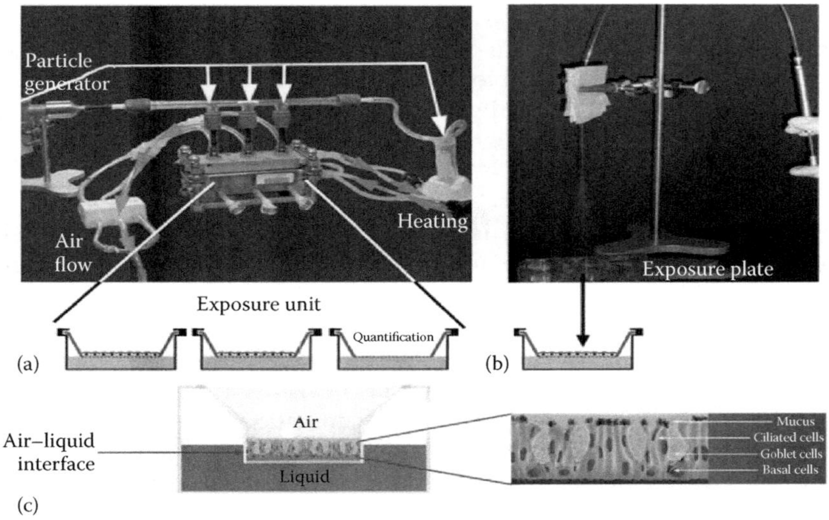

FIGURE 4.7
Set-up for aerosol-exposure at the air-liquid interface based on the VitroCell® (Vitrocell Systems GmbH, Waldkirch, Germany) exposure system (a) and by manually generated aerosols (MicroSprayer®, b). (a) Cells are cultured on transwells and exposed in the compartments of the exposure unit, thermostabilized by a water bath (c) MucilAir model based on multiple cells coculture. (From Fröhlich, Eleonore, and Sharareh Salar-Behzadi. 2014. Toxicological Assessment of Inhaled Nanoparticles: Role of in Vivo, ex Vivo, in Vitro, and in Silico Studies. *International Journal of Molecular Sciences* 15 (3):4795; Baxter, A., S. Thain, A. Banerjee, L. Haswell, A. Parmar, G. Phillips and E. Minet 2015. Targeted omics analyses, and metabolic enzyme activity assays demonstrate maintenance of key mucociliary characteristics in long term cultures of reconstituted human airway epithelia. *Toxicology in Vitro* 29(5):864–875.)

4.9.2.1.4 MucilAir™ (3D airway model)

Epithelix developed an in-vitro 3D cell model of human airway epithelium. MucilAir™ is an in-vitro model consisting of human basal, goblet, and ciliated cells cultured at an ALI featuring a fully differentiated and functional respiratory epithelium. The cells can be obtained from a wide array of individuals according to the need for the experiments. The cells are procured either from a single or from a pool of donors. The cells from individuals suffering from pathologies such as asthma, cystic fibrosis, allergic rhinitis, and COPD are obtained for this cell model. Pseudostratified columnar epithelium, possession of cilia beating, mucus production, and cytokine and chemokine expression (2A13, 2B6, 2F1) are some of the characteristics of a MucilAir™ model. This model mimics upper respiratory tract of the human lung (Movia, Di Cristo et al. 2017). The model displays in-vivo characteristics including tight junctions, metabolic activity, mucus production, and beating cilia. It has been utilized to evaluate suitability for repeated dose testing of inhaled chemicals, as airway surface liquid markers and the gene expression of enzymes metabolizing xenobiotics are similar in MucilAir™ and a normal lung.

The toxicity of cerium oxide nanoparticles was assessed with the aforementioned model. The nanoparticle inhalation study performed reported that the MucilAir 3D airway model predicted local toxicity and systemic availability better as compared to monolayer (2D cell) model. They reported that the monolayer cell generally provides an overestimated value for the toxicity of particles (Frieke Kuper, Gröllers-Mulderij et al. 2015).

4.9.2.1.5 Lung-on-a-Chip

One advanced in-vitro model is lung-on-a-chip, wherein the endothelial cells present on the lung lining are placed very close to blood vessels across a porous flexible boundary. Media is circulated through the blood vessels to mimic blood flow, while air is circulated on the other side. Mechanical stretching is generated by a cyclic vacuum which mimics breathing (Huh, Matthews et al. 2010). Lung-on-a-chip has been used for the screening of chemotherapeutic drug oncogene modeling, infectious disease research, and inflammatory diseases of the lung. It is a promising way of studying various lung diseases. A detailed review of its utilities is available elsewhere (Konar, Devarasetty et al. 2016).

4.9.2.2 In-Vitro Drug Release Studies

The dialysis tubing technique is one the methods employed in determining the release of the drug from the optimized formulation. In this technique, a phosphate buffered saline at pH 7.4 or any other release medium (such as lung simulated fluid) which mimics lung conditions can be used. The formulation, or the free drug, is added to the dialysis bag, sealed, and inserted in the medium. The whole system is subjected to a constant motion created by a magnetic stirrer. According to the requirement, the sample is extracted and a similar amount of fresh medium is added to maintain sink conditions. The extracted sample is further measured to determine the amount of drug released into the medium from the formulation. The released drug provides us with an insight into the release profile of the formulation (Feng, Zhang et al. 2014; Kaur, Garg et al. 2016).

4.9.2.2.1 Particle Deposition by Cascade Impactors

Cascade impactors (CIs) are widely used to determine size-based particle deposition patterns for formulations delivered through the pulmonary route. As CIs work on constant flow rates, they cannot be considered as in-vitro lung simulators, as they lack the variable airflow rate observed in a functional lung. However, they provide invaluable data regarding the aerodynamic diameter and the deposition patterns of particulate drug carriers in the respiratory tract. Various types of CIs are available and a few of the most commonly used ones are listed below in Table 4.1 (Dunber, Hickey et al. 1998). Inertial impaction is the mechanism by which a cascade impactor operates. A cascade impactor is generally made with an induction port with one port for the attachment of mouthpiece, entrance cone, and stages. An Anderson impactor is one of the most commonly used impactors for evaluating inhalation products. It has 8 stages arranged in such a way that particles with different diameters (10 µm for stage 0, to 0.4 µm for stage 7) will be collected on a particular stage on passing an aerosol stream. This is known as inertia-based separation and is the working principle of most of the CIs. Fine particle doses, fine particle fractions, and MMAD are determined by calculation of the drug deposited on the individual stages. Different flow rates of 28.3 L/min, 30 L/min, or 60 L/min are according to the formulation requirement and the disorder (or disease) being targeted. Different CIs, such as the multistage impinger or the Marple-Miller impactor, have also been used to determine these parameters. A combined effort by various pharmaceutical companies has led to the development of the next-generation impactor, and its flexible and productive nature has made it a popular CI (Marple, Olson et al. 2003; Patil-Gadhe, Kyadarkunte et al. 2014; USP 2017).

To negate the "particle bounce effect," the USP and the other pharmacopeias suggest coating the plates with silicone oil, thus trapping all of the particles. It should be noted that this particle bounce effect can be reduced but not completely eliminated. To obtain a meaningful result from the use of CIs, the jet velocity, dose, and the type of impactor should be optimized.

In-vitro models provide reproducible data at a fraction of the cost. Different models provide different levels of in-vivo–in-vitro correlations. However, in the current situation, in-vitro models do not provide a linear in-vivo correlation or cross-species correlation, as in-vitro setup cannot mimic the 40 different cells present in the lungs. It is widely used as it offers a probable insight on in-vivo nanoparticle fate and as a means to reduce animal testing. It is important to note here that the risk assessment using in-vitro models is completely dependent on the capability of the model to mimic the in-vivo

TABLE 4.1

Reported Calibrated Cut-Off Diameters (D_{50}, μm) of Various Cascade Impactors on the Basis of Inertial Impaction

Apparatus Flow Rate (L/min)	Twin Stage Liquid Impinger	USPB	Multistage Impinger				Anderson Viable Impactor	Anderson Non-Viable Impactor			Next-Generation Impactor			Marple-Miller Impactor		
Stage / Apparatus flow rate (L/min)	60	60	30	60	80	100	28.3	28.3	60	90	30	60	100	30	60	90
Pre-seperator	–	–	–	–	–	–	–	–	–	–	15	13	10	–	–	–
–1	–	–	–	–	–	–	–	–	8.6	8	–	–	–	–	–	–
0	6.4	9.8	–	–	–	–	7.1	9	6.5	6.5	–	–	–	–	–	–
1	–	–	16.9	13.3	11.8	10.4	4.7	5.8	4.4	5.2	11	7.8	6	10	10	8
2	–	–	9.3	6.7	5.6	4.6	3.3	4.7	3.3	3.5	6.6	4.6	3.6	5	5	4
3	–	–	4.5	3.2	2.7	2.4	2.1	3.3	2	2.6	3.9	2.7	2.1	2.5	2.5	2
4	–	–	2.5	1.7	1.4	1.2	1.1	2.1	1.1	1.7	2.3	1.6	1.2	1.25	1.25	1
5	–	–	–	–	–	–	0.65	1.1	0.54	1	1.4	0.96	0.72	0.63	0.63	0.63
6	–	–	–	–	–	–	–	0.7	0.25	0.43	0.84	0.57	0.42	–	–	–
7	–	–	–	–	–	–	–	0.4	–	–	0.51	0.33	0.23	–	–	–
MOC	–	–	–	–	–	–	–	–	–	–	0.31	0.13	0.055	–	–	–

Source: Dunber, Craig A., Anthony J. Hickey, and Peter Holzner. 1998. Dispersion and characterization of pharmaceutical dry powder aerosols. *KONA Powder and Particle Journal* 16:7–45.

situation, which can be ensured by utilizing different cell cocultures and microfluidic devices (Fröhlich and Salar-Behzadi 2014).

4.9.2.3 **Ex Vivo *Lung Tissue Models***

Apart from the in-vitro and in-vivo models, *ex vivo* lung tissue models are extensively used, as they provide vital information for analyzing in-vitro–in-vivo correlations and drug transport mechanisms across lung tissue. *Ex vivo* experiments simulate the in-vivo biological environment while reducing the complications associated with the use of the whole body (Beck-Broichsitter, Gauss et al. 2009). Additionally, they provide information which would be lost due to moral and ethical considerations governing the use of intact subjects. Isolated perfused lung (IPL) and precision cut lung slices (PCLS) are two established *ex vivo* models (Beck-Broichsitter, Gauss et al. 2009; Beck-Broichsitter, Schmehl et al. 2011).

4.9.2.3.1 Isolated Perfused Lung Murine

This model is used for the preparation of IPL. Mice are small in size and process of isolation is difficult. Hence they are rarely used for IPL. Apart from rats and rabbits, guinea pigs are also commonly used for *ex vivo* models. IPL models require a complete isolated lung of the selected animal physiologically resembling the artificial system. Buffer systems such as Krebs-Ringer or Krebs-Henseleit are used for encasing IPL and maintaining the temperature at 37°C. Maintaining 12–15 mL/min of perfusate flow, and equilibrating the perfused solution with carbon dioxide and oxygen ensures proper functioning of lung tissue. The elimination of the first-pass effect influence and retention of physiological properties of lung tissue are some of the advantages of using IPL model. Due to the aforementioned reasons, it is more identical to in-vivo systems compared to in-vitro lung models. The challenges of using IPL are as follows: high precision and skill are required for the removal of the intact lungs from an animal, for maintenance of its physiological conditions. Beck-Broichsitter et al. studied the uptake in an IPL *ex vivo* model of nebulized 5(6)-carboxyfluorescein (CF)-loaded polymeric nanoparticles. The amount of CF transported into the lungs was analyzed by absorption of CF nanoparticle in the perfusate solution. A comparison of absorption and uptake of CF in nanoparticle and CF in solution in the rabbit isolated perfused lung model was carried out by the authors, and a high concentration of CF was observed in the perfusate of CF solution compared to the nanoparticles. Particle size and nebulization performance were analyzed; to which the authors concluded that nebulization doesn't affect the particle size and polydispersity index of the nanoparticles (Beck-Broichsitter, Schmehl et al. 2011).

4.9.2.3.2 Precision Cut Lung Slices

PCLS is prepared using rats and mice (murine model); the lungs are loaded with a low melting agarose solution. The gelation process is enabled by maintaining cold conditions for agarose-filled lungs; these lungs are then sliced according to the requirements (thickness) by a tissue slicer. The slicing process is carried out in a cold cell culture medium, in order to maintain the viability of the slices. The slicing process is carried out in a cold cell culture medium, in order to maintain the viability of the slices and to confirm removal of agarose. The cutting of slices at a specific position can also be carried in order to obtain alveoli and pulmonary vessel slices, which can be used to study the effect of the drug on the contraction–relaxation intensity of pulmonary vessels. As the slices are procured from actual lung tissue, they maintain their receptor mechanisms and physiological properties and are viable for three days (Morin, Baste et al. 2013).

4.9.2.4 In-Vivo Animal Models

In-vivo experiments are crucial in the determination of the safe dosage of the formulation during preclinical trials, which can further provide insight in phase 1 clinical trials. Currently, there is no substitute for drug absorption and disposition studies carried in the in-vivo models. Rodents like guinea pigs, mice, and rats are commonly used animals in laboratory settings. Nose/head-only exposure or lung exposure (inhalation/intratracheal instillation) and whole-body exposure are some of the common routes of inhalation exposure. The amount of material required is a determining factor as per testing material guidelines. Inhalation pharmacokinetics, device efficiency, and formulation sometimes require larger and much more expensive animals, such as sheep, rabbits, monkeys, or pigs. Various techniques include intratracheal instillation and whole-animal exposure (Patil and Sarasija 2012a).

4.9.2.4.1 Surgical Intratracheal Instillation

In this technique, the trachea is exposed, and an incision is made between the tracheal rings through an endotracheal tube. A microsyringe is used for instillation of 10–200 µl of aqueous solution or suspension of the formulation through bifurcation. Intratracheal instillation requires a personnel expert in that technique. This also provides controlled dosing, but it takes a toll on local tissue, and causes uneven distribution of the formulation observed in the animal's lungs. However, it is not a preferred method of choice as it is not a physiologically relevant method. This is due to the inability to dose multiple times, the usage of anesthetized animals, and probability of injury. Intratracheal delivery of curcumin acetate nanocrystals revealed particle size–dependent, antipulmonary arterial hypertension activity (Hu, Yang et al. 2017). Several investigators have reported discrepancies in particle distribution and efficacy when same formulations are administered to animals via instillation and inhalation (Vogel, Rivera et al. 1996). This method has been replaced by a non-invasive method that uses aerosolizers.

4.9.2.4.2 Nonsurgical Intratracheal Instillation

Noninvasive devices have been developed to deliver liquids and dry powders directly to the lungs of rats and mice. After visualization of the trachea, the dose is administered using devices such as a Penn-Century® Liquid Microsprayer®. The MMAD is of these droplets is around ~16–22 μm (Suarez, Garcia-Contreras et al. 2001).

4.9.2.4.3 Aerosolizers

They consist of a whole-body exposure system or nose-only exposure system (Figure 4.8). These chambers allow the drugs to be administered directly to conscious animals and mimic a physiological condition (Dorato 1990). Whole-body chambers are fabricated of different materials, such as glass polyester or Teflon®. It is suggested that in case of whole-body exposure systems, the drug dose can enter the body via other routes such as a percutaneous or oral route (Nahar, Gupta et al. 2013). Rodents are classified as nose breathers (Harkema, Carey et al. 2006) and breathe only through their nose when conscious. Hence nose-only or head-only apparatuses would be preferred over a whole-body apparatus. This involves exposing the animal directly to an aerosol in a chamber where the nose of the animal is anchored for the inhalation of the aerosol. The animal placed in such a way in a tube, that the nose is positioned to the hole or extension of aerosol producing chamber and the aerosol generated is then directed to the nose of the animal. At the other end of the tube, a restraint is attached which prevents the leakage of aerosol and as well as any backing out movement of the animal. A partial closing of the restraint present portion helps avoid overproduction of moisture and heat inside the tube. These systems allow for the collection of multiple blood samples, thus decreasing

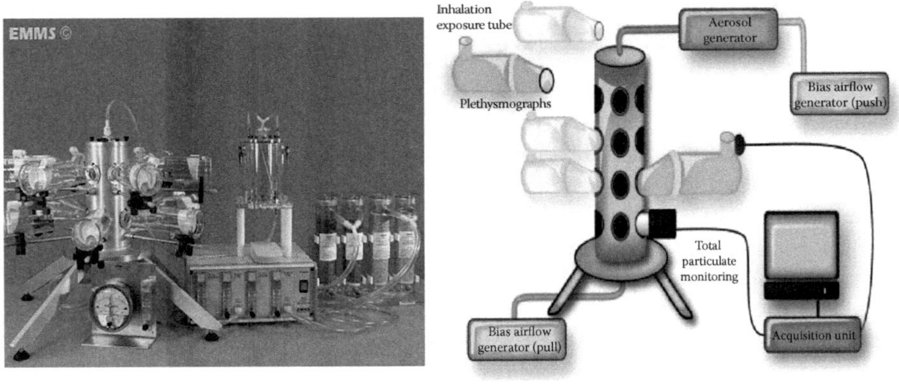

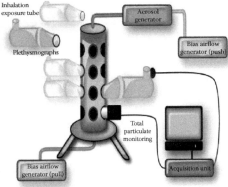

FIGURE 4.8
Apparatus for simultaneous wet aerosol delivery to multiple animals and a schematic layout describing the apparatus. Image obtained from http://www.electromedsys.com/inhalation.html. This needs to be here or I shall have to add an reference in the text next to figure 8.

the inter-animal variability and reduce the number of animals used. One disadvantage of these systems is that the aerosol deposited in the nasal passage is ultimately transported to the gastrointestinal tract (GIT) via mucociliary clearance and might give false positive results in pharmacokinetic experiments. Companies such as CH Technologies (Westwood, NJ), Intox products (Edgewood, NM), TSE GmbH (Badhamburg, Germany), and ADG Developments (Herts, UK) provide various types of nose-only chambers on a commercial basis. Another advantage of the nose-only systems is that they are amenable to small samples, thus making them suitable for testing liposomes (Mainelis, Seshadri et al. 2013), polymeric microparticles (Garcia-Contreras, Sethuraman et al. 2006), and porous nanoparticle aggregate particles (Sung, Padilla et al. 2009).

Balb/c mice subjected to aerosolized solid lipid nanoparticles (SLN) for a time span of 16 days revealed no significant signs of inflammation at a deposit dose of 200 µg. Cytotoxicity studies were carried by probing the broncho-alveolar lavage (BAL) and cytokines along with a histopathological examination of lungs revealed the amenability of the murine inhalational model for testing of nanoparticulate systems (Nassimi, Schleh et al. 2010). In another study, the antifibrotic activity of oldaterol was evaluated in a bleomycin-induced fibrotic mice model using a whole-body exposure chamber equipped with a jet nebulizer.

A detailed review with regards to the advantages and disadvantages of intratracheal and nose-only exposure models is available elsewhere (Sakagami 2006; Cryan, Sivadas, and Garcia-Contreras 2007).

4.10 In-Vitro–In-Vivo Correlations for Pulmonary Drug Delivery

The FDA defines IVIVC as A predictive mathematical model describing the relationship between an in-vitro property of a dosage form and an in-vivo response. The aim behind developing in-vitro–in-vivo correlations (IVIVCs) is to predict therapeutically efficient characteristics, such as bioavailability and release kinetics of formulation, by comparing them with in-vitro tests. This creates the relationship between the in-vitro properties of the drug and the drug response in-vivo. IVIVCs form a firm base in developing a quality product, also while scaling up the formulation. Predictions that provide a reliable correlation are the goal of the pharmaceutical sciences. Although the correlations are being used in preclinical testing, a significant alteration of inhaler design has not been approved by the regulatory authorities (Wachtel, Ertunc et al. 2008).

In the case of inhaled products, since there is no established in-vitro model, the correlation becomes challenging. It is quite difficult to establish a correlation between in-vivo particle deposition and drug absorption with in-vitro tissue culture studies. Recent tissue culture methods that have been reported include a combination of in-vitro drug dissolution and particle deposition in a thin layer of liquid (Agu and Ugwoke 2011). These new methods help in establishing a correlation between in-vitro data and in-vivo particle deposition and efficacy as they mimic the in-vivo delivery of the drug in cell culture (Agu and Ugwoke 2011). Gamma scintigraphy or the charcoal block pharmacokinetic method can be used to evaluate particle deposition in the whole lung (Silkstone, Dennis et al. 2002). For measurement of a deposition in the regional lung, a 3D imaging technique called single photon emission computed tomography (Eberl, Chan et al. 2006), or positron emission tomography may be employed (Dolovich and Bailey 2012).

However, establishing this correlation is not easy, as in-vitro data scarcely has any predictive power for in-vivo studies (BORGSTROM and Clark 2002). IVIVCs aid in a better understanding of the various manufacturing and process elements that influence the critical quality attributes (de Matas, Shao et al. 2007). IVIVCs also corroborate biowaivers; they can be also be substituted for in-vivo bioequivalence studies thereby reducing the number of in-vivo studies that need to be conducted (de Matas, Shao et al. 2007). Publications successfully establish there is a linear correlation between an inhalable dose from a nebulizer (Silkstone, Dennis et al. 2002), fine particle fraction (FPF) from a DPI, and the pharmacokinetics of lung bioavailability. Moreover, establishing the correlation between aerosol performance with lung deposition and in-vivo bioavailability can be taxing due to variation in patient characteristics. There are various patient related and physiological parameters which affect various characteristics like safety, efficacy, and drug deposition of the inhaled formulation (de Matas, Shao et al. 2007). Therefore, it is necessary to generate models for IVIVCs which take these factors into account. The linear mathematical model which uses the regression equation cannot account for this variability (Agu and Ugwoke 2011). To overcome these limitations, an artificial neural network (ANN) model was developed which was successful in establishing various complex as well as nonlinear relationships (de Matas, Shao et al. 2007). ANN is based on artificial intelligence technology. ANN is capable of producing models which take all the variable factors into account and can predict the correlation between various critical product features. Based on performance characteristics, an ANN model was generated which predicted the relative bioavailability of salbutamol in the lung based on urinary excretion 30 minutes post inhalation in a much better manner than linear correlation model when applied to the same data (de Matas, Shao et al. 2007). To form an IVIVC for inhalations, computer-based in-vitro models are being used. One of the models developed was with liposomal inhalation delivery in isolated, perfused, and ventilated lung models to measure absorbance in the lung (Barakat, Kramer et al. 2015).

Since the oropharyngeal geometry differs within the population, a combination of results obtained from different models can be a quality parameter to consider for in-vivo performance analysis. Also, the flow rates have intrapopulation variability. Similar to geometrical in-vitro models, if different flow rate models are developed and then combined qualitatively, the in-vitro models can serve as a tool to establish a better IVIVC for drugs delivered via inhalation. Mimicking the different respiratory patterns in patients for in-vitro evaluation can correlate the in-vivo results (Forbes, Bäckman et al. 2015).

In order to improve IVIVCs, it is necessary to incorporate features that mimic patient and clinically relevant conditions, such as humidified air and valve spacers in the in-vitro tests. Such data would help gain accurate in-vivo assessments. Similarly, in-vitro data analysis and interpretation should be done using appropriate metrics. Developing in-vitro studies that are orientated towards clinical use will lead to enhancement of IVIVCs.

4.11 Chemistry, Manufacturing, and Controls Consideration for Scale-Up

4.11.1 Toxicological Studies

Toxicological studies for inhaled formulations are different than oral or systemically delivered drugs. The dose generation methods and delivery device simulations are some of the differentiating factors for inhalation delivery systems. In animal studies, the atmosphere for inhalation delivery is created by exposing the animals to the environment in which dose of the drug is released. Some important parameters which are needed to be considered here are the extent of exposure, the time period of the exposure particle size of exposed drug, and nasal deposition. The aim of toxicological studies is to ensure the safety of the drug before administering it to the patients. Inhalation toxicological studies stand at a unique place because of the different techniques used in toxicological testing.

The dose administered, dose exposed, and the distribution throughout the length of the respiratory tract are of important consideration in toxicological studies. Each process is essentially done by histopathological examination of organs in animals after exposing them to suitable conditions.

The dose-response data obtained in toxicologic studies is used to interpret the no-observed-adverse-effect level (NOAEL). It is measured in the 1–10 safety dose range which is then used for phase I clinical trials.

4.11.2 PK-PD Considerations

Pharmacokinetics for pulmonary delivered drugs is different than that of locally and systemically administered drugs (Figure 4.9). Adsorption of the drug depends on the desired site of action, and distribution characteristics that are dependent on the cells to which the drug will be exposed, as well as

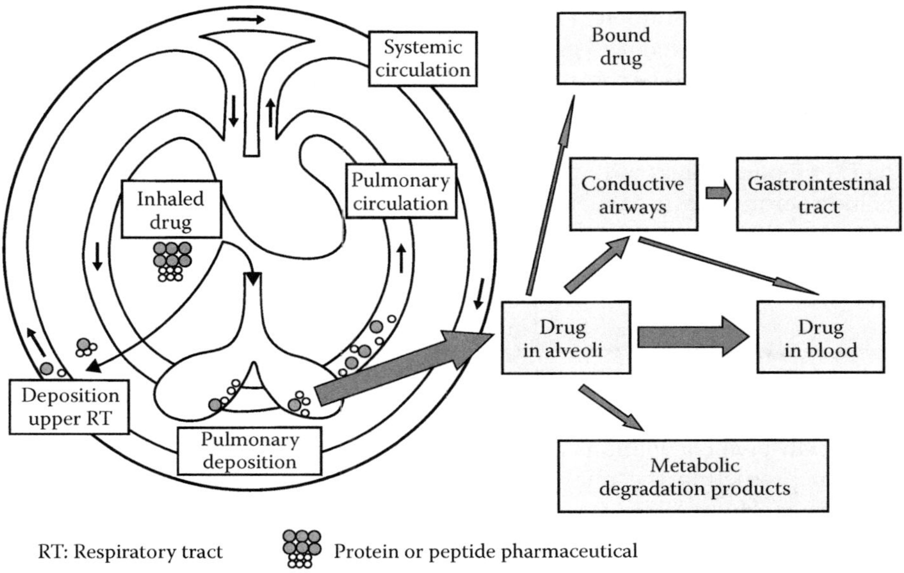

FIGURE 4.9
Pharmacokinetics of drugs via the pulmonary route. The major routes of distribution and metabolism are displayed. Adapted with permission from R. Siekmeier and G. Scheuch. 2008. Systemic treatment by inhalation of macromolecules—Principles, problems, and examples, *Journal of Physiology and Pharmacology: an Official Journal of the Polish Physiological Society* 59 Suppl 6(6): 53–79.

the properties of the delivery system and drug. The lungs have a low level of enzymes compared to the liver, and therefore metabolism is by different CYPs than the liver's. The most common mechanisms of clearance from the lungs are mucociliary clearance and alveolar macrophage clearance, as mentioned earlier. Drug duration within the lung and the available drug at the site of action are dependent upon these factors.

4.11.3 Setting up QTPP

The quality target product profile (QTPP) is the summary of characteristics such as stability, administration route, the bioavailability of the dosage form, efficacy, etc. The QTPP includes all these characteristics for the desired product to achieve and maintain those parameters in the final product. The final product developed should reflect the parameters summarized in the QTPP once it is manufactured. Quality by design (QbD) has a significant contribution to the development of the QTPP.

QTPP defines the purpose of the product for the consumer or end user. QTPP is an essential step in the QbD approach. By defining the QTPP for the product, as well as materials to be used, the output or response is controlled.

Once the critical parameters in the process and formulation are defined, the process risk assessment is performed in order to establish the risk parameters which are further controlled to maintain the required QTPP in the final formulation.

For inhalation formulations depending on the device used for delivery, the QTPP parameters will be changed. For example, the QTPP for DPIs will include aerodynamic particle size distribution, in-vivo performance, uniformity in delivered dose, local or systemic delivery, patient compliance, etc.

4.12 Summary

In-vitro-in-vivo correlations and *ex vivo* lung models provide a robust rationale for predicting the pharmacokinetic parameters for pulmonary delivered drugs. However, these are still incomplete and evolving. Enhancement in the activity of the drug, the increased solubility of the drug, and the increased encapsulation efficiency of the delivery system due to different modifications in the nanoparticle-mediated drug delivery system have developed a strong perspective to enhance and alter the release profile of the drug. However, although many nanoparticulate systems are considered efficient and safe for pulmonary drug delivery, there is still a further need for investigation of their actions across the pulmonary barrier. The recent successful phase trials of liposomal systems for pulmonary administration have appraised the future of nanotechnology and ascertains its feasibility in pulmonary delivery. Due to its multiple advantages, the usage of the pulmonary route is favorable because of the elimination of first-pass effect, and the large surface area provides a faster absorption of drugs. In addition to this, vast technological development in the design of delivery systems, a better understanding of drug design related parameters (such as particle size, the charge on the particle, or the surface presence of proteins or ligands) can help in maximizing the therapeutic advantages of these systems in treatment of various lung diseases by using pulmonary delivery as a major route. In order to harmonize the development, regulatory bodies such as the FDA, USP, and expert groups have devoted much attention to developing and introducing the appropriate test methods.

Rising interest in an intranasal route as one of the drug delivery systems demands reliable in-vitro, *ex vivo*, and in-vivo assessment tools and techniques for the clinical outcomes. Recently remarkable research has been carried out, and is still in development for the different models, such as various cell cultural models, excise tissue models for the nasal transport, and the permeability assessment of the drug molecules. Nevertheless, the importance of *ex vivo* models in drug delivery, certain important factors like the viability of the particular tissues and variation in activity of various nasal cavity enzymes into different species, should be taken into consideration during data interpretation of experiments and extrapolation of the data. In-vitro models are also complimentary tools for in-vivo nasal delivery and very useful into the toxicity screening and facilitate absorption improvement techniques. Conclusively, drug delivery through the nasal route is exceptionally qualified, and the tools/techniques described in Chapter 4 assist fundamental research on the permeability of drugs, toxicity, and formulation development for better clinical outcomes.

Acknowledgments

Dr. Chougule acknowledges the support of the National Institute of General Medical Science of the National Institutes of Health under award number SC3GM109873. Dr. Mahavir Chougule acknowledges Hawai'i Community Foundation, Honolulu, HI 96813, USA, for research support on lung cancer (LEAHI FUND for Pulmonary Research Award; ID# 15ADVC-74296). The Chougule group would like to acknowledge the 2013 George F. Straub Trust and Robert C. Perry Fund of the Hawai'i Community Foundation, Honolulu, HI 96813, USA, for research support on lung cancer. The authors acknowledge the Department of Pharmaceutics and Drug Delivery, School of Pharmacy, University of Mississippi, University, MS, USA for providing start-up financial support to Dr. Chougule's lab.

References

Adjei, Akwete, Dean Sundberg, James Miller, and Alexander Chun. 1992. "Bioavailability of leuprolide acetate following nasal and inhalation delivery to rats and healthy humans." *Pharmaceutical Research* 9 (2):244–249.

Agu, Remigius U. and Michael I. Ugwoke. 2011. "In vitro and in vivo testing methods for respiratory drug delivery." *Expert Opinion on Drug Delivery* 8 (1):57–69. doi: 10.1517/17425247.2011.543896.

Agu, Remigius Uchenna, Mark Jorissen, Tom Willems, Patrick Augustijns, Renaat Kinget, and Norbert Verbeke. 2001. "In-vitro nasal drug delivery studies: Comparison of derivatised, fibrillar and polymerised collagen matrix-based human nasal primary culture systems for nasal drug delivery studies." *Journal of Pharmacy and Pharmacology* 53 (11):1447–1456.

Alam, Sanjar, Zeenat I. Khan, Gulam Mustafa, Manish Kumar, Fakhrul Islam, Aseem Bhatnagar, and Farhan J. Ahmad. 2012. "Development and evaluation of thymoquinone-encapsulated chitosan nanoparticles for nose-to-brain targeting: A pharmacoscintigraphic study." *International Journal of Nanomedicine* 7:5705.

Alfaro-Moreno, E., T. S. Nawrot, B. M. Vanaudenaerde, M. F. Hoylaerts, J. A. Vanoirbeek, B. Nemery, and P. H. Hoet. 2008. "Co-cultures of multiple cell types mimic pulmonary cell communication in response to urban PM10." *Eur Respir J* 32 (5):1184–1194. doi: 10.1183/09031936.00044008.

Ali, Mohamed Ehab, and Alf Lamprecht. 2014. "Spray freeze drying for dry powder inhalation of nanoparticles." *European Journal of Pharmaceutics and Biopharmaceutics* 87 (3):510–517.

Ando, Taeko, Yoshie Maitani, Tomonaga Yamamoto, Kozo Takayama, and Tsuneji Nagai. 1998. "Nasal insulin delivery in rabbits using soybean-derived sterylglucoside and sterol mixtures as novel enhancers in suspension dosage forms." *Biological and Pharmaceutical Bulletin* 21 (8):862–865.

Audus, Kenneth L., Ronnda L. Bartel, Ismael J. Hidalgo, and Ronald T. Borchardt. 1990. "The use of cultured epithelial and endothelial cells for drug transport and metabolism studies." *Pharmaceutical Research* 7 (5):435–451.

Bahadar, Haji, Faheem Maqbool, Kamal Niaz, and Mohammad Abdollahi. 2016. "Toxicity of Nanoparticles and an Overview of Current Experimental Models." *Iranian Biomedical Journal* 20 (1):1–11. doi: 10.7508/ibj.2016.01.001.

Bahadur, Shiv, and Kamla Pathak. 2012. "Physicochemical and physiological considerations for efficient nose-to-brain targeting." *Expert Opinion on Drug Delivery* 9 (1):19–31.

Bailey, Mark M., and Cory J. Berkland. 2009. "Nanoparticle formulations in pulmonary drug delivery." *Medicinal Research Reviews* 29 (1):196–212.

Barakat, Abdulwahab, Johannes Kramer, Carvalho de Souza Cristiane, and Claus-Michael Lehr. 2015. "In vitro—in vivo correlation: Shades on some non-conventional dosage forms." *Dissolution Technologies* 22 (2):19–23.

Baroody, F. M. 2007. "Nasal and paranasal sinus anatomy and physiology." *Clin Allergy Immunol* 19:1–21.

Baxter, A., S. Thain, A. Banerjee, L. Haswell, A. Parmar, G. Phillips, and E. Minet (2015). "Targeted omics analyses, and metabolic enzyme activity assays demonstrate maintenance of key mucociliary characteristics in long term cultures of reconstituted human airway epithelia." *Toxicology in Vitro* 29(5): 864–875.

Bechgaard, Erik, Sveinbjörn Gizurarson, Lisbeth Jørgensen, and Rikke Larsen. 1992. "The viability of isolated rabbit nasal mucosa in the Ussing chamber, and the permeability of insulin across the membrane." *International Journal of Pharmaceutics* 87 (1–3):125–132.

Beck-Broichsitter, Moritz, Julia Gauss, Claudia B. Packhaeuser, Kerstin Lahnstein, Thomas Schmehl, Werner Seeger, Thomas Kissel, and Tobias Gessler. 2009. "Pulmonary drug delivery with aerosolizable nanoparticles in an ex vivo lung model." *International Journal of Pharmaceutics* 367 (1):169–178. doi: https://doi.org/10.1016/j.ijpharm.2008.09.017.

Beck-Broichsitter, Moritz, Thomas Schmehl, Werner Seeger, and Tobias Gessler. 2011. "Evaluating the controlled release properties of inhaled nanoparticles using isolated, perfused, and ventilated lung models." *Journal of Nanomaterials* 2011:3.

Behl, C. R., H. K. Pimplaskar, A. P. Sileno, and V. D. Romeo. 1998. "Effects of physicochemical properties and other factors on systemic nasal drug delivery." *Advanced Drug Delivery Reviews* 29 (1):89–116.

Belgamwar, Veena S., Dheeraj S. Chauk, Hitendra S. Mahajan, Snehal A. Jain, Surendra G. Gattani, and Sanjay J. Surana. 2009. "Formulation and evaluation of in situ gelling system of dimenhydrinate for nasal administration." *Pharmaceutical Development and Technology* 14 (3):240–248.

Bi, Ru, Wei Shao, Qun Wang, and Na Zhang. 2008. "Spray-freeze-dried dry powder inhalation of insulin-loaded liposomes for enhanced pulmonary delivery." *Journal of Drug Targeting* 16 (9):639–648.

Bivas-Benita, M., S. Romeijn, H. E. Junginger, and G. Borchard. 2004. "PLGA-PEI nanoparticles for gene delivery to pulmonary epithelium." *European Journal of Pharmaceutics and Biopharmaceutics* 58 (1):1–6. doi: 10.1016/j.ejpb.2004.03.008.

BORGSTROM, Andy Clark Lars, and A Clark. 2002. "In vitro testing of pharmaceutical aerosols and predicting lung deposition from in vitro measurements." In *Drug Delivery to the Lung*, 105–139. Marcel Dekker, Inc. New York.

Brittebo, Eva B. 1982. "Metabolism of progesterone by the nasal mucosa in mice and rats." *Basic & Clinical Pharmacology & Toxicology* 51 (5):441–445.

Brittebo, Eva B. 1985. "Localization of oestradiol in the rat nasal mucosa." *Basic & Clinical Pharmacology & Toxicology* 57 (4):285–290.

Brittebo, Eva B. 1987. "Metabolic activation of phenacetin in rat nasal mucosa: dose-dependent binding to the glands of Bowman." *Cancer Research* 47 (5):1449–1456.

Castell, J. V., M. T. Donato, and M. J. Gomez-Lechon. 2005. "Metabolism and bioactivation of toxicants in the lung. The in vitro cellular approach." *Experimental and Toxicologic Pathology* 57 Suppl 1:189–204.

Chan, Hak-Kim, and Nora Y. K. Chew. 2003. "Novel alternative methods for the delivery of drugs for the treatment of asthma." *Advanced Drug Delivery Reviews* 55 (7):793–805.

Chattopadhyay, Pratibhash, Boris Y. Shekunov, D. Yim, David Cipolla, Brooks Boyd, and Stephen Farr. 2007. "Production of solid lipid nanoparticle suspensions using supercritical fluid extraction of emulsions (SFEE) for pulmonary delivery using the AERx system." *Advanced Drug Delivery Reviews* 59 (6):444–453.

Chen, Alex M., Min Zhang, Dongguang Wei, Dirk Stueber, Oleh Taratula, Tamara Minko, and Huixin He. 2009. "Co-delivery of Doxorubicin and Bcl-2 siRNA by Mesoporous Silica Nanoparticles Enhances the Efficacy of Chemotherapy in Multidrug Resistant Cancer Cells." *Small (Weinheim an der Bergstrasse, Germany)* 5 (23):2673–2677. doi: 10.1002/smll.200900621.

Chen, Shu-Chih, Kristine Eiting, Kunyuan Cui, Alexis Kays Leonard, Daniel Morris, Ching-Yuan Li, Ken Farber, Anthony P. Sileno, Michael E. Houston, and Paul H. Johnson. 2006. "Therapeutic utility of a novel tight junction modulating peptide for enhancing intranasal drug delivery." *Journal of Pharmaceutical Sciences* 95 (6):1364–1371.

Cheng, Yung Sung. 2014. "Mechanisms of Pharmaceutical Aerosol Deposition in the Respiratory Tract." *AAPS PharmSciTech* 15 (3):630–640. doi: 10.1208/s12249-014-0092-0.

Chien, Y. W. and S. F. Chang. 1987. "Intranasal drug delivery for systemic medications." *Crit Rev Ther Drug Carrier Syst* 4 (2):67–194.

Chow, Albert H. L., Henry H. Y. Tong, Pratibhash Chattopadhyay, and Boris Y. Shekunov. 2007. "Particle engineering for pulmonary drug delivery." *Pharmaceutical Research* 24 (3):411–437.

Chung, Francis Y. and Maureen D. Donovan. 1996. "Nasal pre-systemic metabolism of peptide drugs: Substance P metabolism in the sheep nasal cavity." *International Journal of Pharmaceutics* 128 (1–2):229–237.

Corace, Giuseppe, Cristina Angeloni, Marco Malaguti, Silvana Hrelia, Paul C Stein, Martin Brandl, Roberto Gotti, and Barbara Luppi. 2014. "Multifunctional liposomes for nasal delivery of the anti-Alzheimer drug tacrine hydrochloride." *Journal of Liposome Research* 24 (4):323–335.

Costantino, Henry R., Laleh Firouzabadian, Ken Hogeland, Chichih Wu, Chris Beganski, Karen G. Carrasquillo, Melissa Córdova, Kai Griebenow, Stephen E. Zale, and Mark A. Tracy. 2000. "Protein spray-freeze drying. Effect of atomization conditions on particle size and stability." *Pharmaceutical Research* 17 (11):1374–1382.

Costantino, Henry R., Lisbeth Illum, Gordon Brandt, Paul H. Johnson, and Steven C. Quay. 2007. "Intranasal delivery: Physicochemical and therapeutic aspects." *International Journal of Pharmaceutics* 337 (1):1–24.

Cremaschi, D., C. Rossetti, M. T. Draghetti, C. Manzoni, and V. Aliverti. 1991. "Active transport of polypeptides in rabbit nasal mucosa: Possible role in the sampling of potential antigens." *Pflügers Archiv European Journal of Physiology* 419 (5):425–432.

Cryan, Sally-Ann, Neeraj Sivadas, and Lucila Garcia-Contreras. 2007. "In vivo animal models for drug delivery across the lung mucosal barrier." *Advanced Drug Delivery Reviews* 59 (11):1133–1151. doi: https://doi.org/10.1016/j.addr.2007.08.023.

de Matas, Marcel, Qun Shao, Victoria Louise Silkstone, and Henry Chrystyn. 2007. "Evaluation of an in vitro in vivo correlation for nebulizer delivery using artificial neural networks." *Journal of Pharmaceutical Sciences* 96 (12):3293–3303.

de Souza Carvalho, Cristiane, Nicole Daum, and Claus-Michael Lehr. 2014. "Carrier interactions with the biological barriers of the lung: Advanced in vitro models and challenges for pulmonary drug delivery." *Advanced Drug Delivery Reviews* 75 (Supplement C):129–140. doi: https://doi.org/10.1016/j.addr.2014.05.014.

Djupesland, Per Gisle. 2013. "Nasal drug delivery devices: Characteristics and performance in a clinical perspective—A review." *Drug Delivery and Translational Research* 3 (1):42–62.

Dolovich, Myrna B. and D. L. Bailey. 2012. "Positron emission tomography (PET) for assessing aerosol deposition of orally inhaled drug products." *J Aerosol Med Pulm Drug Deliv* 25 Suppl 1: S52–71.

Dolovich, Myrna B., N. M. 1993. *Allergy: Principles and Practice*. St Louis.

Dorato, M. A. 1990. "Overview of inhalation toxicology." *Environmental Health Perspectives* 85:163–170.

Duddu, Sarma P., Steven A. Sisk, Yulia H. Walter, Thomas E. Tarara, Kevin R. Trimble, Andrew R. Clark, Michael A. Eldon, Rebecca C. Elton, Matthew Pickford, and Peter H. Hirst. 2002. "Improved lung delivery from a passive dry powder inhaler using an engineered PulmoSphere® powder." *Pharmaceutical Research* 19 (5):689–695.

Dunber, Craig A., Anthony J. Hickey, and Peter Holzner. 1998. "Dispersion and characterization of pharmaceutical dry powder aerosols." *KONA Powder and Particle Journal* 16:7–45.

Eberl, Stefan, Hak-Kim Chan, and Evangelia Daviskas. 2006. "SPECT imaging for radioaerosol deposition and clearance studies." *Journal of Aerosol Medicine* 19 (1):8–20.

Edwards, D. A., J. Hanes, G. Caponetti, J. Hrkach, A. Ben-Jebria, M. L. Eskew, J. Mintzes, D. Deaver, N. Lotan, and R. Langer. 1997. "Large porous particles for pulmonary drug delivery." *Science* 276 (5320):1868–1871.

Edwards, David A., Robert S. Langer, Rita Vanbever, Jeffrey Mintzes, Jue Wang, and Donghao Chen. 2003. Preparation of novel particles for inhalation. Google Patents.

El-Sherbiny, Ibrahim M., Nancy M. El-Baz, and Magdi H. Yacoub. 2015. "Inhaled nano- and microparticles for drug delivery." *Global Cardiology Science and Practice* 2.

Elhissi, Abdelbary. 2017. "Liposomes for pulmonary drug delivery: The role of formulation and inhalation device design." *Current Pharmaceutical Design* 23 (3):362–372.

Enlow, Elizabeth M., J. Christopher Luft, Mary E. Napier, and Joseph M. DeSimone. 2011. "Potent engineered PLGA nanoparticles by virtue of exceptionally high chemotherapeutic loadings." *Nano Letters* 11 (2):808–813.

Feng, R., Z. Zhang, Z. Li, and G. Huang. 2014. "Preparation and in vitro evaluation of etoposide-loaded PLGA microspheres for pulmonary drug delivery." *Drug Delivery* 21 (3): 185–192

Feng, Ruihua, Zhiyue Zhang, Zhongwen Li, and Guihua Huang. 2014. "Preparation and in vitro evaluation of etoposide-loaded PLGA microspheres for pulmonary drug delivery." *Drug Delivery* 21 (3):185–192.

Forbes, Ben, Per Bäckman, David Christopher, Myrna Dolovich, Bing V. Li, and Beth Morgan. 2015. "In Vitro Testing for Orally Inhaled Products: Developments in Science-Based Regulatory Approaches." *The AAPS Journal* 17 (4):837–852. doi: 10.1208/s12248-015-9763-3.

Foster, K. A., M. L. Avery, M. Yazdanian, and K. L. Audus. 2000. "Characterization of the Calu-3 cell line as a tool to screen pulmonary drug delivery." *International Journal of Pharmaceutics* 208 (1–2):1–11.

Frieke Kuper, C., Mariska Gröllers-Mulderij, Thérèse Maarschalkerweerd, Nicole M. M. Meulendijks, Astrid Reus, Frédérique van Acker, Esther K. Zondervan-van den Beuken, Mariëlle E. L. Wouters, Sabina Bijlsma, and Ingeborg M. Kooter. 2015. "Toxicity assessment of aggregated/agglomerated cerium oxide nanoparticles in an in vitro 3D airway model: The influence of mucociliary clearance." *Toxicology In Vitro* 29 (2):389–397. doi: https://doi.org/10.1016/j.tiv.2014.10.017.

Fröhlich, Eleonore, and Sharareh Salar-Behzadi. 2014. "Toxicological Assessment of Inhaled Nanoparticles: Role of in Vivo, ex Vivo, in Vitro, and in Silico Studies." *International Journal of Molecular Sciences* 15 (3):4795.

Gaber, Nazar Noureen, Yusrida Darwis, Kok-Khiang Peh, and Yvonne Tze-Fung Tan. 2006. "Characterization of polymeric micelles for pulmonary delivery of beclomethasone dipropionate." *Journal of Nanoscience and Nanotechnology* 6 (9–1):3095–3101.

Gavini, Elisabetta, Anne Bee Hegge, G. Rassu, V. Sanna, Cecilia Testa, G. Pirisino, Jan Karlsen, and Paolo Giunchedi. 2006. "Nasal administration of carbamazepine using chitosan microspheres: In vitro/in vivo studies." *International Journal of Pharmaceutics* 307 (1):9–15.

Gavini, Elisabetta, Giovanna Rassu, Vanna Sanna, Massimo Cossu, and Paolo Giunchedi. 2005. "Mucoadhesive microspheres for nasal administration of an antiemetic drug, metoclopramide: In-vitro/ex-vivo studies." *Journal of Pharmacy and Pharmacology* 57 (3):287–294.

Gehr, Peter, Marianne Bachofen, and Ewald R. Weibel. 1978. "The normal human lung: Ultrastructure and morphometric estimation of diffusion capacity." *Respiration Physiology* 32 (2):121–140. doi: https://doi.org/10.1016/0034-5687(78)90104-4.

Gilani, Kambiz, Abdolhossien Rouholamini Najafabadi, Mohammadali Barghi, and Morteza Rafiee-Tehrani. 2005. "The effect of water to ethanol feed ratio on physical properties and aerosolization behavior of spray dried cromolyn sodium particles." *Journal of Pharmaceutical Sciences* 94 (5):1048–1059.

Gizurarson, Sveinbjörn. 2015. "The effect of cilia and the mucociliary clearance on successful drug delivery." *Biological and Pharmaceutical Bulletin* 38 (4):497–506.

Graff, Candace L. and Gary M. Pollack. 2015. "Nasal Drug Administration: Potential for Targeted Central Nervous System Delivery." *Journal of Pharmaceutical Sciences* 94 (6):1187–1195.

Haque, Shadabul, Shadab Md, Mohammad Fazil, Manish Kumar, Jasjeet Kaur Sahni, Javed Ali, and Sanjula Baboota. 2012. "Venlafaxine loaded chitosan NPs for brain targeting: Pharmacokinetic and pharmacodynamic evaluation." *Carbohydrate Polymers* 89 (1):72–79.

Harkema, Jack R., Stephan A. Carey, and James G. Wagner. 2006. "The Nose Revisited: A Brief Review of the Comparative Structure, Function, and Toxicologic Pathology of the Nasal Epithelium." *Toxicologic Pathology* 34 (3):252–269. doi: 10.1080/01926230600713475.

Harris, A. S. 1993. "Review: Clinical opportunities provided by the nasal administration of peptides." *Journal of Drug Targeting* 1 (2):101–16. doi: 10.3109/10611869308996066.

Henry, R. J., N. Ruano, D. Casto, and R. H. Wolf. 1998. "A pharmacokinetic study of midazolam in dogs: Nasal drop vs. atomizer administration." *Pediatric Dentistry* 20 (5):321–326.

Hidalgo, I. J., T. J. Raub, and R. T. Borchardt. 1989. "Characterization of the human colon carcinoma cell line (Caco-2) as a model system for intestinal epithelial permeability." *Gastroenterology* 96 (3):736–749.

Hoang, Vu Dang, Agu Remigius Uchenna, Jorissen Mark, Kinget Renaat, and Verbeke Norbert. 2002. "Characterization of human nasal primary culture systems to investigate peptide metabolism." *International Journal of Pharmaceutics* 238 (1):247–256.

Hu, Xiao, Fei-Fei Yang, Xiao-Lan Wei, Guang-Yin Yao, Chun-Yu Liu, Ying Zheng, and Yong-Hong Liao. 2017. "Curcumin Acetate Nanocrystals for Sustained Pulmonary Delivery: Preparation, Characterization and In Vivo Evaluation." *Journal of Biomedical Nanotechnology* 13 (1):99–109.

Huh, D., B. D. Matthews, A. Mammoto, M. Montoya-Zavala, H. Y. Hsin, and D. E. Ingber. 2010. "Reconstituting organ-level lung functions on a chip." *Science* 328 (5986):1662–1668. doi: 10.1126/science.1188302.

Hussain, Anwar A., Shinichiro Hirai, and Rima Bawarshi. 1981. "Nasal absorption of natural contraceptive steroids in rats—Progesterone absorption." *Journal of Pharmaceutical Sciences* 70 (4):466–467.

Hussain, Munir A., and Bruce J. Aungst. 1994. "Nasal mucosal metabolism of an LH-RH fragment and inhibition with boroleucine." *International Journal of Pharmaceutics* 105 (1):7–10.

Illum, Lisbeth. 2000. "Transport of drugs from the nasal cavity to the central nervous system." *European Journal of Pharmaceutical Sciences* 11 (1):1–18.

Illum, Lisbeth. 2002. "Nasal drug delivery: New developments and strategies." *Drug Discovery Today* 7 (23):1184–1189.

Illum, Lisbeth. 2003. "Nasal drug delivery—Possibilities, problems and solutions." *Journal of Controlled Release* 87 (1):187–198.
Illum, Lisbeth. 2012. "Nasal drug delivery—Recent developments and future prospects." *Journal of Controlled Release* 161 (2):254–263.
Jacobs, Claudia, and Rainer Helmut Müller. 2002. "Production and Characterization of a Budesonide Nanosuspension for Pulmonary Administration." *Pharmaceutical Research* 19 (2):189–194. doi: 10.1023/a:1014276917363.
Jadhav, Kisan R., Manoj N. Gambhire, Ishaque M. Shaikh, Vilarsrao J. Kadam, and Sambjahi S. Pisal. 2007. "Nasal drug delivery system-factors affecting and applications." *Current Drug Therapy* 2 (1):27–38.
Javia, Ankit, and Hetal Thakkar. 2017. "Intranasal delivery of tapentadol hydrochloride–loaded chitosan nanoparticles: Formulation, characterisation and its in vivo evaluation." *Journal of Microencapsulation* 1–15.
Jawahar, N. and Gowtham Reddy. 2012. "Nanoparticles: A novel pulmonary drug delivery system for tuberculosis." *Journal of Pharmaceutical Sciences and Research* 4 (8):1901–1906.
Kataoka, Kazunori, Atsushi Harada, and Yukio Nagasaki. 2001. "Block copolymer micelles for drug delivery: Design, characterization and biological significance." *Advanced Drug Delivery Reviews* 47 (1):113–131.
Kaur, Ranjot, Tarun Garg, Basant Malik, Umesh Datta Gupta, Pushpa Gupta, Goutam Rath, and Amit Kumar Goyal. 2016. "Development and characterization of spray-dried porous nanoaggregates for pulmonary delivery of anti-tubercular drugs." *Drug Delivery* 23 (3):872–877.
Kelly, James T., Bahman Asgharian, and Brian A. Wong. 2005. "Inertial particle deposition in a monkey nasal mold compared with that in human nasal replicas." *Inhalation Toxicology* 17 (14):823–830.
Kelly, Jennifer Y., and Joseph M. DeSimone. 2008. "Shape-specific, monodisperse nano-molding of protein particles." *Journal of the American Chemical Society* 130 (16):5438–5439.
Kissel, Thomas, Juergen Drewe, Siegfried Bantle, Andreas Rummelt, and Christoph Beglinger. 1992. "Tolerability and absorption enhancement of intranasally administered octreotide by sodium taurodihydrofusidate in healthy subjects." *Pharmaceutical Research* 9 (1):52–57.
Kissel, Thomas and Ute Werner. 1998. "Nasal delivery of peptides: An in vitro cell culture model for the investigation of transport and metabolism in human nasal epithelium." *Journal of Controlled Release* 53 (1):195–203.
Klein, Sebastian G., Tommaso Serchi, Lucien Hoffmann, Brunhilde Blömeke, and Arno C. Gutleb. 2013. "An improved 3D tetraculture system mimicking the cellular organisation at the alveolar barrier to study the potential toxic effects of particles on the lung." *Particle and Fibre Toxicology* 10 (1):31.
Konar, Dipasri, Mahesh Devarasetty, Didem V. Yildiz, Anthony Atala, and Sean V. Murphy. 2016. "Lung-On-A-Chip Technologies for Disease Modeling and Drug Development." *Biomedical Engineering and Computational Biology* 7 (Suppl 1):17.
Kubo, Hiroyuki, Ken-Ichi Hosoya, Hideshi Natsume, Kenji Sugibayashi, and Yasunori Morimoto. 1994. "In vitro permeation of several model drugs across rabbit nasal mucosa." *International Journal of Pharmaceutics* 103 (1):27–36.
Kumar, Manoj, Ravi Shankar Pandey, Kartik Chandra Patra, Sunil Kumar Jain, Muarai Lal Soni, Jawahar Singh Dangi, and Jitender Madan. 2013. "Evaluation of neuropeptide loaded trimethyl chitosan nanoparticles for nose to brain delivery." *International Journal of Biological Macromolecules* 61:189–195.

Kumar, Mukesh, Kamla Pathak, and Ambikanandan Misra. 2009. "Formulation and characterization of nanoemulsion-based drug delivery system of risperidone." *Drug Development and Industrial Pharmacy* 35 (4):387–395.

Labiris, N. R. and M. B. Dolovich. 2003. "Pulmonary drug delivery. Part I: Physiological factors affecting therapeutic effectiveness of aerosolized medications." *British Journal of Clinical Pharmacology* 56 (6):588–599.

Lambros, Maria Polikandritou, David W. A. Bourne, Syed Ali Abbas, and David L. Johnson. 1997. "Disposition of aerosolized liposomal amphotericin B." *Journal of Pharmaceutical Sciences* 86 (9):1066–1069.

Lancaster, R. W., H. Singh, and A. L. Theophilus. 2000. "Apparatus and process for preparing crystalline particles." *WO World IPO* 38811.

Leach, W. Thomas, Dale T. Simpson, Tibisay N. Val, Efemona C. Anuta, Zhongshui Yu, Robert O. Williams, and Keith P. Johnston. 2005. "Uniform encapsulation of stable protein nanoparticles produced by spray freezing for the reduction of burst release." *Journal of Pharmaceutical Sciences* 94 (1):56–69.

Lupo, C., L. Lodi, M. Canonaco, A. Valenti, and F. Dessi-Fulgheri. 1986. "Testosterone metabolism in the olfactory epithelium of intact and castrated male rats." *Neuroscience Letters* 69 (3):259–262.

Luppi, Barbara, Federica Bigucci, Giuseppe Corace, Alice Delucca, Teresa Cerchiara, Milena Sorrenti, Laura Catenacci, Anna Maria Di Pietra, and Vittorio Zecchi. 2011. "Albumin nanoparticles carrying cyclodextrins for nasal delivery of the anti-Alzheimer drug tacrine." *European Journal of Pharmaceutical Sciences* 44 (4):559–565.

Maa, Yuh-Fun and Phuong-Anh Nguyen. 2001. Method of spray freeze drying proteins for pharmaceutical administration. Google Patents.

Maa, Yuh-Fun, Phuong-Anh Nguyen, Theresa Sweeney, Steven J. Shire, and Chung C. Hsu. 1999. "Protein inhalation powders: Spray drying vs spray freeze drying." *Pharmaceutical Research* 16 (2):249–254.

Maa, Yuh-Fun and Steven J. Prestrelski. 2000. "Biopharmaceutical powders particle formation and formulation considerations." *Current Pharmaceutical Biotechnology* 1 (3):283–302.

Mahajan, Hitendra S., Milind S. Mahajan, Pankaj P. Nerkar, and Anshuman Agrawal. 2014. "Nanoemulsion-based intranasal drug delivery system of saquinavir mesylate for brain targeting." *Drug Delivery* 21 (2):148–154.

Maitani, Yoshie, Kouji Nakamura, Hiroshi Suenaga, Katsuo Kamata, Kozo Takayama, and Tsuneji Nagai. 2000. "The enhancing effect of soybean-derived sterylglucoside and β-sitosterol β-D-glucoside on nasal absorption in rabbits." *International Journal of Pharmaceutics* 200 (1):17–26.

Mainelis, G., S. Seshadri, O. B. Garbuzenko, T. Han, Z. Wang and T. Minko. 2013. "Characterization and Application of a Nose-Only Exposure Chamber for Inhalation Delivery of Liposomal Drugs and Nucleic Acids to Mice." *Journal of Aerosol Medicine and Pulmonary Drug Delivery* 26 (6): 345–354.

Mall, Marcus A. 2008. "Role of cilia, mucus, and airway surface liquid in mucociliary dysfunction: Lessons from mouse models." *Journal of Aerosol Medicine and Pulmonary Drug Delivery* 21 (1):13–24.

Marple, Virgil A., Bernard A. Olson, Kumaragovindham Santhanakrishnan, Jolyon P. Mitchell, Sharon C. Murray, and Buffy L. Hudson-Curtis. 2003. "Next generation pharmaceutical impactor (a new impactor for pharmaceutical inhaler testing). Part II: Archival calibration." *Journal of Aerosol Medicine* 16 (3):301–324.

Morin, J. P., J. M. Baste, A. Gay, C. Crochemore, C. Corbiere, and C. Monteil. 2013. "Precision cut lung slices as an efficient tool for in vitro lung physio-pharmacotoxicology studies." *Xenobiotica* 43 (1):63–72. doi: 10.3109/00498254.2012.727043.

Mosén, Kristina, Kjell Bäckström, Kyrre Thalberg, Torben Schaefer, Henning G. Kristensen, and Anders Axelsson. 2005. "Particle formation and capture during spray drying of inhalable particles." *Pharmaceutical Development and Technology* 9 (4):409–417.

Movia, Dania, Luisana Di Cristo, Roaa Alnemari, Joseph E. McCarthy, Hanane Moustaoui, Marc Lamy de la Chapelle, Jolanda Spadavecchia, Yuri Volkov, and Adriele Prina-Mello. 2017. "The curious case of how mimicking physiological complexity in in vitro models of the human respiratory system influences the inflammatory responses. A preliminary study focused on gold nanoparticles." *Journal of Interdisciplinary Nanomedicine.*

Müller, R. H., C. Jacobs, and O. Kayser. 2001. "Nanosuspensions as particulate drug formulations in therapy: Rationale for development and what we can expect for the future." *Advanced Drug Delivery Reviews* 47 (1):3–19.

Muralidharan, P., D. Hayes, Jr., and H. M. Mansour. 2015. "Dry powder inhalers in COPD, lung inflammation and pulmonary infections." *Expert Opin Drug Deliv* 12 (6):947–962. doi: 10.1517/17425247.2015.977783.

Mygind, Niels and Ronald Dahl. 1998. "Anatomy, physiology and function of the nasal cavities in health and disease." *Advanced Drug Delivery Reviews* 29 (1):3–12.

Nahar, K., N. Gupta, R. Gauvin, S. Absar, B. Patel, V. Gupta, A. Khademhosseini, and F. Ahsan. 2013. "In vitro, in vivo and ex vivo models for studying particle deposition and drug absorption of inhaled pharmaceuticals." *Eur J Pharm Sci* 49 (5):805–818. doi: 10.1016/j.ejps.2013.06.004.

Nakamura, K., Y. Maitani, and K. Takayama. 2002. "The enhancing effect of nasal absorption of FITC-dextran 4,400 by β-sitosterol β-D-glucoside in rabbits." *Journal of Controlled Release* 79 (1):147–155.

Nassimi, M., C. Schleh, H. D. Lauenstein, R. Hussein, H. G. Hoymann, W. Koch, G. Pohlmann, N. Krug, K. Sewald, S. Rittinghausen, A. Braun, and C. Muller-Goymann. 2010. "A toxicological evaluation of inhaled solid lipid nanoparticles used as a potential drug delivery system for the lung." *Eur J Pharm Biopharm* 75 (2):107–116. doi: 10.1016/j.ejpb.2010.02.014.

Östh, Karin, Mattias Paulsson, Erik Björk, and Katarina Edsman. 2002. "Evaluation of drug release from gels on pig nasal mucosa in a horizontal Ussing chamber." *Journal of Controlled Release* 83 (3): 377–388.

Palakodaty, Srinivas, Peter York, and John Pritchard. 1998. "Supercritical fluid processing of materials from aqueous solutions: The application of SEDS to lactose as a model substance." *Pharmaceutical Research* 15 (12):1835–1843.

Paliwal, Rishi, R Jayachandra Babu, and Srinath Palakurthi. 2014. "Nanomedicine scale-up technologies: Feasibilities and challenges." *AAPS PharmSciTech* 15 (6):1527–1534.

Park, Ju-Hwan, Hyo-Eon Jin, Dae-Duk Kim, Suk-Jae Chung, Won-Sik Shim, and Chang-Koo Shim. 2013. "Chitosan microspheres as an alveolar macrophage delivery system of ofloxacin via pulmonary inhalation." *International Journal of Pharmaceutics* 441 (1):562–569. doi: https://doi.org/10.1016/j.ijpharm.2012.10.044.

Patel, Vishal R., and Y. K. Agrawal. 2011. "Nanosuspension: An approach to enhance solubility of drugs." *Journal of Advanced Pharmaceutical Technology & Research* 2 (2):81–87. doi: 10.4103/2231-4040.82950.

Patil-Gadhe, Arpana, Abhay Kyadarkunte, Milind Patole, and Varsha Pokharkar. 2014. "Montelukast-loaded nanostructured lipid carriers: Part II Pulmonary drug delivery and in vitro–in vivo aerosol performance." *European Journal of Pharmaceutics and Biopharmaceutics* 88 (1):169–177. doi: https://doi.org/10.1016/j.ejpb.2014.07.007.

Patil, J. S., and S. Sarasija. 2012a. "Pulmonary drug delivery strategies: A concise, systematic review." *Lung India: Official Organ of Indian Chest Society* 29 (1):44–49. doi: 10.4103/0970-2113.92361.

Patil, J. S., and S. Sarasija. 2012b. "Pulmonary drug delivery strategies: A concise, systematic review." *Lung India: Official Organ of Indian Chest Society* 29 (1):44.

Patton, John S. 1996. "Mechanisms of macromolecule absorption by the lungs." *Advanced Drug Delivery Reviews* 19 (1):3–36. doi: https://doi.org/10.1016/0169-409X(95)00113-L.

Patton, John S. and Peter R. Byron. 2007. "Inhaling medicines: Delivering drugs to the body through the lungs." *Nature Reviews. Drug Discovery* 6 (1):67.

Pawar, Dilip, Sharad Mangal, Roshan Goswami, and K. S. Jaganathan. 2013. "Development and characterization of surface modified PLGA nanoparticles for nasal vaccine delivery: Effect of mucoadhesive coating on antigen uptake and immune adjuvant activity." *European Journal of Pharmaceutics and Biopharmaceutics* 85 (3):550–559.

Peter, Hagen Georg. 1998. "Cell culture sheets to study nasal peptide metabolism: The human nasal RPMI 2650 cell line model."

Pezron, Isabelle, Ashim K. Mitra, Sridhar Duvvuri, and Giridhar S. Tirucherai. 2002. "Prodrug strategies in nasal drug delivery." *Expert Opinion on Therapeutic Patents* 12 (3):331–340.

Platz, Robert M., John S. Patton, Linda Foster, and Mohammed Eljamal. 2003. Spray drying of macromolecules to produce inhaleable dry powders. Google Patents.

Rahhal, T. B., C. A. Fromen, E. M. Wilson, M. P. Kai, T. W. Shen, J. C. Luft, and J. M. DeSimone. 2016. "Pulmonary Delivery of Butyrylcholinesterase as a Model Protein to the Lung." *Molecular Pharmaceutics* 13 (5):1626–1635.

Rang, Humphrey, Maureen Dale, James Ritter, and Rod Flower. 2007. *Pharmacology*. 6th ed: Churchill Livingstone Elsevier Publications.

Rau, J. L. 2005. "The inhalation of drugs: Advantages and problems." *Respiratory Care* 50 (3):367–382.

Rehman, Mahboob, Boris Y. Shekunov, Peter York, David Lechuga-Ballesteros, Danforth P. Miller, Trixie Tan, and Paul Colthorpe. 2004. "Optimisation of powders for pulmonary delivery using supercritical fluid technology." *European Journal of Pharmaceutical Sciences* 22 (1):1–17.

Rhidian, Rhys, and Ben Greatorex. 2015. "Chest pain in the recovery room, following topical intranasal cocaine solution use." *BMJ Case Reports* 2015:bcr2015209698.

Rogers, True L., Keith P. Johnston, and Robert O. Williams III. 2001. "Solution-based particle formation of pharmaceutical powders by supercritical or compressed fluid CO2 and cryogenic spray-freezing technologies." *Drug Development and Industrial Pharmacy* 27 (10):1003–1015.

Rojanarat, Wipaporn, Narumon Changsan, Ekawat Tawithong, Sirirat Pinsuwan, Hak-Kim Chan, and Teerapol Srichana. 2011. "Isoniazid proliposome powders for inhalation—Preparation, characterization and cell culture studies." *International Journal of Molecular Sciences* 12 (7):4414–4434.

Rolland, Jason P., Benjamin W. Maynor, Larken E. Euliss, Ansley E. Exner, Ginger M. Denison, and Joseph M. DeSimone. 2005. "Direct fabrication and harvesting of monodisperse, shape-specific nanobiomaterials." *Journal of the American Chemical Society* 127 (28):10096–10100.

Rothen-Rutishauser, Barbara, Fabian Blank, Christian Mühlfeld, and Peter Gehr. 2008. "In vitro models of the human epithelial airway barrier to study the toxic potential of particulate matter." *Expert Opinion on Drug Metabolism & Toxicology* 4 (8):1075–1089. doi: 10.1517/17425255.4.8.1075.

Rothen-Rutishauser, Barbara M., Stephen G. Kiama, and Peter Gehr. 2005. "A three-dimensional cellular model of the human respiratory tract to study the interaction with particles." *American Journal of Respiratory Cell and Molecular Biology* 32 (4):281–289.

Ruffin, R. E., M. B. Dolovich, R. K. Wolff, and M. T. Newhouse. 1978. "The Effects of Preferential Deposition of Histamine in the Human Airway 1–4." *American Review of Respiratory Disease* 117 (3):485–492.

Russo, Paola, Cecilia Sacchetti, Irene Pasquali, Ruggero Bettini, Gina Massimo, Paolo Colombo, and Alessandra Rossi. 2006. "Primary microparticles and agglomerates of morphine for nasal insufflation." *Journal of Pharmaceutical Sciences* 95 (12):2553–2561.

Sakagami, Masahiro. 2006. "In vivo, in vitro and ex vivo models to assess pulmonary absorption and disposition of inhaled therapeutics for systemic delivery." *Advanced Drug Delivery Reviews* 58 (9):1030–1060.

Samson, Géraldine, Alicia García de la Calera, Sophie Dupuis-Girod, Frédéric Faure, Evelyne Decullier, Gilles Paintaud, Céline Vignault, Jean-Yves Scoazec, Christine Pivot, and Henri Plauchu. 2012. "Ex vivo study of bevacizumab transport through porcine nasal mucosa." *European Journal of Pharmaceutics and Biopharmaceutics* 80 (2):465–469.

Sarkar, Mohamadi A. 1992. "Drug metabolism in the nasal mucosa." *Pharmaceutical Research* 9 (1):1–9.

Schipper, Nicolaas G. M., J. Coos Verhoef, and Frans W. H. M. Merkus. 1991. "The nasal mucociliary clearance: Relevance to nasal drug delivery." *Pharmaceutical Research* 8 (7):807–814.

Schmidt, M. Christiane, Hagen Peter, Steffen R. Lang, Günter Ditzinger, and Hans P. Merkle. 1998. "In vitro cell models to study nasal mucosal permeability and metabolism." *Advanced Drug Delivery Reviews* 29 (1):51–79.

Schön, Peter, Georgios Ctistis, Wouter Bakker, and Gregor Luthe. 2017. "Nanoparticular surface-bound PCBs, PCDDs, and PCDFs—A novel class of potentially higher toxic POPs." *Environmental Science and Pollution Research* 24 (14):12758–12766. doi: 10.1007/s11356-016-6211-6.

Seju, U., A. Kumar, and K. K. Sawant. 2011. "Development and evaluation of olanzapine-loaded PLGA nanoparticles for nose-to-brain delivery: In vitro and in vivo studies." *Acta Biomaterialia* 7 (12):4169–4176.

Sharma, Deepak, Rakesh Kumar Sharma, Navneet Sharma, Reema Gabrani, Sanjeev K. Sharma, Javed Ali, and Shweta Dang. 2015. "Nose-to-brain delivery of PLGA-diazepam nanoparticles." *AAPS PharmSciTech* 16 (5):1108–1121.

Shekunov, Boris Y., Pratibhash Chattopadhyay, Henry H. Y. Tong, and Albert H. L. Chow. 2007. "Particle size analysis in pharmaceutics: Principles, methods and applications." *Pharmaceutical Research* 24 (2):203–227.

Shekunov, B. Y., B. Chattopadhyay, A. Gibson, and C. Lehmkuhl. 2006. "Influence of spray-freezing parameters on particle size and morphology of insulin." Proceedings of the Conference on Respiratory Drug Delivery, Boca Raton, Florida.

Shekunov, B. Y., B. Chattopadhyay, and J. Seitzinger. 2004. "Production of respirable particles using spray-freeze-drying with compressed CO2." Proceedings of the Conference on Respiratory Drug Delivery, Palm Springs, California.

Silkstone, V. L., J. H. Dennis, C. A. Pieron, and H. Chrystyn. 2002. "An Investigation of In Vitro/In Vivo Correlations for Salbutamol Nebulized by Eight Systems." *Journal of Aerosol Medicine* 15 (3):251–259. doi: 10.1089/089426802760292591.

Singh, Reena M. P., Anil Kumar, and Kamla Pathak. 2013. "Thermally triggered mucoadhesive in situ gel of loratadine: β-cyclodextrin complex for nasal delivery." *AAPS PharmSciTech* 14 (1):412–424.

Steele, Vernon E., and Julia T. Arnold. 1985. "Isolation and long-term culture of rat, rabbit, and human nasal turbinate epithelial cells." *In Vitro Cellular & Developmental Biology-Plant* 21 (12):681–687.

Suarez, S., L. Garcia-Contreras, D. Sarubbi, E. Flanders, D. O'Toole, J. Smart, and A. J. Hickey. 2001. "Facilitation of pulmonary insulin absorption by H-MAP: Pharmacokinetics and pharmacodynamics in rats." *Pharm Res* 18 (12):1677–1684.

Sung, J. C., D. J. Padilla, L. Garcia-Contreras, J. L. Verberkmoes, D. Durbin, C. A. Peloquin, K. J. Elbert, A. J. Hickey, and D. A. Edwards. 2009. "Formulation and pharmacokinetics of self-assembled rifampicin nanoparticle systems for pulmonary delivery." *Pharm Res* 26 (8):1847–1855.

Taylor, K. M., and J. M. Newton. 1992. "Liposomes for controlled delivery of drugs to the lung." *Thorax* 47 (4):257.

Tewes, Frederic, Oliviero L. Gobbo, Carsten Ehrhardt, and Anne Marie Healy. 2016. "Amorphous calcium carbonate based-microparticles for peptide pulmonary delivery." *ACS Applied Materials & Interfaces* 8 (2):1164–1175.

Tiwari, Shailja, Amit K. Goyal, Neeraj Mishra, Bhuvaneshwar Vaidya, Abhinav Mehta, Devyani Dube, and Suresh P. Vyas. 2009. "Liposome in situ gelling system: Novel carrier based vaccine adjuvant for intranasal delivery of recombinant protein vaccine." *Procedia in Vaccinology* 1 (1):148–163.

Türker, Selcan, Erten Onur, and Yekta Ózer. 2004. "Nasal route and drug delivery systems." *Pharmacy World and Science* 26 (3):137–142.

Ugwoke, Michael I., Remigius U. Agu, Norbert Verbeke, and Renaat Kinget. 2005. "Nasal mucoadhesive drug delivery: Background, applications, trends and future perspectives." *Advanced Drug Delivery Reviews* 57 (11):1640–1665.

Ugwoke, Michael Ikechukwu, Remigius Uchenna Agu, Mark Jorissen, Patrick Augustijns, Raf Sciot, Norbert Verbeke, and Renaat Kinget. 2000. "Nasal toxicological investigations of Carbopol 971P formulation of apomorphine: Effects on ciliary beat frequency of human nasal primary cell culture and in vivo on rabbit nasal mucosa." *European Journal of Pharmaceutical Sciences* 9 (4):387–396.

Upadhyay, Pratik, Jatin Trivedi, Kilambi Pundarikakshudu, and Navin Sheth. 2017. "Direct and enhanced delivery of nanoliposomes of anti schizophrenic agent to the brain through nasal route." *Saudi Pharmaceutical Journal* 25 (3):346–358.

USP. 2017. "USP< 601> Aerosols, nasal sprays, metered-dose inhalers and dry powder inhalers." *USP 40-NF 35.*

Vogel, P., V. R. Rivera, M. L. Pitt, and M. A. Poli. 1996. "Comparison of the pulmonary distribution and efficacy of antibodies given to mice by intratracheal instillation or aerosol inhalation." *Laboratory Animal Science* 46 (5):516–523.

Vyas, Tushar K., Aliasgar Shahiwala, Sudhanva Marathe, and Ambikanandan Misra. 2005. "Intranasal drug delivery for brain targeting." *Current Drug Delivery* 2 (2):165–175.

Wachtel, H., O. Ertunc, C. Koksoy, and A. Delgado. 2008. "Aerodynamic optimization of Handihaler and Respimat: The roles of computational fluid dynamics and flow visualization." *Respiratory Drug Delivery* 1 (2008):165–174.

Wadell, Cecilia, Erik Björk, and Ola Camber. 1999. "Nasal drug delivery–evaluation of an in vitro model using porcine nasal mucosa." *European Journal of Pharmaceutical Sciences* 7 (3):197–206.

Wanner, Adam, Matthias Salathé, and Thomas G O'Riordan. 1996. "Mucociliary clearance in the airways." *American Journal of Respiratory and Critical Care Medicine* 154 (6):1868–1902.

Wanning, Stefan, Richard Süverkrüp, and Alf Lamprecht. 2015. "Pharmaceutical spray freeze drying." *International Journal of Pharmaceutics* 488 (1):136–153.

Wengst, Annette, and Stephan Reichl. 2010. "RPMI 2650 epithelial model and three-dimensional reconstructed human nasal mucosa as in vitro models for nasal permeation studies." *European Journal of Pharmaceutics and Biopharmaceutics* 74 (2):290–297.

Werner, Ute, and Thomas Kissel. 1995. "Development of a human nasal epithelial cell culture model and its suitability for transport and metabolism studies under in vitro conditions." *Pharmaceutical Research* 12 (4):565–571.

Wheatley, M. A., J. Dent, E. B. Wheeldon, and P. L. Smith. 1988. "Nasal drug delivery: An in vitro characterization of transepithelial electrical properties and fluxes in the presence or absence of enhancers." *Journal of Controlled Release* 8 (2):167–177.

Wilczewska, A. Z., K. Niemirowicz, K. H. Markiewicz, and H. Car. 2012. "Nanoparticles as drug delivery systems." *Pharmacology Reports* 64 (5):1020–1037.

Williams III, Robert O., Keith P. Johnston, Timothy J. Young, True L. Rogers, Melisa K. Barron, Zhongshui Yu, and Jiahui Hu. 2005. Process for production of nanoparticles and microparticles by spray freezing into liquid. Google Patents.

Wlaz, P., S. Knaga, K. Kasperek, A. Wlaz, E. Poleszak, G. Jezewska-Witkowska, S. Winiarczyk, E. Wyska, T. Heinekamp, and C. Rundfeldt. 2015. "Activity and Safety of Inhaled Itraconazole Nanosuspension in a Model Pulmonary Aspergillus fumigatus Infection in Inoculated Young Quails." *Mycopathologia* 180 (1–2):35–42. doi: 10.1007/s11046-015-9885-2.

Xu, Jing, Dominica H. C. Wong, James D. Byrne, Kai Chen, Charles Bowerman, and Joseph M. DeSimone. 2013. "Future of the particle replication in nonwetting templates (PRINT) technology." *Angewandte Chemie International Edition* 52 (26):6580–6589.

Yamamoto, Tomonaga, Yoshie Maitani, Taeko Ando, Koichi ISOWA, Kozo TAKAYAMA, and Tsuneji NAGAI. 1998. "High absorbency and subchronic morphologic effects on the nasal epithelium of a nasal insulin powder dosage form with a soybean-derived sterylglucoside mixture in rabbits." *Biological and Pharmaceutical Bulletin* 21 (8):866–870.

Yang, Chun, Hongwu Gao, and Ashim K Mitra. 2001. "Chemical stability, enzymatic hydrolysis, and nasal uptake of amino acid ester prodrugs of acyclovir." *Journal of Pharmaceutical Sciences* 90 (5):617–624.

Yang, Yan, Nimisha Bajaj, Peisheng Xu, Kimberly Ohn, Michael D. Tsifansky, and Yoon Yeo. 2009. "Development of highly porous large PLGA microparticles for pulmonary drug delivery." *Biomaterials* 30 (10):1947–1953.

Yu, Zhongshui, True L. Rogers, Jiahui Hu, Keith P. Johnston, and Robert O. Williams. 2002. "Preparation and characterization of microparticles containing peptide produced by a novel process: Spray freezing into liquid." *European Journal of Pharmaceutics and Biopharmaceutics* 54 (2):221–228.

5

Ocular Drug Delivery Systems

Shubhini A. Saraf, Jovita Kanoujia, Samipta Singh, and Shailendra K. Saraf

CONTENTS

5.1 Introduction

Quality of life can be directly affected by vision impairment or ocular diseases. A survey from 39 countries revealed that about 285 million people suffer from visual impairment. Among these, 65% are over 50 years old, and 82% of blind patients are over 50 (Pascolini and Mariotti 2012). Ocular drug delivery (ODD) is one of the most challenging and competitive fields for pharmaceutical researchers (Gaudana et al. 2008). Delivery of an active pharmaceutical ingredient to the eye has been a great challenge due to the complex biochemistry, anatomy, and physiology of this organ (Kanoujia, Kushwaha, and Saraf 2014; Dey and Mitra 2005). Despite significant improvements in ODD, much is to be accomplished for maintaining the requisite quantity of drugs at the site of action for an extended period of time (Saettone and Salminen 1995). Ideally, the drug delivery system must be able to increase ocular drug absorption, have excellent corneal penetration, and infrequent administration with a low toxicity (Keister et al. 1991).

Chapter 5 offers an insight into conventional ODD systems. It also presents a brief overview of novel drug delivery systems in ocular diseases with special reference to the evaluation of these systems through in-vitro studies, toxicological studies, and in-vivo animal pharmacokinetics. The focus is on both novel products and differentiated generics.

General procedures for evaluation of any ocular dosage form include drug content, drug release, ocular irritation, stability of dosage form, and so on. These have been discussed in the section "general evaluation parameters" mentioned below. The specific evaluation methods have been mentioned alongside the brief overview of respective ODD categories, and can be viewed holistically along with the general methods.

5.2 General Evaluation Parameters

5.2.1 Sterility Testing

Sterility testing is performed to assure the sterilization of ocular drug delivery systems and no microbial growth must be observed in any ocular formulation. Sterility testing is done by a direct inoculation method or membrane filtration method, as per procedures mentioned in the respective pharmacopoeias. For the membrane filtration method, membrane filters of 50 mm with a pore size no greater than 0.45 μm are used. A suitable amount of the sample (10–100 ml) is transferred to the membrane filter. The membranes are collected, washed with an appropriate quantity of a suitable liquid, and then incubated. For direct inoculation, the contents are transferred to the culture media and are incubated for no more than 5 days. In both the cases, these are compared with the control (positive and negative control). The positive control needs the routine use of a known solution containing fixed number of microorganisms (IP 2010).

5.2.2 Stability Studies

Stability studies are carried out according to the International Conference on Harmonisation of Technical Requirements for Registration of Pharmaceuticals for Human Use guidelines by storing replicates of ocular formulations; packed in the container in which formulation is proposed to be marketed (often sealed glass vials in an inert atmosphere and away from light); in a stability chamber at 4°C/25% relative humidity (RH), 25°C/60% RH, and 40°C/75% RH. Samples are withdrawn at predefined time intervals and the product is checked for particle size, drug content, percent entrapment efficiency (EE), etc. These have been discussed in the subsequent sections. Ocular formulations are also evaluated for their physicochemical characteristics (Matthews and Wall 2000).

5.2.3 Evaluation of Rate and Extent of Delivery in Different Segments of Eye

5.2.3.1 In-Vitro Release Study

The in-vitro release study of the ocular drug delivery system (ODDS) is carried out to establish the extent of delivery of a drug to or across the biological tissue. It is an important in-vitro tool to provide information regarding tissue permeability, drug solubility, and to predict and establish a correlation between in-vitro and in-vivo studies. This study authenticates the optimum thermodynamic activity, optimization of formulation, and batch to batch uniformity, and is useful for bioequivalence studies. Several methods such as the bottle method, modified rotating basket method, diffusion method using Franz cell, modified rotating paddle apparatus, or the method using flow-through devices are employed for in-vitro release estimation (Satya, Suria, and Muthu 2011). The most commonly used media are phosphate buffer (pH 7.4) or simulated tear fluid (STF), which are used for dissolution studies. A dialysis membrane or diffusion cell membrane acts as ocular barrier for in-vitro release testing to study the release pattern of a drug from the formulation over a period of time. Analytical techniques such as UV-Vis spectroscopy and high pressure liquid chromatography (HPLC) are the most frequently used equipment to find out the quantity of active pharmaceutical ingredient (API) released in media (Aburahma and Mahmoud 2011).

5.2.3.2 In-Vitro Permeation Study

The in-vitro drug permeation study is a very popular study for the quantification of rate and extent of drug transferred across the tissue. A modified diffusion cell is used for the in-vitro drug permeation study to mimic the anatomical and physiological conditions of the ocular tissues. The excised tissue (cornea) from animals like rabbits, goats, sheep, buffaloes, and sometimes humans is employed to investigate the permeability of API to the targeted site. The excised cornea is mounted onto diffusion cells, which are mostly placed side by side having the donor chamber and the receptor chamber maintained at 34°C. The amount of drug at the target site is determined by using UV-Vis spectrophotometry and HPLC, among other quantitative techniques (Noomwong et al. 2011; Khan et al. 2008; Balguri et al. 2017).

5.2.3.3 In-Vitro Cell Culture Models

In-vitro cell-based models are advantageous in terms of simplicity, efficiency, being relatively inexpensive, and reproducible. They can be used to evaluate a number of experimental parameters (Hornof, Toropainen, and Urtti 2005).

5.2.3.3.1 Corneal Models

- *Primary Corneal Models*: The outermost layer of anterior segment of the eye is known as corneal epithelium. Corneal epithelial cells isolated from rabbits are most frequently used to prepare a primary cell culture model. This model is very useful in studying drug permeation and active transport. Models of human primary epithelial cells are also used to study the drug absorption, ocular toxicity, and ocular irritation (Curren and Harbell 2002).
- *Immortalized Corneal Models*: Various models of immortalized corneal epithelial cells are prepared from the hamsters, rats, rabbits, and humans. Statens Seruminstitut rabbit corneal cells (SIRC), HCE (SkinEthic), EpiOcular (MatTek), and Clonetics (Lonza) are the immortalized corneal cell models which are used in the studies of drug metabolism, drug transport, ocular sensitivity, corrosion, transepithelial permeability studies, and ocular irritation (Araki-Sasaki et al. 1995; Offord et al. 1999).

5.2.3.3.2 Conjunctival Models

Conjunctival epithelial cells, mostly derived from rabbit primary cells, are widely used (Yang, Kim, and Lee 2000). Recently, immortalized rat cells and primary bovine conjunctival cells have been used to develop conjunctival models (Civiale et al. 2003). Primary human conjunctival culture models are also utilized for conjunctival tissue transplantation efficacy (Scuderi et al. 2002).

5.2.3.3.3 Retinal Models

- *Retinal Pigment Epithelium*: Frogs, rats, bovines, and chicks are employed to develop primary retinal pigment epithelium (RPE) cell cultures for the measurement of barrier functions and tight junctions. Immortalized RPE cell cultures are used for toxicity studies, gene delivery, and polarity studies of proteins (Mannermaa et al. 2010).
- *Retinal Capillary Endothelium*: Primary isolated bovine retinal capillary endothelial cells (BRCEC) is the retinal capillary endothelium (RCE) used for permeability study. Immortalized rat retinal capillary endothelium is used to investigate barrier properties (Gillies, Su, and Naidoo 1995).

5.2.3.3.4 Ocular Disease Models

The mechanism of ocular diseases at molecular level and screening of ophthalmic drug candidates can be done via an ocular disease model. Cell culture models of RPE cells are most frequently used to explore the pathology and physiology of the disease and age associated macular degeneration (Forest, Johnson, and Clegg 2015). These models are experimentally controlled systems and give reproducible results compared to those obtained from animal models. The variety of cell culture models of diseases, such as glaucoma, consist of mixed retinal cells, neuronal-like cell lines, retinal ganglion cells, and transformed retinal cells used in neuroprotective therapy (Levkovitch-Verbin 2004).

5.2.3.3.5 3D Models of the Eye

In the study conducted by Postnikoff et al., human papillomavirus-immortalized HCE cells were cultured on a curved Millicell-HA membrane (mixed cellulose esters; Millipore, Billerica, MA). This 3D culture state led to suitable stratified, curved, epithelial models appropriate for estimation of cytotoxicity and for biocompatibility testing of contact lenses (Postnikoff et al. 2014).

5.3 Overview of Conventional ODD Systems

An overview of various conventional ODD systems, along with their evaluation methods, has been provided in brief in the subsequent sections.

5.3.1 Ophthalmic Solutions and Suspensions

Ophthalmic solutions (eye drops) are the most popular form for topical administration and lead to good patient compliance, immediate action, and ease of administration. Eye drops contain a drug in dissolved state and are usually adsorbed by the corneal route (cornea, aqueous humor, intraocular tissue), and the conjunctival route (conjunctiva, sclera, choroid, retina, vitreous body). Eye drops (solutions) are capable of treating corneal diseases, iris diseases, and glaucoma (Weijtens et al. 2002; Baudouin et al. 2010). Suspensions are described as a dispersion of an insoluble drug in an aqueous solvent system, containing the appropriate dispersing and suspending agent. They have enhanced residence time at the ocular surface depending upon the drug particle size in suspension (Robinson and Section 1980).

5.3.1.1 Characterization of Ophthalmic Solutions and Suspensions

- *Uniformity of Volume/Weight*: This can be determined by taking 10 filled containers, removing the contents, and weighing/measuring the cumulative content. The average net weight/volume of the contents of the 10 containers should not be less than the labeled amount.
- *Clarity*: Clarity can be determined by visual as well as instrumental methods. The visual method entails observation against a black or a white background with suitable light to detect particulate matter, if any, in the formulation. The instrumentation involves principles of either light scattering or electrical resistance to obtain particle count and particle distribution along with its size. This can also be done with the help of a particle size analyzer. Sometimes, projection of image on a large screen is also employed to visualize particles (Ahmed, Khan, and Shaikh 2017; Makwana, Patel, and Parmar 2016).
- *pH*: The pH of the formulations are measured by a digital pH meter. pH and buffering capacity are of paramount importance as they are linked to irritation as well as stability of the formulations.
- *Tonicity*: Ophthalmic preparations are generally preferred to be isotonic with 0.9% w/v NaCl. However, the limits of 0.5%–5% w/v of sodium chloride are also considered to be safe and non-irritating to the eye.
- *Viscosity*: Various types of thickening agents or viscous liquids can be added to the formulations to increase the time of contact. Viscosity can be measured with the help of various types of viscometers available and should normally be in the range of 15–25 centipoises.
- *Particle Size*: Counting cells or microscopes are used for determination of particle size of ophthalmic suspensions.

5.3.2 Ointments and Emulsions

The ointments are beneficial in terms of longer contact time, reduced nasolacrimal drainage, minimization of tear dilution, higher effective concentration, and greater storage stability. However, ointments have the drawback of causing blurred vision due to film formation when applied into the eye (Sieg and

Robinson 1979). Ophthalmic emulsions give better solubility and bioavailability of drugs. The oil-in-water (o/w) type of emulsion is very common and extensively preferred over the water-in-oil (w/o) type of emulsion in ocular delivery due to its better ocular tolerance and lesser irritation. Emulsions improve drug corneal permeation, precorneal residence time, enhance ocular bioavailability, and provide sustained drug release (Vandamme 2002; Liang et al. 2008).

5.3.2.1 Characterization of Ophthalmic Ointments and Emulsions

- *Particle Size*: Microscopy is used for determination of particle size in ointments.
- *Leakage Test*: The leakage test is carried out on ointment tubes by placing the ointment tubes in a horizontal position on absorbent paper. Leakage, if any, is visually noted (Gad 2008).
- *Globule Size, Zeta Potential, and Emulsion Stability*: Emulsions are evaluated for these parameters. Details have been mentioned in the subsequent sections dealing with novel ODD systems (Mafi et al. 2014).

5.4 Novel Ocular Drug Delivery Systems

5.4.1 *In Situ* Gelling Systems

In situ gelling systems are polymeric solutions which convert from sol (solution) phase to gel phase under the effect of physiological or environmental stimuli. These environmental stimuli include temperature, ions, pH, and UV irradiation–induced gelation. The *in situ* gelling systems enhance viscosity and reduce the drainage of drug from cornea, in response to stimuli, which increases the bioavailability of the drug (Rajoria and Gupta 2012).

5.4.1.1 Characterization of In Situ *Gelling Systems*

- *Physical Appearance*: Color, consistency, homogeneity, and phase separation is observed in the *in situ* gelling systems by visual observation (Mohan, KanduKuri, and Allenki 2009).
- *pH Determination*: A small quantity of gel (e.g., 1 g) is dissolved in distilled water (100 ml) and the pH is measured by a pH meter. pH of the ocular *in situ* gel formulation should be between 6.5–7.4 for ODDS.

- *Drug Content*: The drug content of *in situ* gelling system is measured by diluting 1 ml of gel in 100 ml of appropriate dissolution media (e.g., distilled water, alcohol). The drug in the gel is also analyzed using a UV-visible spectrophotometer or HPLC at the absorption maxima of the drug (Abraham et al. 2009).
- *Viscosity Determination*: Viscosity of *in situ* gelling systems plays a very important role in determining the residence period of API in the eye. Viscosity is determined using different viscometers. A Brookfield digital viscometer is very common equipment used for viscosity measurement. By using different spindles and an angular velocity run (10–100 rpm/min), viscosity can be measured. Factors affecting viscosity such as pressure, temperature, and sample size should be kept constant during the experiment (Patel et al. 2010).
- *Gelling Capacity*: The gelling capacity of *in situ* gels is used to determine the capability of the prepared formulation to gel over the surface of eye. The test sample is placed in the vial/test tube containing STF and is observed visually. The environmental stimuli including temperature, ions, pH, and UV irradiation–induced gelation can be determined to study the gelling capacity. Color (1% Congo red solution in water) can be added to the formulation for better visual appearance of the formed gel. The gelling capacity of the test samples are evaluated on the basis of firmness and time for which the produced gel remains as such (Kanoujia et al. 2012).
- *Gel Strength*: The force in grams required to go through the *in situ* gel is known as the gel strength. A texture analyzer is used to determine the gel strength using a "gelling strength test" mode at the predetermined speed, acquisition rate, and trigger force. An aluminum probe (7.6 cm diameter) is used to penetrate in to the gel contained in beaker at room temperature (Walewijk, Cooper-White, and Dunstan 2008).
- *Spreadability*: The spreadability of gels is determined by a texture analyzer or laboratory method by applying the test sample between two glass slides. Pre-weight is added by introducing 1000 g weight for 5 min. Now weight (50 g) is added to the pan. The spreadability can be defined as the time requisite to detach the two glass slides (Lardy et al. 2000).

5.4.2 Mucoadhesive Gels

Mucoadhesive gels are based on noncovalent bonding of mucoadhesive polymers with conjunctival mucin. Mucoadhesive polymers are classically high molecular weight polymers having numerous hydrophilic functional groups, for example hydroxyl- carboxyl-, sulphate, and amide. Carboxy methyl cellulose (CMC), carbopols, sodium alginate, polyacrylamide, dextran, etc. are examples of mucoadhesive polymers. The mucoadhesive gels provide an intimate link with the corneal layer and enhance the ocular bioavailability of a drug (Hui and Robinson 1985; Mitra 2003).

5.4.2.1 Characterization of Mucoadhesive Gels

Physical appearance, pH, drug content, and viscosity of the mucoadhesive gels are determined as described for *in situ* gelling systems (Ranch et al. 2017).

- *Mucoadhesion Test*: Mucoadhesion testing of the sample discs is carried out using a texture analyzer or a two-arm balance. In a texture analyzer, gel is applied to the cylindrical probe with the help of a double-sided adhesive tape. The probe is allowed to move downward to attach with the soaked tissue (e.g., cornea) at a specified force for specified time. Afterwards, the probe is consequently withdrawn at a specified test speed using the texture analyzer. The highest force requisite to detach the probe from the ocular tissue is known as the mucoadhesive strength (Ponchel et al. 1987). In the two-arm balance method, the outer surface of a beaker (50 ml) is attached with a biological membrane using acrylate adhesive and placed in another beaker (100 ml). The appropriate buffer is placed into the 100 ml beaker up to the upper surface of the biological membrane. A small quantity (1 ml or 1 gm) of gel is sandwiched between the bottom of stainless steel pan and 50 ml beaker. A predetermined pre-load weight is placed over the pan for a predetermined time to contact the gel with the biological membrane. After the predetermined time, weight is added to another pan until the pan is detached from the membrane. The weight in grams necessary to detach the pan from the membrane is considered as the bioadhesive strength (Kanoujia et al. 2012).

5.4.3 Contact Lens

Contact lenses are defined as thin and curved plastic disks which are intended to envelop the cornea. The contact lenses loaded with a drug are placed in to eye, where they adhere to the eye surface due to the existing surface tension. The polymers extensively used for fabrication of contact lenses are silicon hydrogel and poly(hydroxyethyl methacrylate). The benefits of using contact lenses are longer residence time, leading to elevated drug flux via the cornea, and less nasolacrimal drainage (Gupta and Aqil 2012). Recently, contact lenses loaded with particles have been investigated for improved loading and sustained release action. For this type of a delivery system, drugs are loaded in colloids such as liposomes, nanoparticles, and microemulsions. Further, these colloids are dispersed in the contact lenses.

5.4.3.1 Characterization of Contact Lenses

- *Contact Lens Hydration*: Good hydration is necessary for efficient oxygen supply to the cornea. The contact lenses are tested for hydration by soaking in distilled water for 24 h at room temperature. The weights of dry and wet lenses are recorded and the hydration in percentage is calculated (Gulsen and Chauhan 2004).
- *Surface Contact Angle*: Surface wettability of the contact lens is essential for efficient and normal functioning of a contact lens. The contact angle or surface wettability measurement is performed using contact angle goniometer via Young's equation (Tranoudis and Efron 2004).
- *Light Transmission*: The transparency of contact lens is measured by UV-Vis spectrophotometer. The contact lenses are soaked in distilled water for 12 h and measurement is carried out at a wavelength of 600 nm (Garhwal et al. 2012).
- *Texture Analysis*: The elasticity and tensile strength of contact lenses is determined by a texture analyzer. The contact lens is fitted on to the clamp of the texture analyzer and expanded until breaking point. Stress and strain values are obtained by Young's equation. The highest force applied on a contact lens by texture analyzer is calculated as tensile strength (Tranoudis and Efron 2004).

5.4.4 Implants

The intraocular implants are designed to provide prolonged and controlled drug release to overcome the problem of low availability and short activity of topical ophthalmic formulations. Minor surgery is always required for intraocular administration of implants. (Sankar et al. 2006) Ocular implants fall under two categories, which are biodegradable implants and non-biodegradable implants (Bourges et al. 2006; Choonara et al. 2010).

5.4.4.1 Characterization of Implants

- *Thickness, Content Uniformity, Weight, and Physical Appearance*: Calibrated micrometer, spectrophotometric method (UV-Vis spectroscopy, HPLC), digital balance, and optical microscopy can be used respectively (Rao, Ramakrishna, and Diwan 2000).
- *Morphology*: Surface morphology of placebo- and drug-loaded implants may be studied by SEM and scanning force microscopy (SFM) (Rao, Ramakrishna, and Diwan 2000).
- *Tensile Strength*: Tensile strength should range from 2.5 MPa to 53.0 GPa, revealing the strength of the 3D network in these films (Li, Kamath, and Dwivedi 2001).

- *Surface pH*: The surface pH of the implants varies from pH 7.0 to 7.4 indicating that the implants do not cause irritation after application (Kumar, Pandit, and Balasubramaniam 2001).
- *Drug Content Uniformity Test*: The concentration of drug in per unit dose is analyzed by UV-Vis spectroscopy or HPLC (Rojas and Ojeda 2009).
- *Percentage Moisture Absorption and Loss*: A rectangular piece of implant is weighed and placed in a glass chamber saturated with a solution of potassium chloride (84% RH). After a specific period, the implants are taken out of the chamber and weighed. The change in weight is considered as the water absorption capacity of the implants (Rao, Ramakrishna, and Diwan 2000).

5.4.5 Microneedles

Microneedles (MN) are an attractive technology in micron scale, that are minimally invasive and efficient to treat the posterior section of the eye (Gomaa et al. 2012). Microneedle-based therapy reduces the problems associated with intravitreal injections such as retinal detachment, hemorrhaging, cataracts, endophthalmitis, and pseudoendophthalmitis. Recently, diverse types of microneedles have been fabricated and evaluated, such as dissolving, hollow, and solid polymeric microneedles (Donnelly, Singh, and Woolfson 2010).

5.4.5.1 Characterization of Microneedles

- *Drug Content Determination*: MN arrays are dissolved in appropriate buffer and an aliquot is analyzed using a UV-Vis spectroscopy, an HPLC, or a fluorescence plate reader.
- *Moisture Content Determination*: A thermogravimetric analyzer is used to determine the moisture content of samples. The moisture content is determined by calculating the percentage of reduction in sample weight.
- *Determination of Fracture Force*: A texture analyzer in compression mode is used for determining the mechanical strength of a microneedle array.
- *Determination of the Insertion Force in Ocular Tissues*: The insertion force of MNs into the ocular tissues is measured using a texture analyzer, with minor modifications (Larrañeta et al. 2016).
- *Determination of Depth of Penetration in Ocular Tissues*: Optical coherence tomography (OCT) is performed to observe and measure the microneedle depth of penetration in the ocular tissues (Donnelly, Singh, and Woolfson 2010).

5.4.6 Ocular Inserts

Ocular inserts are defined as solid devices placed in the cul-de-sac of the eye. They are advantageous in terms of higher bioavailability of a drug, accurate dosing, increased contact time with the ocular surface, and better patient compliance (Sankar et al. 2006).

5.4.6.1 Characterization of Ocular Inserts

- *Thickness*: A screw gauge is used to measure the thickness at different places of inserts and average of the measurements is considered to be the thickness of inserts (Sankar et al. 2006).
- *Weight Uniformity*: The weight of each insert is observed on a digital balance and the average weight is calculated (Sankar et al. 2006).
- *Drug Content*: Inserts are cut into small and equal size pieces and dissolved in appropriate media (buffer pH 7.4). The obtained solution is diluted and analyzed using a UV-visible spectrophotometer (Ubaidulla et al. 2007).
- *Folding Endurance*: Inserts are repeatedly folded at the same place till breaking and the number of times the inserts could be folded till breaking is known as the folding endurance (Ubaidulla et al. 2007; Sakar et al. 2006).
- *Percentage Moisture Uptake*: The moisture uptake in percentage is calculated as given in Section 4.4.1.
- *Surface pH*: Inserts are allowed to swell in distilled water and the pH is measured using a digital pH meter (Pandey et al. 2011).

5.4.7 Collagen Shields, Gels, Hydrogels, and Sponges

Collagen shields are disc shaped, soluble protein templates, often made of animal collagen, to be placed on the cornea for the increment in ocular wound healing. They can be medicated or non-medicated. These are hydrophilic and can be saturated with drug solutions before use (Silbiger and Stern 1992). As the name suggests, they shield the corneal epithelium from external factors such as dust particles, constant movement of eyelids, etc. They have now evolved to behave like drug reservoirs. Gels, hydrogels, and sponges can also be fabricated from collagen for ocular use.

5.4.7.1 Characterization of Collagen Shields, Gels, Hydrogels, Sponges

They are characterized and evaluated through biocompatibility studies using cell cultures or immunogenicity studies, which will be discussed in Section 5.6. Inflammatory responses can be recorded in-vivo in guinea pigs, which will be discussed in Section 7.6 (Lee, Singla, and Lee 2001).

5.5 Nanosystem-Based Ocular Drug Delivery

Many approaches have been investigated by researchers to conquer topical ocular drug delivery challenges. Nanosystems such as lipid nanoparticles, nanoparticles, niosomes, liposomes, nanosuspensions, nanoemulsions, and dendrimers have been investigated for ODD. Nanosystems are beneficial in terms of low irritation, the controlled release of drug, satisfactory bioavailability, and the ability to adhere to the ocular surface and facilitate the delivery of drugs to both anterior and posterior segments (Yellepeddi and Palakurthi 2016; Patel et al. 2013).

5.5.1 Liposomes and Niosomes

Liposomes are lipid vesicles, with a size range of 0.08 to 10.00 µm, composed of one or additional phospholipid bilayers surrounding an aqueous core (Nanjawade, Manvi, and Manjappa 2007). The potential of liposomes in ODD is tremendous due to biocompatibility, the ability to encapsulate both lipophilic and hydrophilic drugs, and a cell membrane–like structure. The bioavailability of a drug molecule increases due to the ability of liposomes to make a close contact with the corneal surface (Kaur et al. 2004). Liposomes with a positive charge exhibit a prolonged and sustained precorneal retention due to strong interactions with the corneal epithelium (Meisner and Mezei 1995; Moustafa et al. 2017). Niosomes are a special kind of bilayered vesicular systems, composed of amphiphilic nonionic surfactants. Niosomes are varied in a size range of 10 to 1000 nm and are capable of encapsulating both, lipophilic, and hydrophilic drugs (Rózsa and Beuerman 1982; Sahoo, Dilnawaz, and Krishnakumar 2008). They are chemically stable, show very low toxicity and improved bioavailability, and also control drug delivery at the targeted ocular site (Cholkar et al. 2013). Niosomes reduce the systemic drainage of a drug which further improves residence time, leading to increased ocular bioavailability.

5.5.1.1 Characterization of Vesicular Drug Delivery Systems (Liposomes/Niosomes)

- *Particle Size, Zeta Potential and Polydispersity Index*: The size measurement of vesicles in ODDS is important, as the vesicles with higher size can irritate the eye upon application. Particle size should not exceed 10 µm for an ophthalmic application. The zeta potential is an important attribute to describe the stability of vesicular formulation. The mean particle size, zeta potential, and polydispersity index for a vesicular delivery system is determined by photon correlation spectroscopy. Laser diffraction particle analyzers, light obscuration particle count testing, dynamic imaging analysis, dynamic light scattering, and Coulter counter techniques are also used to study the vesicle size and polydispersity index (Taha et al. 2014).

- *Morphology*: The morphology assessment of drug-loaded vesicles can be carried out using scanning electron microscopy and transmission electron microscopy. The size, structure (monolayer, bilayer, or multilayer), shape, and distribution can be easily described by morphological study (Abdelbary et al. 2017).
- *Entrapment Efficiency (EE%)*: Niosomes and liposomes are separated from the free drug using ultra centrifugation technique at a high rpm (20000 rpm for 1 h). The quantity of entrapped drug is determined by lyses of vesicles. The concentration of the drug in the supernatant is determined by UV-Vis spectroscopy and HPLC. The amount of free drug in the aqueous phase is subtracted from the total drug content and then divided by the total drug content. This is converted into a percentage calculation to obtain the percentage entrapment efficiency (Abdelbary et al. 2017; Taha et al. 2014).
- *Corneal Deposition and Permeation*: Details will be discussed in Section 7.1.

5.5.2 Microemulsion and Nanoemulsions

Microemulsions are defined as clear and stable (thermodynamically) dispersions of two immiscible liquids (oil and water), stabilized by using surfactants. Microemulsions provide the advantage of higher thermodynamic stability, improved solubility, and improved corneal permeation. The selection of organic phase, aqueous phase, and surfactant/cosurfactant systems are the important factors for the fabrication of a stable microemulsions (Vandamme 2002; Kapoor and Chauhan 2008). Nanoemulsions are known as oil-in-water (o/w) emulsions, with standard droplet size varying from 50–1000 nm. Nanoemulsions are kinetically stable systems which provide numerous advantages like a higher capacity to dissolve lipophilic drugs and hydrophilic drugs, good spreadability, stability, improved bioavailability, and improved permeability of the drug across the cornea. Surfactants (for example, phospholipids, polysorbates, stearylamine, oleylamine, etc.) are used for the formulation of nanoemulsions. Creaming, coalescence, flocculation, and sedimentation do not occur in these formulations since they are thermodynamically stable.

5.5.2.1 Characterization of Micro/Nanoemulsions

- *Globule Size, Polydispersity Index, and Drug Content*: Globule size and polydispersity index (PDI) is determined by photon correlation spectroscopy at room temperature and drug content is determined by UV-Vis spectroscopy (Kumar and Sinha 2014).

- *Viscosity, pH, and Conductivity*: The viscosity of dispersions (undiluted) is determined by using a viscometer and a suitable spindle at room temperature. The pH meter and conductivity meter are used to determine the pH and conductivity of the dispersions (Zhu and Chauhan 2008; Moulik and Rakshit 2006).
- *Surface Morphology and Structure*: Transmission electron microscopy is most commonly used for structural and morphological examination of drug-loaded dispersions (Kumar and Sinha 2014).
- *Stability Evaluation*: Thermodynamic stability of the dispersions is determined by centrifugation, the freeze–thaw cycle, and the heating–cooling cycle (Kumar and Sinha 2014).
- *Physicochemical Interactions*: Different techniques such as nuclear magnetic resonance, Fourier transform infrared spectroscopy (FTIR), and differential scanning calorimetry (DSC) are commonly used to study the extent of interaction between a drug and excipients (Moghimipour, Salimi, and Zadeh 2013).

5.5.3 Nanosuspensions

Nanosuspensions are described as colloidal dispersions containing submicron particles of drugs stabilized by polymers or surfactants. These have emerged as a promising approach for the delivery of lipophilic drugs. Nanosuspensions are fabricated by high-pressure homogenization, pearl milling, and precipitation techniques. They offer numerous advantages, such as sterilization, ease of eye drop formulation, increased precorneal residence time, less irritation potential, and enhanced ocular bioavailability of lipophilic drugs. Prednisolone, dexamethasone, and hydrocortisone have been widely investigated by researchers in nanosuspension form. Polymers like, Pluronic F108 solution, Eudragit RS 100, methacrylates, etc. are generally used for the fabrication of nanosuspensions (Patravale and Kulkarni 2004; Sanders et al. 1983; Mandal, Alexander, and Riga 2010).

5.5.3.1 Characterization of Nanosuspensions

- *Particle Size and Zeta Potential Measurement*: These tests are described in the section named characterization of vesicles (Mezei and Meisner 2003; Joshi 1994).
- *Morphological Study*: Scanning electron microscopy and transmission electron microscopy are used for morphological characterization. 10–20 µl of sample is placed on a parafilm. This is taken on to a grid, stained, dried, and the image is then recorded (Balguri et al. 2017).

- *Drug Entrapment Efficiency*: The ultracentrifugation technique (indirect method) is used to determine the entrapment efficiency of nanosuspensions (Greenwald and Kleinmann 2008).
- *Powder X-Ray Diffractometry*: Powder X-ray diffractometry (PXRD) is used to study the crystalline nature of the formulation (Mudgil and Pawar 2013).
- *Drug–Polymer Interaction*: FTIR is mostly used to study the drug polymer interaction (Abboud and Massoud 1972).

5.5.4 Nanoparticles and Polymeric Micelles

Nanoparticles are polymeric particulate delivery systems with diameters ranging from 10 to 1000 nm. In this delivery system, a drug may be dispersed, encapsulated, conjugated, or adsorbed (Patel, Shastri, et al. 2010; Harmia et al. 1986). The nanoparticulate system can be employed as a choice to alleviate the problem of irritation and toxicity caused by liposomes and dendrimers. They are capable of showing sustained drug delivery without repeated drug administration, due to different mechanisms such as diffusion, dissolution, and erosion of the polymer matrix (Pignatello et al. 2002). Drug-loaded nanoparticles are able to efficiently deliver to both anterior as well as posterior ocular tissues (Parveen et al. 2010). Tailored nanoparticles, such as mucoadhesive nanoparticles or coating of nanoparticles with mucoadhesive polymers, can be easily prepared to improve precorneal residence time. Hence, the nanoparticles show an immense potential to deliver ophthalmic drugs.

Polymeric micelles (PMs) are self-assembled micelles prepared by diblock/multiblock amphiphilic copolymers (Ahmad et al. 2014; Mandal et al. 2017). The advantages of PMs are ascribed to their small size, easy preparation techniques, high drug encapsulation capability, and enhanced bioavailability of therapeutic drugs at the site of action (Civiale et al. 2009). The polymers majorly used for the designing of PMs are polyoxyethylene-polyoxypropylene (POE/POP) block copolymers. Also, acrylic acid, poly(ethylene glycol)-hexylsubstituted poly (lactides), N-isopropylacrylamide, vinyl pyrrolidone, polyhydroxyehtyl aspartamide, and surfactants (ionic, nonionic, and zwitterionic) are well known for the fabrication of micelles (Mandal et al. 2017).

5.5.4.1 Characterization of Particulate Systems (Nanoparticles/Polymeric Micelles)

Entrapment efficiency, particle size, zeta potential, particle morphology, and X-ray diffractometry (XRD) have been described in Section 5.3.1 (Gupta et al. 2000; Yousry et al. 2017).

5.5.5 Dendrimers

Dendrimers are defined as nanosized, star-shaped, and vastly branched polymeric systems (Fischer and Vögtle 1999). The highly branched structure of dendrimers permits the inclusion of a wide array of therapeutic moieties (hydrophobic or hydrophilic). The properties of dendrimers such as shape, molecular size, density, dimensions, polarity, solubility, and flexibility can be modified easily because they are artificially synthesized. Poly(amidoamine) (PAMAM) and polypropylenimine dendrimers are extensively used in ODD.

5.5.5.1 Characterization of Dendrimers

- *FTIR and DSC*: FTIR spectroscopy and DSC are used for the analysis of drug-dendrimer complexes, indicate the binding of a drug with the dendrimers (Agnihotri and Aminabhavi 2007; Li et al. 2008).
- *Drug Content*: Drug content is determined by UV-Vis spectroscopy to estimate the amount of a drug complexed or incorporated in the dendrimers (Ma et al. 2007).
- *Rheology*: The measurement of viscosity of dendrimers can be carried out using the cone and plate viscometer. The change in viscosity after instillation of dendrimers and its mixing with lachrymal fluid has also been evaluated. The viscosity change is measured after diluting the dendrimers with simulated lachrymal fluid, with and without mucin (15% w/w). The rheological data is analyzed using the equation:

$$\eta_b = \eta_t - \eta_m - \eta_p$$

  where η_b is the viscosity of the component due to bioadhesion, η_t is the viscosity of the test preparation or mucin system, η_m and η_p are the viscosity coefficients of mucin, and the bioadhesive polymer, respectively (Vandamme and Brobeck 2005).

5.6 Safety and Toxicity Evaluation: In-Vitro, In-Vivo and *Ex Vivo* Methods

Toxicology testing is very important for ophthalmic formulations to evaluate the risk linked with the formulations after application to the eye. Toxicity evaluation can be done by in-vitro, in-vivo, and *ex vivo* methods.

5.6.1 In-Vitro Tests

In-vitro testing using cell based models are beneficial in term of being inexpensive, simple, and less time consuming. These methods allow the understanding of toxicity mechanisms at the cellular level. The toxicity testing can be done at multiple endpoints, concentration, exposure time, and the methods provide wide range of information regarding toxicity (Takahashi et al. 2008). In-vitro tests are considered to be superior to in-vivo tests, as animal testing is avoided.

Cytotoxicity estimation is frequently carried out on monolayered culture cells, using different assay methods, on exposure to a test formulation. Cytotoxicity is determined by using different assay methods, which include protein measurements by Coomassie brilliant blue, crystal violet assay, thymidine incorporation, Lowry reagent assay, 3-(4,5-dimethylthiazol-2-yl)-2,5-diphenyltetrazolium bromide assays (MTT assays), lactate dehydrogenase leakage (LDH), fluorescein leakage (FL) trypan blue exclusion, florescent staining by propidium iodide, and neutral red uptake/release tests (Huhtala et al. 2008).

Corneal epithelial models such as SIRC cells, human corneal cells, and rabbit corneal epithelium (RCE) are utilized for ocular toxicity studies. In-vitro assays and models provide useful data, in agreement with in-vivo studies, allowing for significant reductions in the numbers of animals used.

5.6.2 In-Vivo Tests

5.6.2.1 In-Vivo Draize Eye Test

The FDA developed the in-vivo Draize test using rabbits for evaluating acute ocular toxicity. In this protocol, the test formulation (0.1 ml liquid or 0.1 gm solid) is applied on the cornea as well as the conjunctival sac of one eye of the conscious rabbit, whereas the untreated eye acts as a control. After 72 h of the test, rabbits are observed for any signs of irritation which include redness, hemorrhage, swelling, edema, cloudiness, discharge, and blindness. The test substance can be categorized on the basis of scoring that varies from non-irritating to severely irritating, attributed to the effect on the conjunctiva, cornea, and iris (Draize, Woodard, and Calvery 1944).

5.6.2.2 In-Vivo Low-Volume Eye Irritation Test

In-vivo low-volume eye irritation test (LVET) is an alternative animal method, recommended by the National Research Council in 1977 (NRC 1977). In this test, only a small quantity (0.01 ml or 0.01 gm) of test formulation is applied to the surface of the cornea but not applied in the conjunctival sac of the rabbit's eye. The low volume of a test substance causes less stress to the test animals and also leads to better prediction of human ocular irritation response (Jester et al. 2001).

5.7 In-Vivo and *Ex Vivo* Evaluation of Intraocular Parameters

5.7.1 Study of Corneal Penetration Process

This is performed through confocal laser scanning microscopy, fluorescence microscopy, and other such sensitive microscopic methods (Liu et al. 2017).

5.7.2 Intraocular Pressure

Intraocular pressure (IOP) is the fluid pressure inside the eye. IOP rises in many pathological conditions of the eye. To assess the pressure, various direct as well as indirect methods are used. Normally, the direct methods utilize canulation and continuous measurement. This procedure suffers from the disadvantage of inflammation and may thereby change IOP (Millar and Pang 2015).

5.7.2.1 Acute Measurement of Intraocular Pressure

- *Microcannulation of the Globe*: A fine hypodermic needle of up to 1 mm in diameter is inserted into the anterior chamber or the posterior chamber, or sometimes the vitreous chamber. This is connected to a thin tubing with a pressure measurement device or a saline column which can be equated to pressure in millimeters of mercury (mmHg).
- *Servo-Null Micropipette Procedure*: A servo-null device measures the IOP by converting it in terms of counter pressure required to nullify resistance (Avila et al. 2001). Since a micropipette of up to 5 micrometers is utilized for cannulating, it is less invasive and causes lesser injury and inflammation of the cornea.
- *Non-invasive Approximation of IOC*: Tonometers are the equipment used to determine IOP in experimental animals like New Zealand white rabbit or rats. Tonometers are standardized to determine IOP in mmHg but are mostly indirect measurements of other variables which are linked to IOP (Stamper 2011). A few rebound-type tonometers are designed for rodents and the portable veterinary MacKay-Marg tonometers are designed for animal studies in dogs, cats, and other animals. The rest (Schiøtz tonometer, Goldmann Applanation tonometer, Perkins and Draeger tonometers, pneumatonometer, etc.) are mostly for human use and since these are non-invasive, they can easily be used for clinical studies. The readings of a tonometer are often calibrated with the help of a cannulation method to get a regressed curve. A few suppliers of the equipment provide the paraphernalia required for a simulated calibration so that the invasive

cannulation calibration method is not required. A correction factor is calculated for the tonometer with the help of the readings obtained by the invasive cannulation method. During animal handling, a calm atmosphere needs to be maintained and a trained animal handler is required. Local or general anesthesia could also be required. Corneal photoelasticity, acoustic oscillations of the eye, sonoelastic Doppler ultrasound, respectively, have all been successfully utilized for tonometric measurements in-vitro as well as in-vivo. Being non-invasive in nature, this method is not continuous.

Measurement of Intraocular Pressure by Telemetry Since IOC follows a circadian rhythm and peaks during the night time, sometimes a continuous method of determination is required. Telemetry is a popular method to determine IOC (Paschalis et al. 2014). Implantable pressure transducers with motion sensors have been used for continuous IOP evaluations. The transducers are surgically inserted in between the tissue and muscles and a catheter is secured to the sclera with the help of transducers. Rabbits are acclimatized for half an hour and then a baseline of IOP recorded. Ocular hypertension can be induced with 5% w/v sterile glucose injection and effect of a drug/dosage form can be monitored on a continuous basis for a required time period, which could include diurnal and nocturnal variability. Many variations of this method have been reported in the literature (McLaren, Brubaker, and FitzSimon 1996).

5.7.3 Aqueous Humor Flow Rate

This can be monitored through ocular fluorophotometry. Tracer substances like fluorescein are administered so that their concentration can be monitored as an estimation of aqueous flow in the relevant compartment of the eye. Initially, basal flow rates are recorded and then treated with test drug. Flow rates are then recorded again through a fluorophotometer (Smith 1991). Gadolinium contrast agent (Gd-DTPA) has also been systemically administered, so that its concentration can be monitored in adult rats through gadolinium-enhanced magnetic resonance imaging (Ho et al. 2014). Gd-DTPA, 234.5 mg/ml, is instilled into SD rat eyes in 0.2% azone as eye drops, and modification of Gd signal envisaged by 7.0 T magnetic resonance imaging (Li et al. 2017).

5.7.4 Experimental Glaucoma

Various agents like alpha-chymotrypsin, betamethasone-21-phosphate, India ink, and even tap water have been used to induce glaucoma. Polybead microspheres have also been used to induce experimental glaucoma, after anesthesia with ketamine. The outcomes were observed through techniques like RT-PCR, western blot, scleral thickness, and histology (Quigley et al.

2015). Uveitic glaucoma has been induced in rats with the help of S-antigen. Ischemia/reperfusion and the optic nerve crush models have also been described (Goldblum and Mittag 2002).

5.7.5 Local Anesthesia of the Cornea

Corneal anesthesia can be induced by the use of a sodium channel blocker or tetrodotoxin, applied with either proparacaine or a chemical permeation enhancer. Normally, an anesthetic is administered and disappearance of blinking reflex and reappearance of blinking reflex is noted when the eyelid is prodded with a von Frey hair (equine or camel hair attached to a glass rod) or even a nylon fiber.

5.7.6 Models of Eye Inflammation

5.7.6.1 Allergic Conjunctivitis Model

Allergic conjunctivitis is an inflammatory condition affecting the ocular surface and is linked with type 1 hypersensitivity reaction. It can be divided into two categories: intermittent and persistent allergic conjunctivitis. Both the problems are mast cell–mediated.

Guinea pigs, rats, and mice are the animals normally used. Murine models are most preferred form of model for studying allergic conjunctivitis. Models may either be IgE mediated or they can be non-IgE mediated. The most commonly used antigen for IgE-mediated ones are ovalbumin and ragweed. Haptens and compound 40/80 are the most common molecules used for non-IgE models (Groneberg et al. 2003).

In a procedure using ragweed extract, the ragweed extract is biotinylated. Ragweed and alum are mixed before immunization. On day 0, while injecting the ragweed and alum mixture in one hind footpad, the animals are anesthetized with methoxyflurane. On day 10, conjunctivitis is generated by applying the ragweed suspension topically into each eye. Two groups of mice (immunized or not immunized with alum alone) are challenged with the same dose of ragweed. Animals are examined for the signs of inflammatory response (conjunctivital redness, chemosis, lid edema, and tearing). Grading is done on a 0–3+ scale. The sums of scores in each of the four categories of inflammatory response are regarded as the total score. Histological studies are also done to observe the changes at a cellular level (Vogel 2002).

The allergic conjunctivitis model has recently been tried by many researchers. The potential of interleukin-28A using ovalbumin induced mouse model was also checked (Chen et al. 2016). Most of the studies have reported the use of ovalbumin, like attenuation of Th2- driven allergic conjunctivitis by superoxide dismutase 3 (Lee et al. 2017), β-1,3-glucan's administration for modulating allergic conjunctivitis (Lee, Kwon, and Joo 2016), and the therapeutic potential of a peptide ZY12 (Xu et al. 2017). A ligand (CCL7) has been found to play a role in mediating hypersensitivity reactions in murine allergic conjunctivitis (Kuo et al. 2017).

In an ovalbumin-induced allergic conjunctivitis model, BALB/c mice are immunized with an intraperitoneal injection of ovalbumin in PBS, also containing aluminum hydroxide and pertussis toxin. Mice are sensitized for two weeks and these are challenged once daily by topical ovalbumin for 13 days. Evaluation involves checking for a hypersensitive response after the ovalbumin challenge. Four parameters are checked: conjunctival chemosis, conjunctival redness, lid swelling, and tearing. Parameters are recorded on score basis (Lee, Kwon, and Joo 2016).

5.7.6.2 Corneal Inflammation Models

Corneal inflammation (also called keratitis) may be a result of either an injury or an infection. Thus, mechanical or chemical injury would mimic this situation. In a procedure, rabbits are anesthetized (sodium thiamylal 15 mg/kg i.v.). Clove oil is injected to produce the corneal inflammatory response. In a period of 24 h, all the animals receives two i.v. injections of tritiated thymidine. The second injection of thymidine is given 24 h before inducing corneal inflammatory response. Therapy is initiated instantly after injecting clove oil intracorneally, using one drop of test or a standard compound every hour for a total of 6 doses. After 1 h, the animals are sacrificed under anesthesia and a 10-ml penetrating corneal button is removed by trephination. The tissue samples are solubilized in a suitable solubilizing substance. The samples are counted in a scintillation counter. The radioactivity in each cornea is documented. The data are expressed as percentage change in radioactivity in comparison to their own untreated controlled eyes. The average of the mean values of these differences is then determined (Vogel 2002).

There are several models for keratitis induced by infection. Silk sutures contaminated with bacteria passed through the rabbit corneal stroma can induce keratitis. In another model, bacteria are injected directly into the cornea. Contaminated lenses have also been utilized. The bacteria under investigation are *Pseudomonas* and *Staphylococcus aureus.* Similar studies have been reported with mice. Keratitis can also be induced genetically through mutations (Marquart 2011).

Pinnock et al. (2017) came up with an *ex vivo* model as an alternative of single and dual infection corneal inflammation. In the method, they infected excised rabbit corneoscleral rims using 10^8 cells of microbes by wounding with a scalpel and exposing cornea to a microbial suspension or by injecting it via intrastromal route. After the inoculation was done, cornea was maintained at 37°C for 24 h and 48 h. After incubation, either the corneas were homogenized (for determining colony-forming units) or they were processed for histological examination. The model also supports dual infection (Pinnock et al. 2017). Zhu et al. (2017) developed an in-vivo mouse model and an *ex vivo* rabbit model for infectious keratitis where they used bioluminescent strains of pathogens which allowed non-invasive assessment of the extent of infection via bioluminescence imaging. The quantification was related to colony forming units. The effectiveness of antimicrobial blue light was then accessed using these models (Zhu et al. 2017).

5.7.6.3 Autoimmune Uveitis

Uveitis has been induced in guinea pigs (Vogel 2002). It may be induced in a variety of species. Lewis rats are a well-established model (Wildner, Diedrichs-Möhring, and Thurau 2008). For the induction of uveitis, retina binding protein/homologous retinas are emulsified in a complete Freund's adjuvant. After the intradermal immunization with Freund's adjuvant, the animals develop pan uveitis in 1–2 weeks. Choroidal/retinal inflammation, retinal vasculitis, loss of visual function, and photoreceptors are the main features of experimental autoimmune uveitis (Bansal et al. 2015). The involvement of T_H17 cells in autoimmune uveitis has also been reported (Amadi-Obi et al. 2007).

5.7.6.4 Endotoxin-Induced Uveitis Model

The endotoxin-induced uveitis (EIU) model can be developed after administration of bacterial endotoxin, like lipopolysaccharide (LPS) (Yadav and Ramana 2013). EIU in rats has been induced by injecting LPS either via the IP route or foot pad route (Rosenbaum et al. 1980). The attenuation of ocular inflammation by gabapentin has also been reported (Anfuso et al. 2017). They used rabbits (male New Zealand albino) for this purpose and the method which they opted is as follows: animals were anaesthetized by intra venous injection of ZOLETIL® and a drop of a local anesthetic was administered to the eye. EIU was induced by injecting LPS intravitreally. An ophthalmic solution containing 0.5% gabapentin in an isotonic buffered solution was prepared and a multiple treatment was carried out in which one instillation was administered 30 min prior to LPS injection and four instillations post the LPS. After 7 h or 24 h of LPS injection, tears were obtained with the help of glass capillary tubes. Then, the animals were sacrificed. Various parts, such as aqueous, cornea, conjunctiva, and iris-ciliary body were collected. Care was taken while obtaining tear samples (10 microliters from each eye) avoiding simulated tear production. Tears were stored at –80°C until analysis (Anfuso et al. 2017). This model was further utilized and it was reported that ocular inflammatory reactions are disturbed in diabetic conditions (Tamura et al. 2005).

5.7.6.5 UV-Induced Uveitis

The ultraviolet B (UVB) lamp is utilized for this purpose and its distance to mouse cornea is kept approximately 15 cm for 90 sec and for 7 days.

Mice are assigned to four groups. One is the blank and the other three are UVB treatment groups, in which one is nonprotected while the other two are protected either with etafilcon A contact lens or with nelficon A contact lens, respectively. All the contact lenses are used once per eye and are discarded daily after ultraviolet radiation. UV radiation is administered to both the eyes of each mouse. After general anesthesia, the eyes of the mice are covered individually with the appropriate contact lenses and then are exposed to UVB and the corneal surface and uveitis was evaluated (Shao et al. 2017).

5.7.6.6 Ocular Inflammation Induced by Paracentesis

This model is based on the rationale that paracentesis stimulates some inflammatory mediators (PGE_2 and $PGF_{2\alpha}$) into the eye's anterior chamber. As per the procedure, rabbits are anesthetized using a xylazine and ketamine mixture. The test compound/vehicle is administered topically and the eyes are taped in order to prevent drying. After an hour of administration, ocular inflammation is induced by paracentesis. Aqueous humor from the rabbit's both eyes is combined to form a "pre" sample. Test compound or vehicle is again administered and the eyes are taped. After 1 h, paracentesis is induced for the second time and again; the aqueous humor from both the eyes is combined to form a "post" sample. Before second paracentesis, the general appearance of the eye is rated. This rating gives indication of the inflammation and correlates with the amount of prostaglandin accumulation. PGs are measured from the "pre" and "post" samples of the aqueous humor by RIA (Vogel 2002).

The utility of a drug in treating trauma-induced ocular inflammation was evaluated. In this trauma-induced model, animals received the dose of a test compound or a vehicle, administered bilaterally. After 45 min of dosing, a single drop of 0.5% propocaine was instilled in the eye, and trauma was elicited by paracentesis within 5 min. Removal of the aqueous humor was accomplished by the puncture of cornea with a 27-g needle. 100 µl of this was diluted with an equal volume of a 2% solution of EDTA in saline (pH 7.4) frozen on dry ice and stored for later analysis (of protein and PGE_2 content) at –70°C. Protein and PGE_2 levels in the aqueous humor samples obtained before and after paracentesis were determined. Thirty minutes after the initial paracentesis, the animals were euthanized with the help of sodium pentobarbital overdose via the marginal ear vein. Post-trauma aqueous humor samples were obtained immediately, and were stored and analyzed (Gamache et al. 2000).

5.7.6.7 Ocular Inflammation by Lens Proteins

Rabbits are anesthetized and the anesthesia is allowed to be maintained every hour. After deep anesthesia of animals, the drug is applied topically to the one of the eyes and the other eye received solvent. After 1 h, the lens protein is intracamerally injected. Contact with the iris is avoided. After 15 min of injecting lens proten, fluorescein is injected through the marginal ear vein at a rate of 1 ml/min. The inflammation in rabbit's eye is measured at various intervals by quantification of the fluorescein leakage into the anterior chamber using a fluorophotometer (Vogel 2002).

5.7.6.8 Proliferative Vitreoretinopathy in Rabbits

Proliferative vitreoretinopathy is a severe eye injury or a complication of retinal detachment. This model is based on the principle that if the cultured fibroblast cells are cultured into the rabbit's vitreous region, they might mimic the condition of proliferative vitreoratinopathy in them. The explants obtained from the dorsal skin of rabbit, after removing hair and subcutaneous fat, are rinsed, finely minced, and placed in petri dishes containing Dulbecco's medium supplemented with heat-inactivated fetal calf serum, garamycin, and fungizone. When the cells started growing, the pieces of tissue and the medium are removed and 2 ml of fresh medium is added. After the cells are grown to convergence, they are passaged into flasks by rinsing with PBS and incubating in trypsin/EDTA. Subsequent to the initial passage, cells are split in a ratio of 1:4 upon reaching confluence. For injection purpose, cells from third to tenth passage are trypsinized using trypsin/ EDTA, centrifuged, washed, and re-suspended in a medium and permitted to stand for some time till cell clumps are settled. The number of cells is determined using a Coulter counter.

Pigmented rabbits are anesthetized. The eyes are dilated using a drop of neo-synephrine (10%), atropine (1%), and mydriacyl (1%). One eye is injected with using an operating microscope. The fibroblast cells suspended in a medium are then injected slowly into the posterior portion of the mid-vitreous. Test drug is injected slowly through and to the same site in the vitreous. Needles have to be withdrawn slowly in order to prevent leakage. Parenthesis is performed to equilibrate the intraocular pressure. An ointment containing antibiotics and atropine is applied at the end of the procedure (Vogel 2002).

There are additional models for proliferative vitreoretinopathy. A model in which rabbits are first anesthetized, their pupils are dilated before the surgery using tropicamide and the rabbits are given antibiotics has also been reported. Proliferative vitreoretinopathy is induced by injecting retinal pigment epithelial cells (suspended in platelet rich plasma) intravitreally. An anterior chamber paracentesis is done and aqueous humor is drained prior to injection (Wang et al. 2014). The MDM2 T309G mutation enhances proliferative vitreoretinopathy in primary human retinal epithelial cells has also been reported (Zhou et al. 2017).

5.7.7 In-Vivo Confocal Microscopy

In-vivo confocal microscopy (IVCM) can be used to observe the changes that occur at the cellular level on the ocular surface. High-resolution images are obtained non-invasively. Ocular diseases, such as herpes, graft rejection, and dry eye syndrome can be studied on the basis of density, kinking, tortuosity, etc. of the corneal innervations (He et al. 2017).

5.7.8 *Ex Vivo* and In-Vitro Tests Recommended by Federal Agencies

The Interagency Coordinating Committee on the Validation of Alternative Methods (ICCVAM) has approved revised regulatory test methods which minimize pain and distress to animals and encourage the use of alternative methods. They recommended five in-vitro methods for the evaluation that have gained much popularity over the last two decades. These methods include the Hen's egg test, chorioallantoic membrane test method, isolated rabbit eye test method, bovine corneal opacity, permeability test method, isolated chicken eye, and cytosensor microphysiometer test method. Reduce, refine, and replace are the global buzzwords when it comes to animal testing. The principles enumerated by ICCVAM should be followed in letter and spirit which recommends using alternatives to animal testing procedures (ICCVAM).

5.7.8.1 Isolated/Enucleated Organ/Organotypic Methods

- *Experimental Cataract Formation*: This evaluation can be performed with isolated goat eyes which are available through any slaughterhouse; goat eye lenses are removed by extracapsular extraction and dipped in tyrrode solution. The group is treated with 0.5 mM hydrogen peroxide solution for induction of cataract. The lenses can be checked for transparency/opacity and biochemical estimation of oxidative stress markers. Ocular organotypic models are secluded systems that are able to maintain the normal biochemical and physiological function of the enucleated eye or cornea for a short duration. Corneal opacity and corneal histology are utilized to detect the changes in isolated tissue on exposure to test formulations. Corneal opacity indicates the swelling, protein denaturation, vacuolation, and damage to the corneal epithelium and corneal stroma (Barile 2010). On the other hand, histology describes the irritation in terms of a slight irritant (causes damage to superficial epithelium), mild irritants (cause damage to stroma), and severe irritants (cause damage to endothelium layer) (Jester et al. 2001).

5.7.8.2 Enucleated Eye Tests

Enucleated eye tests (EETs) were proposed by Burton et al. (1981) using isolated rabbit eyes (IREs) for test ocular toxicity. Mostly, chemical/synthetic intermediates API, raw materials, cleaners, detergents, solvents, soaps, and surfactants are used in an isolated rabbit eyes for ocular toxicity study (ICCVAM 2010c). Porcine corneas, bovine corneas, and chicken enucleated eye tests (CEETs) are frequently used to test the ocular irritation. ICCVAM gave the recommendations for both, IREs and isolated chicken eye test methods (ICCVAM). The toxic effects of formulations are measured by observing change in surface of tissue, swelling, fluorescein retention, and opacity (Prinsen 1996; OCED 2013b).

5.7.8.3 Non-Ocular Organotypic Models

The Hen's egg test/Huhner-embryonen test on chorioallantoic membrane (HET-CAM) assay has been proposed (Luepke 1985; Luepke and Kemper 1986). In this test, the test formulation (0.2–0.3 ml liquid, 0.1–0.3 gm solid) is applied to the CAM surface and observation of time required for any change in CAM morphology (hemorrhage, vasoconstriction, and/or coagulation) is recorded, scored, and categorized. The breed of hen, number of replicates, relative humidity, and incubation times may affect the toxicity results (Gettings et al. 1998; Steiling et al. 1999). ICCVAM has evaluated several HET-CAM methods and has given recommendations. Before initiating non-regulatory, validation, or optimization HET-CAM studies, investigators are encouraged to visit the ICCVAM website to ensure that the test method protocol is most recent (ICCVAM 2012).

Conclusion

In conclusion, ocular therapeutics are extremely complex and are vital to the quality of life of a human being. The research being done on the topic is also substantial and rapid strides have been made in this area. Due to the insights gained about drug transporters and the mechanism of drug absorption as well as the anatomy and physiology of the eye, advanced and novel drug delivery systems are now available which have alleviated the problems of frequent instillation of medicament, low visibility, poor bioavailability, systemic side effects, inaccurate dosing, and so on. With proper in-vivo, *ex vivo*, and in-vitro evaluation methods available, improvement in ocular therapy can be expected and unanswered questions related to macular degeneration and loss of vision and blindness may be addressed.

References

Abboud, I. and W. H. Massoud. 1972. "Effect of blephamide in blepharitis." *Bull Ophthalmol Soc Egypt* no. 65 (69):539–43.

Abdelbary, Ahmed, Heba F. Salem, Rasha A. Khallaf, and Ahmed M. A. Ali. 2017. "Mucoadhesive niosomal in situ gel for ocular tissue targeting: In vitro and in vivo evaluation of lomefloxacin hydrochloride." *Pharmaceutical Development and Technology* no. 22 (3):409–417.

Abraham, Sindhu, Sharon Furtado, S. Bharath, B. V. Basavaraj, R. Deveswaran, and V. Madhavan. 2009. "Sustained ophthalmic delivery of ofloxacin from an ion-activated in situ gelling system." *Pakistan Journal of Pharmaceutical Sciences* no. 22 (2).

Aburahma, Mona Hassan, and Azza Ahmed Mahmoud. 2011. "Biodegradable ocular inserts for sustained delivery of brimonidine tartarate: Preparation and in vitro/in vivo evaluation." *Aaps Pharmscitech* no. 12 (4):1335–1347.

Agnihotri, Sunil A. and Tejraj M. Aminabhavi. 2007. "Chitosan nanoparticles for prolonged delivery of timolol maleate." *Drug Development and Industrial Pharmacy* no. 33 (11):1254–1262.

Ahmad, Zaheer, Afzal Shah, Muhammad Siddiq, and Heinz-Bernhard Kraatz. 2014. "Polymeric micelles as drug delivery vehicles." *Rsc Advances* no. 4 (33):17028–17038.

Ahmed, Zahid Zaheer, Furquan Nazimuddin Khan, and Darakhshan Afreen Shaikh. 2017. "Reverse engineering and formulation by QBD of olopatadine hydrochloride ophthalmic solution." *Journal of Pharmaceutical Investigation.* doi: 10.1007/s40005-017-0312-1.

Amadi-Obi, Ahjoku, Cheng-Rong Yu, Xuebin Liu, Rashid M. Mahdi, Grace Levy Clarke, Robert B. Nussenblatt, Igal Gery, Yun Sang Lee, and Charles E. Egwuagu. 2007. "TH17 cells contribute to uveitis and scleritis and are expanded by IL-2 and inhibited by IL-27/STAT1." *Nat Med* no. 13 (6):711–718. doi: http://www.nature.com/nm/journal/v13/n6/suppinfo/nm1585_S1.html.

Anfuso, Carmelina D., Melania Olivieri, Annamaria Fidilio, Gabriella Lupo, Dario Rusciano, Salvatore Pezzino, Caterina Gagliano, Filippo Drago, and Claudio Bucolo. 2017. "Gabapentin Attenuates Ocular Inflammation: In vitro and In vivo Studies." *Frontiers in Pharmacology* no. 8:173. doi: 10.3389/fphar.2017.00173.

Araki-Sasaki, Kaoru, Yuichi Ohashi, Tetsuo Sasabe, Kozaburo Hayashi, Hitoshi Watanabe, Yasuo Tano, and Hiroshi Handa. 1995. "An SV40-immortalized human corneal epithelial cell line and its characterization." *Investigative Ophthalmology & Visual Science* no. 36 (3):614–621.

Avila, M. Y., D. A. Carre, R. A. Stone, and M. M. Civan. 2001. "Reliable measurement of mouse intraocular pressure by a servo-null micropipette system." *Invest Ophthalmol Vis Sci* no. 42 (8):1841–1846.

Balguri, S. P., G. R. Adelli, K. Y. Janga, P. Bhagav, and S. Majumdar. 2017. "Ocular disposition of ciprofloxacin from topical, PEGylated nanostructured lipid carriers: Effect of molecular weight and density of poly (ethylene) glycol." *International Journal of Pharmaceutics* no. 529 (1–2):32–43. doi: 10.1016/j.ijpharm.2017.06.042.

Bansal, S., V. A. Barathi, D. Iwata, and R. Agrawal. 2015. "Experimental autoimmune uveitis and other animal models of uveitis: An update." *Indian Journal of Ophthalmology* no. 63 (3):211–218. doi: 10.4103/0301-4738.156914.

Barile, Frank A. 2010. "Validating and Troubleshooting Ocular In Vitro Toxicology Tests." *Journal of Pharmacological and Toxicological Methods* no. 61 (2):136–145. doi: 10.1016/j.vascn.2010.01.001.

Baudouin, Christophe, Antoine Labbé, Hong Liang, Aude Pauly, and Françoise Brignole-Baudouin. 2010. "Preservatives in eyedrops: The good, the bad and the ugly." *Progress in Retinal and Eye Research* no. 29 (4):312–334.

Bourges, J. L., C. Bloquel, Aurélien Thomas, F. Froussart, A. Bochot, F. Azan, Robert Gurny, D. BenEzra, and F. Behar-Cohen. 2006. "Intraocular implants for extended drug delivery: Therapeutic applications." *Advanced Drug Delivery Reviews* no. 58 (11):1182–1202.

Burton, A. B. G., M. York, and R. S. Lawrence. 1981. "The in vitro assessment of severe eye irritants." *Food and Cosmetics Toxicology* no. 19:471–480. doi: https://doi.org/10.1016/0015-6264(81)90452-1.

Chen, J., J. Zhang, R. Zhao, J. Jin, Y. Yu, W. Li, W. Wang, H. Zhou, and S. B. Su. 2016. "Topical Application of Interleukin-28A Attenuates Allergic Conjunctivitis in an Ovalbumin-Induced Mouse Model." *Investigative Ophthalmology & Visual Science* no. 57 (2):604–610. doi: 10.1167/iovs.15-18457.

Cholkar, Kishore, Sulabh P. Patel, Aswani Dutt Vadlapudi, and Ashim K. Mitra. 2013. "Novel strategies for anterior segment ocular drug delivery." *Journal of Ocular Pharmacology and Therapeutics* no. 29 (2):106–123.

Choonara, Yahya E., Viness Pillay, Michael P. Danckwerts, Trevor R. Carmichael, and Lisa C. Du Toit. 2010. "A review of implantable intravitreal drug delivery technologies for the treatment of posterior segment eye diseases." *Journal of Pharmaceutical Sciences* no. 99 (5):2219–2239.

Civiale, C., M. Licciardi, G. Cavallaro, G. Giammona, and M. G. Mazzone. 2009. "Polyhydroxyethylaspartamide-based micelles for ocular drug delivery." *International Journal of Pharmaceutics* no. 378 (1):177–186.

Civiale, Claudine, Grazia Paladino, Clara Marino, Francesco Trombetta, Teodoro Pulvirenti, and Vincenzo Enea. 2003. "Multilayer primary epithelial cell culture from bovine conjunctiva as a model for in vitro toxicity tests." *Ophthalmic Research* no. 35 (3):126–136.

Curren, Rodger D., and John W. Harbell. 2002. "Ocular safety: A silent (in vitro) success story." *ATLA-NOTTINGHAM* no. 30:69–74.

Dey, Surajit and Ashim K. Mitra. 2005. *Transporters and Receptors in Ocular Drug Delivery: Opportunities and Challenges*. Taylor & Francis.

Donnelly, Ryan F., Thakur Raghu Raj Singh, and A. David Woolfson. 2010. "Microneedle-based drug delivery systems: Microfabrication, drug delivery, and safety." *Drug Delivery* no. 17 (4):187–207.

Draize, John H., Geoffrey Woodard, and Herbert O. Calvery. 1944. "Methods for the study of irritation and toxicity of substances applied topically to the skin and mucous membranes." *Journal of Pharmacology and Experimental Therapeutics* no. 82 (3):377–390.

Fischer, Marco and Fritz Vögtle. 1999. "Dendrimers: From design to application—A progress report." *Angewandte Chemie International Edition* no. 38 (7):884–905.

Forest, David L., Lincoln V. Johnson, and Dennis O. Clegg. 2015. "Cellular models and therapies for age-related macular degeneration." *Disease Models & Mechanisms* no. 8 (5):421–427. doi: 10.1242/dmm.017236.

Gad, Shayne Cox. 2008. *Pharmaceutical Manufacturing Handbook: Production and Processes*. Wiley.

Gamache, Daniel A., Gustav Graff, Milton T. Brady, Joan M. Spellman, and John M. Yanni. 2000. "Nepafenac, a Unique Nonsteroidal Prodrug with Potential Utility in the Treatment of Trauma-Induced Ocular Inflammation: I. Assessment of Anti-Inflammatory Efficacy." *Inflammation* no. 24 (4):357–370. doi: 10.1023/a:1007049015148.

Garhwal, Rahul, Sally F. Shady, Edward J. Ellis, Jeanne Y. Ellis, Charles D. Leahy, Stephen P. McCarthy, Kathryn S. Crawford, and Peter Gaines. 2012. "Sustained ocular delivery of ciprofloxacin using nanospheres and conventional contact lens materials." *Investigative Ophthalmology & Visual Science* no. 53 (3):1341–1352.

Gaudana, Ripal, J. Jwala, Sai H. S. Boddu, and Ashim K. Mitra. 2008. "Recent Perspectives in Ocular Drug Delivery." *Pharmaceutical Research* no. 26 (5):1197. doi: 10.1007/s11095-008-9694-0.

Gettings, S. D., R. A. Lordo, P. I. Feder, and K. L. Hintze. 1998. "A comparison of low volume, Draize and in vitro eye irritation test data. III. Surfactant-based formulations." *Food and Chemical Toxicology* no. 36 (3):209–231.

Gillies, Mark C., Tao Su, and Daya Naidoo. 1995. "Electrical resistance and macromolecular permeability of retinal capillary endothelial cells in vitro." *Current Eye Research* no. 14 (6):435–442.

Goldblum, D. and T. Mittag. 2002. "Prospects for relevant glaucoma models with retinal ganglion cell damage in the rodent eye." *Vision Research* no. 42 (4):471–478.

Gomaa, Yasmine A., Martin J. Garland, Fiona McInnes, Labiba K. El-Khordagui, Clive Wilson, and Ryan F. Donnelly. 2012. "Laser-engineered dissolving microneedles for active transdermal delivery of nadroparin calcium." *European Journal of Pharmaceutics and Biopharmaceutics* no. 82 (2):299–307.

Greenwald, Yoel and Guy Kleinmann. 2008. "Use of collagen shields for ocular-surface drug delivery." *Expert Review of Ophthalmology* no. 3 (6):627–633.

Groneberg, D. A., L. Bielory, A. Fischer, S. Bonini, and U. Wahn. 2003. "Animal models of allergic and inflammatory conjunctivitis." *Allergy* no. 58 (11):1101–1113.

Gulsen, Derya and Anuj Chauhan. 2004. "Ophthalmic drug delivery through contact lenses." *Investigative Ophthalmology & Visual Science* no. 45 (7):2342–2347.

Gupta, Ajay Kumar, Sumit Madan, D. K. Majumdar, and Amarnath Maitra. 2000. "Ketorolac entrapped in polymeric micelles: Preparation, characterisation and ocular anti-inflammatory studies." *International Journal of Pharmaceutics* no. 209 (1):1–14. doi: https://doi.org/10.1016/S0378-5173(00)00508-1.

Gupta, Himanshu and Mohammed Aqil. 2012. "Contact lenses in ocular therapeutics." *Drug Discovery Today* no. 17 (9):522–527.

Harmia, T., J. Kreuter, P. Speiser, T. Boye, R. Gurny, and A. Kubi. 1986. "Enhancement of the myotic response of rabbits with pilocarpine-loaded polybutylcyanoacrylate nanoparticles." *International Journal of Pharmaceutics* no. 33 (1–3):187–193.

He, Jingliang, Yoko Ogawa, Shin Mukai, Yumiko Saijo-Ban, Mizuka Kamoi, Miki Uchino, Mio Yamane, Nobuhiro Ozawa, Masaki Fukui, Takehiko Mori, Shinichiro Okamoto, and Kazuo Tsubota. 2017. "In Vivo Confocal Microscopy Evaluation of Ocular Surface with Graft-Versus-Host Disease-Related Dry Eye Disease." *Scientific Reports* no. 7 (1):10720. doi: 10.1038/s41598-017-10237-w.

Ho, L. C., I. P. Conner, C. W. Do, S. G. Kim, E. X. Wu, G. Wollstein, J. S. Schuman, and K. C. Chan. 2014. "In vivo assessment of aqueous humor dynamics upon chronic ocular hypertension and hypotensive drug treatment using gadolinium-enhanced MRI." *Investigative Ophthalmology & Visual Science* no. 55 (6):3747–3757. doi: 10.1167/iovs.14-14263.

Hornof, Margit, Elisa Toropainen, and Arto Urtti. 2005. "Cell culture models of the ocular barriers." *European Journal of Pharmaceutics and Biopharmaceutics* no. 60 (2):207–225.

Huhtala, A., L. Salminen, H. Tähti, and H. Uusitalo. 2008. "Corneal models for the toxicity testing of drugs and drug releasing materials." *Topics in Multifunctional Biomaterials and Devices* no. 1 (2):1–23.

Hui, Ho-Wah and Joseph R Robinson. 1985. "Ocular delivery of progesterone using a bioadhesive polymer." *International Journal of Pharmaceutics* no. 26 (3):203–213.

ICCVAM. *ICCVAM- Recommended Test Method Protocols.* National Toxicology Program, US Department of Health and Human Services. Available from https://ntp.niehs.nih.gov/pubhealth/evalatm/test-method-evaluations/protocols/index.html.

ICCVAM. 2010c. ICCVAM Recommended Test Method Protocol Isolated Rabbit Eye Test Method. edited by N.I.o.E.H. Sciences (Ed.). National Institute of Environmental Health Sciences, Triangle Park, NC: NIH Publication.

ICCVAM. 2012. ICCVAM-Recommended Protocol for Using the Isolated Chicken Eye (ICE) Test Method. National Institute of Environmental Health Sciences, Triangle Park, NC: NIH Publication.

IP. 2010. *Indian Pharmacopoeia*. 3 vols. Vol. 1: The Controller of Publications, New Delhi; Minister of Health and Family Welfare, India.

Jester, J. V., L. Li, A. Molai, and J. K. Maurer. 2001. "Extent of initial corneal injury as a basis for alternative eye irritation tests." *Toxicology In Vitro* no. 15 (2):115–130.

Joshi, Abhay. 1994. "Microparticulates for ophthalmic drug delivery." *Journal of Ocular Pharmacology and Therapeutics* no. 10 (1):29–45.

Kanoujia, Jovita, Kanchan Sonker, Manisha Pandey, Koshy M Kymonil, and Shubhini A Saraf. 2012. "Formulation and characterization of a novel pH-triggered in-situ gelling ocular system containing Gatifloxacin." *International Current Pharmaceutical Journal* no. 1 (3):43–49.

Kanoujia, Jovita, Priya Singh Kushwaha, and Shubhini A Saraf. 2014. "Evaluation of gatifloxacin pluronic micelles and development of its formulation for ocular delivery." *Drug Delivery and Translational Research* no. 4 (4):334–343.

Kapoor, Yash, and Anuj Chauhan. 2008. "Ophthalmic delivery of Cyclosporine A from Brij-97 microemulsion and surfactant-laden p-HEMA hydrogels." *International Journal of Pharmaceutics* no. 361 (1):222–229.

Kaur, Indu P., Alka Garg, Anil K. Singla, and Deepika Aggarwal. 2004. "Vesicular systems in ocular drug delivery: An overview." *International Journal of Pharmaceutics* no. 269 (1):1–14.

Keister, J. C., E. R. Cooper, P. J. Missel, J. C. Lang, and D. F. Hager. 1991. "Limits on optimizing ocular drug delivery." *Journal of Pharmaceutical Sciences* no. 80 (1):50–53.

Khan, Shagufta, Asgar Ali, Dilesh Singhavi, and Pramod Yeole. 2008. "Controlled ocular delivery of acyclovir through rate controlling ocular insert of eudragit: a technical note." *AAPS PharmSciTech* no. 9 (1):169–173.

Kumar, M. T., J. K. Pandit, and J. Balasubramaniam. 2001. "Novel therapeutic approaches for uveitis and retinitis." *Journal of Pharmacy & Pharmaceutical Sciences: A Publication of the Canadian Society for Pharmaceutical Sciences, Societe Canadienne des Sciences Pharmaceutiques* no. 4 (3):248–254.

Kumar, Rakesh and V. R. Sinha. 2014. "Preparation and optimization of voriconazole microemulsion for ocular delivery." *Colloids and Surfaces B: Biointerfaces* no. 117:82–88.

Kuo, Chuan-Hui, Andrea M. Collins, Douglas R. Boettner, YanFen Yang, and Santa J. Ono. 2017. "Role of CCL7 in Type I Hypersensitivity Reactions in Murine Experimental Allergic Conjunctivitis." *Journal of Immunology Author Choice* no. 198 (2):645–656. doi: 10.4049/jimmunol.1502416.

Lardy, F., B. Vennat, M. P. Pouget, and A. Pourrat. 2000. "Functionalization of hydrocolloids: Principal component analysis applied to the study of correlations between parameters describing the consistency of hydrogels." *Drug Development and Industrial Pharmacy* no. 26 (7):715–721.

Larrañeta, Eneko, Sarah Stewart, Steven J. Fallows, Lena L. Birkhäuer, Maeliosa T. C. McCrudden, A. David Woolfson, and Ryan F. Donnelly. 2016. "A facile system to evaluate in vitro drug release from dissolving microneedle arrays." *International Journal of Pharmaceutics* no. 497 (1):62–69.

Lee, Chi H., Anuj Singla, and Yugyung Lee. 2001. "Biomedical applications of collagen." *International Journal of Pharmaceutics* no. 221 (1):1–22. doi: https://doi.org/10.1016/S0378-5173(01)00691-3.

Lee, H. J., B. M. Kim, S. Shin, T. Y. Kim, and S. H. Chung. 2017. "Superoxide dismutase 3 attenuates experimental Th2-driven allergic conjunctivitis." *Clinical Immunology* no. 176:49–54. doi: 10.1016/j.clim.2016.12.010.

Lee, H. S., J. Y. Kwon, and C. K. Joo. 2016. "Topical Administration of beta-1,3-Glucan to Modulate Allergic Conjunctivitis in a Murine Model." *Investigative Ophthalmology & Visual Science* no. 57 (3):1352–1360. doi: 10.1167/iovs.15-17914.

Levkovitch-Verbin, H. 2004. "Animal models of optic nerve diseases." *Eye (Lond)* no. 18 (11):1066–1074. doi: 10.1038/sj.eye.6701576.

Li, Jing, Kalpana Kamath, and Chandradhar Dwivedi. 2001. "Gellan Film as an Implant for Insulin Delivery." *Journal of Biomaterials Applications* no. 15 (4):321–343. doi: 10.1106/R3TF-PT7W-DWN0-1RBL.

Li, L., Y. Yuan, L. Chen, M. Li, P. Ji, J. Gong, Y. Zhao, and H. Zhang. 2017. "Gadolinium-enhanced 7.0 T magnetic resonance imaging assessment of the aqueous inflow in rat eyes in vivo." *Exp Eye Res* no. 162:18–26. doi: 10.1016/j.exer.2017.06.019.

Li, Ying, Da-Jian Yang, Shi-Lin Chen, Si-Bao Chen, and Albert Sun-Chi Chan. 2008. "Comparative physicochemical characterization of phospholipids complex of puerarin formulated by conventional and supercritical methods." *Pharmaceutical Research* no. 25 (3):563–577.

Liang, H., F. Brignole-Baudouin, L. Rabinovich-Guilatt, Z. Mao, L. Riancho, M.-O. Faure, J.-M. Warnet, G. Lambert, and C. Baudouin. 2008. "Reduction of quaternary ammonium-induced ocular surface toxicity by emulsions: An in vivo study in rabbits." *Investigative Ophthalmology & Visual Science* no. 49 (13):2357–2357.

Liu, C., Q. Lan, W. He, C. Nie, C. Zhang, T. Xu, T. Jiang, and S. Wang. 2017. "Octa-arginine modified lipid emulsions as a potential ocular delivery system for disulfiram: A study of the corneal permeation, transcorneal mechanism and anti-cataract effect." *Colloids Surf B Biointerfaces* no. 160:305–314. doi: 10.1016/j.colsurfb.2017.08.037.

Luepke, N. P. 1985. "Hen's egg chorioallantoic membrane test for irritation potential." *Food Chem Toxicol* no. 23 (2):287–291.

Luepke, N. P. and F. H. Kemper. 1986. "The HET-CAM test: An alternative to the draize eye test." *Food and Chemical Toxicology* no. 24 (6):495–496. doi: https://doi.org/10.1016/0278-6915(86)90099-2.

Ma, Minglu, Yiyun Cheng, Zhenhua Xu, Peng Xu, Haiou Qu, Yujie Fang, Tongwen Xu, and Longping Wen. 2007. "Evaluation of polyamidoamine (PAMAM) dendrimers as drug carriers of anti-bacterial drugs using sulfamethoxazole (SMZ) as a model drug." *European Journal of Medicinal Chemistry* no. 42 (1):93–98.

Mafi, Roozbeh, Cameron Gray, Robert Pelton, Howard Ketelson, and James Davis. 2014. "On formulating ophthalmic emulsions." *Colloids and Surfaces B: Biointerfaces* no. 122 (Supplement C):7–11. doi: https://doi.org/10.1016/j.colsurfb.2014.06.039.

Makwana, S. B., V. A. Patel, and S. J. Parmar. 2016. "Development and characterization of in-situ gel for ophthalmic formulation containing ciprofloxacin hydrochloride." *Results in Pharma Sciences* no. 6 (Supplement C):1–6. doi: https://doi.org/10.1016/j.rinphs.2015.06.001.

Mandal, Abhirup, Rohit Bisht, Ilva D. Rupenthal, and Ashim K. Mitra. 2017. "Polymeric micelles for ocular drug delivery: From structural frameworks to recent preclinical studies." *Journal of Controlled Release* no. 248 (Supplement C):96–116. doi: https://doi.org/10.1016/j.jconrel.2017.01.012.

Mandal, B., K. S. Alexander, and A. T. Riga. 2010. "Sulfacetamide loaded Eudragit(R) RL100 nanosuspension with potential for ocular delivery." *Journal of Pharmacy and Pharmaceutical Sciences* no. 13 (4):510–523.

Mannermaa, Eliisa, Mika Reinisalo, Veli-Pekka Ranta, Kati-Sisko Vellonen, Heidi Kokki, Anni Saarikko, Kai Kaarniranta, and Arto Urtti. 2010. "Filter-cultured ARPE-19 cells as outer blood–retinal barrier model." *European Journal of Pharmaceutical Sciences* no. 40 (4):289–296.

Marquart, Mary E. 2011. "Animal Models of Bacterial Keratitis." *Journal of Biomedicine and Biotechnology* no. 2011:680642. doi: 10.1155/2011/680642.

Matthews, Brian R. and G. Michael Wall. 2000. "Stability storage and testing of ophthalmic products for global registration." *Drug Development and Industrial Pharmacy* no. 26 (12):1227–1237.

McLaren, J. W., R. F. Brubaker, and J. S. FitzSimon. 1996. "Continuous measurement of intraocular pressure in rabbits by telemetry." *Investigative Ophthalmology & Visual Science* no. 37 (6):966–975.

Meisner, Dale and Michael Mezei. 1995. "Liposome ocular delivery systems." *Advanced Drug Delivery Reviews* no. 16 (1):75–93.

Mezei, M. and D. Meisner. 2003. *Biopharmaceutics of Ocular Drug Delivery* P. (ed) Boca Raton, FL, USA: CRC Press.

Millar, J. Cameron and Iok-Hou Pang. 2015. "Non-continuous measurement of intraocular pressure in laboratory animals." *Experimental Eye Research* no. 141 (Supplement C):74–90. doi: https://doi.org/10.1016/j.exer.2015.04.018.

Mitra, Ashim K. 2003. *Ophthalmic Drug Delivery System*. 2nd ed. New York, United States: Marcel Dekker Inc.

Moghimipour, Eskandar, Anayatollah Salimi, and Behzad Sharif Makhmal Zadeh. 2013. "Effect of the Various Solvents on the In Vitro Permeability of Vitamin B 12 through Excised Rat Skin." *Tropical Journal of Pharmaceutical Research* no. 12 (5):671–677.

Mohan, Eaga Chandra, Jagan Mohan KanduKuri, and Venkatesham Allenki. 2009. "Preparation and evaluation of in-situ-gels for ocular drug delivery." *Journal of Pharmacy Research 2009, 2 (6), 1089* no. 1094.

Moulik, Satya Priya and Animesh Kumar Rakshit. 2006. "Physicochemisty and applications of microemulsions." *Journal of Surface Science and Technology* no. 22 (3/4):159.

Moustafa, M. A., Y. S. R. Elnaggar, W. M. El-Refaie, and O. Y. Abdallah. 2017. "Hyalugel-integrated liposomes as a novel ocular nanosized delivery system of fluconazole with promising prolonged effect." *International Journal of Pharmaceutics* no. 534 (1–2):14–24. doi: 10.1016/j.ijpharm.2017.10.007.

Mudgil, Meetali, and Pravin K. Pawar. 2013. "Preparation and In Vitro/Ex Vivo Evaluation of Moxifloxacin-Loaded PLGA Nanosuspensions for Ophthalmic Application." *Scientia Pharmaceutica* no. 81 (2):591–606. doi: 10.3797/scipharm .1204-16.

Nanjawade, Basavaraj K., F. V. Manvi, and A. S. Manjappa. 2007. "Retracted: In situ-forming hydrogels for sustained ophthalmic drug delivery." *Journal of Controlled Release* no. 122 (2):119–134.

Noomwong, Pawinee, Wantanee Ratanasak, Assadang Polnok, and Narong Sarisuta. 2011. "Development of acyclovir-loaded bovine serum albumin nanoparticles for ocular drug delivery." *International Journal of Drug Delivery* no. 3 (4):669.

NRC. 1977. *Principles and Procedures for Evaluating the Toxicity of Household Substances.* National Academy of Sciences, Washington, DC.

OCED. 2013b. Test No. 438: Isolated Chicken Eye Test Method for Identifying i) Chemicals Including Serious Eye Damage and ii) Chemicals Not Requiring Classification for Eye Irritation or Serious Eye Damage. Paris: OCED Publishing.

Offord, Elizabeth A., Najam A. Sharif, Katherine Mace, Yvonne Tromvoukis, Elisa A. Spillare, Ornella Avanti, William E. Howe, and A. M. Pfeifer. 1999. "Immortalized human corneal epithelial cells for ocular toxicity and inflammation studies." *Investigative Ophthalmology & Visual Science* no. 40 (6):1091–1101.

Pandey, Prasoon, Aakash Singh Panwar, Pankaj Dwivedi, Priya Jain, Ashish Agarwal, and Dheeraj Jain. 2011. "Design and Evaluation of Ocular Inserts For Controlled Drug Delivery of Acyclovir." *International Journal of Pharmaceutical and Biological Archieves* no. 2(4).

Parveen, S., M. Mitra, S. Krishnakumar, and S. K. Sahoo. 2010. "Enhanced antiproliferative activity of carboplatin loaded chitosan-alginate nanoparticles in a retinoblastoma cell line." *Acta Biomater* no. 6:3120–3131.

Paschalis, Eleftherios I., Fabiano Cade, Samir Melki, Louis R. Pasquale, Claes H. Dohlman, and Joseph B. Ciolino. 2014. "Reliable intraocular pressure measurement using automated radio-wave telemetry." *Clinical Ophthalmology (Auckland, N.Z.)* no. 8:177–185. doi: 10.2147/OPTH.S54753.

Pascolini, D. and S. P. Mariotti. 2012. "Global estimates of visual impairment: 2010." *Br J Ophthalmol* no. 96 (5):614–618. doi: 10.1136/bjophthalmol-2011-300539.

Patel, Ashaben, Kishore Cholkar, Vibhuti Agrahari, and Ashim K. Mitra. 2013. "Ocular drug delivery systems: An overview." *World Journal of Pharmacology* no. 2 (2):47.

Patel, P., D. Shastri, P. Shelat, and A. Shukla. 2010. "Ophthalmic drug delivery system: Challenges and approaches." *Systematic Reviews in Pharmacy* no. 1 (2):113.

Patel, R. P., B. Dadhani, R. Ladani, A. H. Baria, and Jigar Patel. 2010. "Formulation, evaluation and optimization of stomach specific in situ gel of clarithromycin and metronidazole benzoate." *International Journal of Drug Delivery* no. 2 (2).

Patravale, V. B. and R. M. Kulkarni. 2004. "Nanosuspensions: A promising drug delivery strategy." *Journal of Pharmacy and Pharmacology* no. 56 (7):827–840.

Pignatello, Rosario, Claudio Bucolo, Piera Ferrara, Adriana Maltese, Antonina Puleo, and Giovanni Puglisi. 2002. "Eudragit RS100® nanosuspensions for the ophthalmic controlled delivery of ibuprofen." *European Journal of Pharmaceutical Sciences* no. 16 (1):53–61.

Pinnock, Abigail, Nagaveni Shivshetty, Sanhita Roy, Stephen Rimmer, Ian Douglas, Sheila MacNeil, and Prashant Garg. 2017. "Ex vivo rabbit and human corneas as models for bacterial and fungal keratitis." *Graefe's Archive for Clinical and Experimental Ophthalmology* no. 255 (2):333–342. doi: 10.1007/s00417-016-3546-0.

Ponchel, Gilles, Frederic Touchard, Dominique Duchêne, and Nikolaos A Peppas. 1987. "Bioadhesive analysis of controlled-release systems. I. Fracture and interpenetration analysis in poly (acrylic acid)-containing systems." *Journal of Controlled Release* no. 5 (2):129–141.

Postnikoff, C. K., R. Pintwala, S. Williams, A. M. Wright, D. Hileeto, and M. B. Gorbet. 2014. "Development of a curved, stratified, in vitro model to assess ocular biocompatibility." *PLoS One* no. 9 (5):e96448. doi: 10.1371/journal.pone.0096448.

Prinsen, M. K. 1996. "The chicken enucleated eye test (CEET): A practical (pre)screen for the assessment of eye irritation/corrosion potential of test materials." *Food Chem Toxicol* no. 34 (3):291–296.

Quigley, H. A., I. F. Pitha, D. S. Welsbie, C. Nguyen, M. R. Steinhart, T. D. Nguyen, M. E. Pease, E. N. Oglesby, C. A. Berlinicke, K. L. Mitchell, J. Kim, J. J. Jefferys, and E. C. Kimball. 2015. "Losartan Treatment Protects Retinal Ganglion Cells and Alters Scleral Remodeling in Experimental Glaucoma." *PLoS One* no. 10 (10):e0141137. doi: 10.1371/journal.pone.0141137.

Rajoria, Gourav and Arushi Gupta. 2012. "In-Situ Gelling System: A novel approach for ocular drug delivery." *AJPTR* no. 2:24–53.

Ranch, Ketan, Hetal Patel, Laxman Chavda, Akshay Koli, Furqan Maulvi, and Rajesh K Parikh. 2017. "Development of in situ Ophthalmic gel of Dexamethasone Sodium Phosphate and Chloramphenicol: A Viable Alternative to Conventional Eye Drops." *Journal of Applied Pharmaceutical Science Vol* no. 7 (03):101–108.

Rao, P. Rama, S. Ramakrishna, and Prakash V. Diwan. 2000. "Drug Release Kinetics from Polymeric Films Containing Propranolol Hydrochloride for Transdermal Use." *Pharmaceutical Development and Technology* no. 5 (4):465–472. doi: 10.1081/PDT-100102030.

Robinson, J. R., and Academy of Pharmaceutical Sciences. Industrial Pharmaceutical Technology Section. 1980. *Ophthalmic Drug Delivery Systems: A Symposium*: The Section.

Rojas, Sánchez F. and Bosch C. Ojeda. 2009. "Recent development in derivative ultraviolet/visible absorption spectrophotometry: 2004–2008: A review." *Analytica Chimica Acta* no. 635 (1):22–44. doi: https://doi.org/10.1016/j.aca.2008.12.039.

Rosenbaum, J. T., H. O. McDevitt, R. B. Guss, and P. R. Egbert. 1980. "Endotoxin-induced uveitis in rats as a model for human disease." *Nature* no. 286 (5773):611–613.

Rózsa, Andrew J. and Roger W. Beuerman. 1982. "Density and organization of free nerve endings in the corneal epithelium of the rabbit." *Pain* no. 14 (2):105–120. doi: https://doi.org/10.1016/0304-3959(82)90092-6.

Saettone, Marco Fabrizio and Lotta Salminen. 1995. "Ocular inserts for topical delivery." *Advanced Drug Delivery Reviews* no. 16 (1):95–106.

Sahoo, Sanjeeb K., Fahima Dilnawaz, and S. Krishnakumar. 2008. "Nanotechnology in ocular drug delivery." *Drug Discovery today* no. 13 (3):144–151.

Sanders, Donald R., Bruce Goldstick, Cheryl Kraff, Robert Hutchins, Melvin S. Bernstein, and Michael A. Evans. 1983. "Aqueous penetration of oral and topical indomethacin in humans." *Archives of Ophthalmology* no. 101 (10):1614–1616.

Sankar, V., A. K. Chandrasekaran, S. Durga, G. Geetha, V. Ravichandran, A. Vijayakumar, S. Raguraman, and G. Geetha. 2006. "Design and evaluation of diclofenac sodium ophthalmic inserts." *The Indian Pharmacist*:98–100.

Satya, D.S., K.P. Suria, and P.P. Muthu. 2011. "Advanced approaches and evaluation of ocular drug delivery systems." *American Journal of PharmTech Research* no. 1 (4):72–92.

Scuderi, Nicolò, Carmine Alfano, Guido Paolini, Cinzia Marchese, and Gianluca Scuderi. 2002. "Transplantation of autologous cultivated conjunctival epithelium for the restoration of defects in the ocular surface." *Scandinavian Journal of Plastic and Reconstructive Surgery and Hand Surgery* no. 36 (6):340–348.

Shao, Yi-Ching, Jyh-Cheng Liou, Chan-Yen Kuo, Yun-Shan Tsai, En-Chieh Lin, Ching-Ju Hsieh, Si-Ping Lin, and Bo-Yie Chen. 2017. "UVB promotes the initiation of uveitic inflammatory injury in vivo and is attenuated by UV-blocking protection." *Molecular Vision* no. 23:219–227.

Sieg, James W. and Joseph R. Robinson. 1979. "Vehicle effects on ocular drug bioavailability III: Shear-facilitated pilocarpine release from ointments." *Journal of Pharmaceutical Sciences* no. 68 (6):724–728.

Silbiger, Jonathan, and George A. Stern. 1992. "Evaluation of corneal collagen shields as a drug delivery device for the treatment of experimental Pseudomonas keratitis." *Ophthalmology* no. 99 (6):889–892.

Smith, S. D. 1991. "Measurement of the rate of aqueous humor flow." *The Yale Journal of Biology and Medicine* no. 64 (1):89–102.

Stamper, R. L. 2011. "A history of intraocular pressure and its measurement." *Optom Vis Sci* no. 88 (1):E16-28. doi: 10.1097/OPX.0b013e318205a4e7.

Steiling, W., M. Bracher, P. Courtellemont, and O. de Silva. 1999. "The HET-CAM, a Useful In Vitro Assay for Assessing the Eye Irritation Properties of Cosmetic Formulations and Ingredients." *Toxicology In Vitro* no. 13 (2):375–384.

Taha, Ehab I., Magda H. El-Anazi, Ibrahim M. El-Bagory, and Mohsen A. Bayomi. 2014. "Design of liposomal colloidal systems for ocular delivery of ciprofloxacin." *Saudi Pharmaceutical Journal* no. 22 (3):231–239.

Takahashi, Y., M. Koike, H. Honda, Y. Ito, H. Sakaguchi, H. Suzuki, and N. Nishiyama. 2008. "Development of the short time exposure (STE) test: An in vitro eye irritation test using SIRC cells." *Toxicology In Vitro* no. 22 (3):760–770. doi: 10.1016/j.tiv.2007.11.018.

Tamura, H., J. Kiryu, K. Miyamoto, K. Nishijima, H. Katsuta, S. Miyahara, F. Hirose, Y. Honda, and N. Yoshimura. 2005. "In vivo evaluation of ocular inflammatory responses in experimental diabetes." *British Journal of Ophthalmology* no. 89 (8):1052–1057. doi: 10.1136/bjo.2004.061929.

Tranoudis, Ioannis and Nathan Efron. 2004. "Tensile properties of soft contact lens materials." *Contact Lens and Anterior Eye* no. 27 (4):177–191.

Ubaidulla, Udhumansha, Molugu V. S. Reddy, Kumaresan Ruckmani, Farhan J. Ahmad, and Roop K. Khar. 2007. "Transdermal therapeutic system of carvedilol: Effect of hydrophilic and hydrophobic matrix on in vitro and in vivo characteristics." *AAPS PharmSciTech* no. 8 (1):E13–E20. doi: 10.1208/pt0801002.

Vandamme, Th F. 2002. "Microemulsions as ocular drug delivery systems: Recent developments and future challenges." *Progress in retinal and eye research* no. 21 (1):15–34.

Vandamme, Th F. and L. Brobeck. 2005. "Poly(amidoamine) dendrimers as ophthalmic vehicles for ocular delivery of pilocarpine nitrate and tropicamide." *Journal of Controlled Release* no. 102 (1):23–38. doi: https://doi.org/10.1016/j.jconrel.2004.09.015.

Vogel, Gerhard H. 2002. *Drug Discovery and Evaluation: Pharmacological Assays* 2nd ed: Springer.

Walewijk, A., Justin J. Cooper-White, and D. E. Dunstan. 2008. "Adhesion measurements between alginate gel surfaces via texture analysis." *Food Hydrocolloids* no. 22 (1):91–96.

Wang, Ying, Zhigang Yuan, Caiyun You, Jindong Han, Haiyan Li, Zhuhong Zhang, and Hua Yan. 2014. "Overexpression p21WAF1/CIP1 in suppressing retinal pigment epithelial cells and progression of proliferative vitreoretinopathy via inhibition CDK2 and cyclin E." *BMC Ophthalmology* no. 14 (1):144. doi: 10.1186/1471-2415-14-144.

Weijtens, Olga, Rik C. Schoemaker, Fred P. H. T. M. Romijn, Adam F. Cohen, Eef G. W. M. Lentjes, and Jan C. van Meurs. 2002. "Intraocular penetration and systemic absorption after topical application of dexamethasone disodium phosphate." *Ophthalmology* no. 109 (10):1887–1891.

Wildner, G., M. Diedrichs-Möhring, and S. R. Thurau. 2008. "Rat Models of Autoimmune Uveitis." *Ophthalmic Research* no. 40 (3–4):141–144.

Xu, C., X. He, W. Liu, Y. Chen, C. Zhou, Z. Duan, Q. Lu, X. Yan, Z. Zhang, and R. Zheng. 2017. "An inhibitor peptide of toll-like receptor 2 shows therapeutic potential for allergic conjunctivitis." *International Immunopharmacology* no. 46:9–15. doi: 10.1016/j.intimp.2017.02.024.

Yadav, U. C. and K. V. Ramana. 2013. "Endotoxin-induced uveitis in rodents." *Methods Mol Biol* no. 1031:155–162. doi: 10.1007/978-1-62703-481-4_18.

Yang, Johnny J., Kwang-Jin Kim, and Vincent H. L. Lee. 2000. "Role of P-glycoprotein in restricting propranolol transport in cultured rabbit conjunctival epithelial cell layers." *Pharmaceutical Research* no. 17 (5):533–538.

Yellepeddi, Venkata Kashyap, and Srinath Palakurthi. 2016. "Recent advances in topical ocular drug delivery." *Journal of Ocular Pharmacology and Therapeutics* no. 32 (2):67–82.

Yousry, Carol, Seham A. Elkheshen, Hanan M. El-laithy, Tamer Essam, and Rania H. Fahmy. 2017. "Studying the influence of formulation and process variables on Vancomycin-loaded polymeric nanoparticles as potential carrier for enhanced ophthalmic delivery." *European Journal of Pharmaceutical Sciences* no. 100 (Supplement C):142–154. doi: https://doi.org/10.1016/j.ejps.2017.01.013.

Zhou, Guohong, Yajiang Duan, Gaoen Ma, Wenyi Wu, Zhengping Hu, Na Chen, Yewlin Chee, Jing Cui, Arif Samad, and Joanne A Matsubara. 2017. "Introduction of the MDM2 T309G Mutation in Primary Human Retinal Epithelial Cells Enhances Experimental Proliferative Vitreoretinopathy." *Investigative Ophthalmology & Visual Science* no. 58 (12):5361–5367.

Zhu, H., I. E. Kochevar, I. Behlau, J. Zhao, F. Wang, Y. Wang, X. Sun, M. R. Hamblin, and T. Dai. 2017. "Antimicrobial Blue Light Therapy for Infectious Keratitis: Ex Vivo and In Vivo Studies." *Investigative Ophthalmology & Visual Science* no. 58 (1):586–593. doi: 10.1167/iovs.16-20272.

Zhu, Heng and Anuj Chauhan. 2008. "Effect of viscosity on tear drainage and ocular residence time." *Optometry & Vision Science* no. 85 (8):E715–E725.

6

Gastroretentive Drug Delivery Systems

Bhupinder Singh, Hetal P. Thakkar, Sanjay Bansal, Sumant Saini, Meena Bansal, and Praveen K. Srivastava

CONTENTS

6.1 Introduction

The development of formulations based on prolongation of gastric residence has been gaining significant attention of the researchers. The stomach is a major organ of the digestive system and plays a significant role in drug absorption too. The gastrointestinal (GI) tract is highly heterogeneous in nature with wide variation in the pH, size of the segment, structure of the membrane, vascularity, and smoothness of inner mucosa, etc. (PubMed Health 2017). The residence time of the ingested matter, including drugs, is also not uniform throughout. The normal gastric residence time, i.e., around 2 h, may not be sufficient for the complete absorption of drugs which are absorbed predominantly from stomach. Moreover, for the drugs acting locally in the stomach for conditions such as hyperacidity, peptic ulcers, or gastric infections, prolongation of the residence time may prove to be advantageous (Hasler and Owyang 2015). Medicinal substances, which cause irritation in the intestine or degraded in the intestinal environment, are preferentially formulated as gastroretentive (GR) dosage forms (Johns Hopkins University 2017). Various strategies such as floating delivery systems, mucoadhesive formulations, high-density systems, and swelling dosage forms are invariably used to achieve enhancement in gastric residence time. Chapter 6 endeavors to discuss the stellar merits and limitations of each approach, their evaluation, and characterization aspects including gastroretention and bioavailability studies. Each of such systems has also been duly illustrated citing laboratory instances, and/or apt literature reports.

6.1.1 Anatomy and Physiology of the Stomach

The GI tract is approximately 9 m long starting from mouth through throat, oesophagus, stomach, small intestine, and large intestine, up to the anus. The stomach serves as the storage and mixing organ for the ingested food. The inner muscular layers of stomach aids in churning and grinding movements (Matta et al. 2015). The stomach is the major organ with capacity to store the ingested matter and its subsequent mixing. The wall of stomach consists of an extra oblique layer of smooth muscle inside the circular layer, which facilitates performance of complex grinding motions, as illustrated in Figure 6.1.

As shown in the figure, the stomach is divided broadly into three regions: the fundus, body, and pylorus, with the cardiac sphincter at the esophageal opening and pyloric sphincter at the GI junction. The stomach is lined by several layers of tissues, the innermost layer is mucosa, followed by the submucosa, the muscularis propria (or muscularis externa), and the serosa, which is the fibrous membrane that covers the outside of the stomach. The main functions of the stomach include the temporary storage of food coming from the esophagus, the mixing and breakdown of food by contraction, and relaxation of the muscle layers for digestion (Kong and Singh 2008). The motility of the stomach plays an important role in emptying of the contents and digestion.

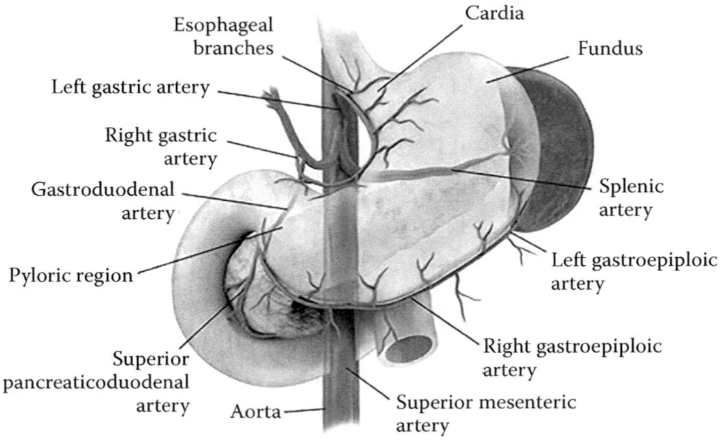

FIGURE 6.1
Anatomy of stomach.

TABLE 6.1

Various Phases of Gastric Emptying and Their Duration

Phase	Description	Duration (min)
Phase I (basal phase)	Noncontractile state	30 to 60
Phase II (pre-burst phase)	Moderate contraction	20 to 40
Phase III (burst phase)	High frequency contractions	10 to 20
Phase IV	Transition phase between Phase III and Phase I	0 to 5

The GIT shows continuous motility, changing with fasting and the fed state. During fasting, the stomach and duodenum exhibit a periodic contractile activity. This process of gastric emptying is characterized by a distinct cycle (about 1.5 to 2.0 h) of electromechanical activity known as the Interdigestive Migrating Myoelectric Complex (IMMC), which can be classified into four consecutive phases as shown in Table 6.1 (Takahashi 2012).

However, in the fed state, the strong contraction of Phase III are replaced by weak amplitude (only 15–25%), propagating contractions and pyloric sphincter is closed. This is the plausible reason for retention of the undigested solid, until the process of digestion goes and fasting peristaltic movement returns. The stomach is also responsible for secretion of various substances such as mucus, enzymes, and hormones. Very few substances are absorbed from the stomach. Solvents, such as ethyl alcohol, and other lipid-soluble compounds, including aspirin and other NSAIDs, are absorbed quite rapidly. Different substances including drugs are absorbed from the different sites of the GI tract. The site of the GI tract from where the absorption takes place predominantly is known as "absorption window". Various factors, which govern the absorption of variegated drugs through a specific absorption window are enumerated in the following sections.

6.1.2 Physicochemical Factors

6.1.2.1 pKa of the Drug

As per the pH-partition hypothesis, the ionization state of a drug depends on its dissociation constant and the pH of the fluid at the absorption site. Thus, weakly acidic drugs (pKa 2.5–7.5), which remain unionized in the acidic medium are predominantly absorbed from the stomach.

6.1.2.2 Solubility

Most drugs are absorbed by passive diffusion in their unionized form. One of the prerequisites for passive diffusion is that the drug should be in the solubilized state. Thus, drugs with higher solubility in the acidic medium are predominantly absorbed from the stomach.

6.1.2.3 Stability

The pH of the GI segment affects the stability of many drugs. The degradation of the drug at a particular site retards its absorption, and hence difference in drug absorption from various regions in the GI tract is observed. Thus, drugs which are stable in the acidic medium show their absorption window usually in the stomach (Lopes et al. 2016).

6.1.2.4 Enzymatic Degradation

Various enzymes present in the particular GI segment can cause drug degradation, resulting in regional variability during drug absorption. Thus, the drugs which are not substrates to the enzymes present in the stomach are absorbed from gastric region.

6.1.3 Physiological Factors

6.1.3.1 Mechanism of Absorption

Absorption of certain drugs can be enhanced by local active and facilitated transport mechanisms present only at a particular site of GI tract.

6.1.3.2 Microbial Degradation

Degradation of certain drugs by microflora is also responsible for regional variability in absorption from the GI tract.

6.1.4 Biochemical Factors

Some drugs are substrates to Cytochrome P450 (CYP3A) enzyme. As the distribution of CYP3A enzyme is heterogeneous in the bowel epithelium, it leads to regional variability in the absorption of drugs.

The summary of the anatomical and physiological differences across various segments of the GI tract is depicted in Table 6.2.

6.2 Concept and Significance of Gastroretention

As discussed previously, the bioavailability is often dependent on the availability of drug in unionized and solubilized state at its "absorption window."

TABLE 6.2

An Overview of Anatomical and Physiological Differences among the Various Segments of Gastrointestinal Tract

Section	Average Length (cm)	Diameter (cm)	Villi	Absorption Mechanism	pH	Major Constituents	Transit Time of Food (h)
Oral cavity	18	10	–	PD, CT	5.2 to 6.8	Amylase, maltase, ptyalin, mucins	Short
Esophagus	25	2.5	–	–	5 to 6	–	Very Short
Stomach	20	15	–	PD, CT	1.2 to 3.5	HCl, rennin, pepsin, lipase, Castle's intrinsic factor	0.25 to 3.00
Duodenum	25	5	+	PD, CT, AT, FT, IP	4.6 to 6.0	Bile, amylase, CYP3A4, maltose, lipase, nuclease	1.0 to 2.0
Jejunum	300	5	++	PD, CT, AT, FT	6.3 to 7.3	Amylase, lactase, maltase, CYP3A5, sucrase	–
Ileum	300	2.5-5.0	++	PD, CT, AT, FT, IP, P	7.6	Lipase, enterokinase, nuclease, nucleotidase	1.0 to 10.0
Cecum	20	2.5-5.0	+	PD, CT, AT, P	7.5 to 8.0	–	Short
Colon	150	5	–	PD, CT	7.9 to 8.0	–	4.0 to 4.5
Rectum	17	2.5	–	PD, CT, P	7.5 to 8.0	–	Inconsistent

Source: OpenStax. 2016. *Digestive System Processes and Regulation. Anatomy and Physiology.* Rice University, USA, BC open textbooks, OpenStax.
Abbreviations: PD: Passive diffusion, CT: convective transport, AT: active transport, FT: facilitated transport, IP: ion pair, P: pinocytosis.

Most of the acidic and very weakly basic drugs, which exist in unionized state at low gastric pH, have the capacity to be absorbed from stomach. For instance, salicylic acid, aspirin, thiopental, secobarbital, and antipyrine are absorbed from the stomach. The gastric retention time is around 2 h and complete absorption may not take place during its residence in stomach. Thus, formulating a GR dosage form for these categories of drugs proves to be advantageous. In the diseases such as hyperacidity, peptic ulcer, or local infections, prolongation of the gastric residence time proves to be beneficial as the drug remains for longer time at the site of action to elicit its effect. For instance, the incidence of gastric tumor was reduced by 74% in mice in case of a 5-flurouracil GR formulation, which in case of conventional tablet was only 25% (Shishu et al. 2007). A drug that degrades in the intestine or cause irritation to the intestinal membrane, but is stable in the stomach, also advocates the formulation of GR dosage forms. For instance, captopril undergoes pH-dependent degradation. It is stable at pH 1.2, but as the pH increases, it starts to degrade. Hence, its GR formulation has been reported (Meka et al. 2008). Solubility is one of the major factors affecting bioavailability. There are number of drugs which exhibit pH-dependent solubility and have a significantly higher solubility in the acidic media of the stomach than the basic intestinal content. The formulation of GR delivery systems for such drugs leads to enhancement of bioavailability, e.g., propranolol, metoprolol, and diazepam. One of the reasons for poor oral bioavailability is the efflux mechanism by the transporters such as P-glycoprotein (P-gp) the levels of which are higher in more distal regions (stomach < jejunum < colon). Drugs that are P-gp substrates thus suffer from poor oral bioavailability. The GR systems decrease the exposure of such drugs to P-gp, and hence increase the oral bioavailability. For example, the propranolol GR system is reported for decreasing its P-gp efflux.

6.2.1 Suitable Drug Candidates for Gastroretention

Table 6.3 lists the most common drugs that are good candidates to be formulated using gastroretention strategies.

6.2.2 Various Strategies for Achieving Gastroretention

Different strategies, such as floating, swelling, inflation, and adhesion (Shah and Agnihotri 2011; Eisenächer et al. 2014) have been applied to increase the gastric retention time (GRT) of dosage forms. These are classified in Figure 6.2 as below.

6.2.2.1 Floating Drug Delivery Systems

A floating drug delivery system (FDDS) is designed such that it has a bulk density less than that of gastric fluids (1.004 g/cc), and it remains

TABLE 6.3

The Most Relevant Drug Candidates Suitable for GR Systems

	Drug(s)	Hurdles, Problem(s)	Therapeutic Indication(s)	(References)
1.	Acyclovir	Poor and inconsistent oral bioavailability owing to low aqueous solubility and lack of site-specific absorption in the gastric region	*Herpes simplex, Varicella zoster,* and *Herpes zoster* infections	(Singh et al. 2016)
2.	Amoxicillin	Local activity	*H. pylori* infection	(Rajinikanth et al. 2007)
3.	Atenolol	Poor absorption from lower GIT and fluctuations in the plasma drug levels	HT	(Singh et al. 2006)
4.	Captopril	Instabilities in the colonic environment, short $t_{1/2}$	HT, CHF	(Meka et al. 2008)
5.	Cefixime	Short $t_{1/2}$	RTI, sinusitis, gonorrhea, otitis media, etc.	(Paul et al. 2011)
6.	Cefuroxime	Site specific absorption and high first-pass metabolism	Gram-negative and gram-positive infections	(Bansal et al. 2016b)
7.	Clarithromycin	Plasma fluctuations and short $t_{1/2}$	*H. pylori* and upper RTI	(Nama et al. 2008)
8.	Furosemide	Narrow absorption window in upper GIT	CHF, CRF, and hepatic cirrhosis	(Darandale and Vavia 2012)
9.	Gabapentin	Short $t_{1/2}$ of elimination and low absorption due to amino acid transport saturable system	Neuralgia	(Gupta and Li 2013)
10.	Hydralazine	Significant first-pass metabolism short half-life	HT and CHF	(Singh et al. 2009)
11.	Itopride	Shorter $t_{1/2}$ (i.e. <6 h) and low absorption window in the stomach and upper part of the small intestine	Functional dyspepsia, anorexia, upper abdominal pain, chronic gastritis	(Bansal et al. 2016a)
12.	Lamivudine	Narrow absorption window and short half life	HAART	(Singh et al. 2012)
13.	Levodopa	Short half-life, narrow absorption window in upper GIT	PD, HT, CHF, angina, and arrhythmias	(Klausner et al. 2003b)
14.	Metformin	Short half-life, narrow absorption window in upper GIT	Type II diabetes mellitus	(Ali et al. 2007)

(Continued)

TABLE 6.3 (CONTINUED)

The Most Relevant Drug Candidates Suitable for GR Systems

	Drug(s)	Hurdles, Problem(s)	Therapeutic Indication(s)	(References)
15.	Metoprolol succinate	Short half-life, narrow absorption window in upper GIT	HT, CHF, angina, and arrhythmias	(Boldhane and Kuchekar 2010)
16.	Metronidazole	Local activity	*Helicobacter pylori* infection	(Ishak et al. 2007)
17.	Nitrofurantoin	Narrow absorption window in upper GIT	Bacterial UTI	(Gröning et al. 1998)
18.	Ofloxacin	Low solubility at alkaline pH	Bacterial UTI and respiratory infections	(Patil et al. 2013)
19.	Ranitidine	Narrow absorption window in upper GIT, short half-life Local activity	Peptic ulcer and reflux oesophagitis	(Rohith et al. 2009)
20.	Rivastigmine	Large fluctuations in the plasma levels, high values of maximal plasma concentration (i.e., C_{max}), and short values of time to reach C_{max} (i.e., t_{max})	AD	(Kapil et al. 2012)
21.	Tramadol	Narrow absorption window and short half life	Analgesic	(Singh et al. 2010)
22.	Verapamil	Low solubility at alkaline pH	HT and tachycardiac disturbances	(Sawicki 2002)
23.	Zidovudine and Lamivudine	High frequency of administration	HAART	(Singh et al. 2017)

Abbreviations: AD: Alzheimer's disease, HT: hypertension, CHF: congestive heart failure, CRF: chronic renal failure, HAART: highly active antiretroviral therapy, PD: Parkinson's disease, RTI: respiratory tract infection, UTI: urinary tract infections.

in buoyant state in gastric fluids without influencing GRT for extended periods of time (Tamizharasi et al. 2011). When the system remains in the flotation state, the drug is released in a slow, continuous, but controlled fashion (Singh et al. 2017). After the release of the drug, the residual system gets emptied from the stomach. This increases GRT, whereby the plasma drug level variability can be controlled. An FDDS owes its buoyancy either to its lower density than the stomach contents or due to the gaseous phase formed inside the system after it comes in contact with the gastric environment (Figure 6.6a). A floating DDS, also known as a hydrodynamically balanced system (HBS), must comply with three major criteria (Bansal et al. 2016b):

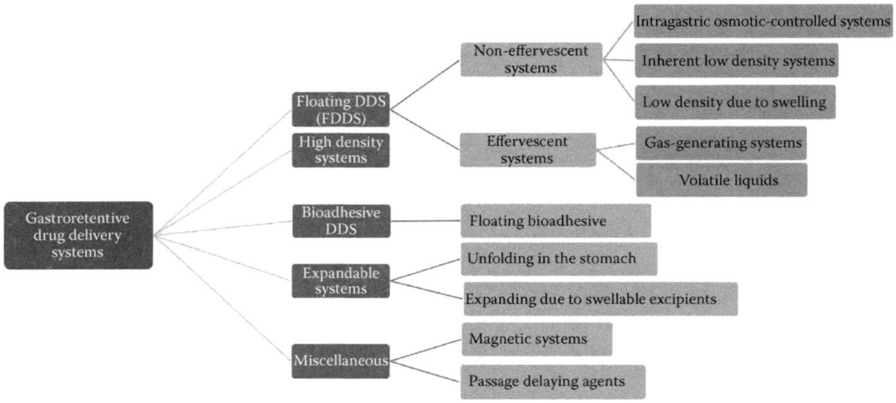

FIGURE 6.2
Strategies for gastroretentive drug delivery systems.

- It must have a suitable architecture to generate a cohesive gelatinous barrier.
- The relative density of the delivery system must be less than gastric contents (1.004–1.010 g/mL).
- It should serve as a reservoir system releasing a drug at controlled rate.

Based on the mechanism of buoyancy, two distinctly different technologies (i.e., noneffervescent and effervescent systems) have been utilized in the development of FDDS.

6.2.2.2 Noneffervescent Systems

The floatation of noneffervescent FDDS can be either because of i) low density due to swelling or ii) inherent low density.

6.2.2.3 Low Density Due to Swelling

This type of system involves the admixture of a drug with a gel, which, after swallowing, swells due to imbibitions of gastric fluid, attaining a bulk density lower than the outer corona. The entrapped air provides the necessary floatation to the dosage forms. The most commonly used polymers include the gel forming or highly swellable cellulose type hydrocolloids, matrix-forming materials and polysaccharides, which also work as bioadhesive polymers such as carbopol and chitosan (Chen et al. 2015).

This technology involves encasing of a drug reservoir in a microporous compartment with apertures along its upper as well as lower walls, as depicted in Figure 6.3.

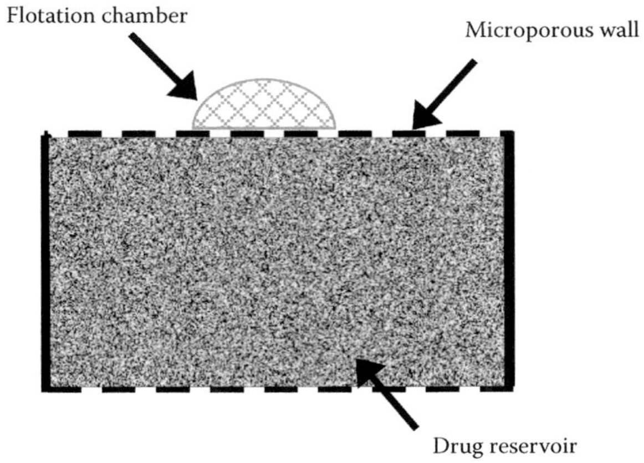

FIGURE 6.3
Reservoir-based gastroretentive drug delivery system.

The peripheral walls are impervious to the gastric fluids. The entrapped air provides the necessary buoyancy in the gastroretentive system. Gastric fluid enters via small orifice in the delivery system and dissolves the drug.

6.2.2.4 Inherent Low-Density Systems

The system initially settles down, and then comes to the brim after a specific lag time, thus poses a plausible risk of premature emptying from the stomach. Therefore, there is an ardent need of a system that floats immediately as soon as it comes in contact with gastric fluids. This can only be accomplished with the provision of a low-density device since its inception. Low-density systems are generally made by air entrapment. Watanabe et al. (1976) prepared a single-unit FDDS with inherent low density, consisting of a hollow core (empty, hard gelatin capsule, polystyrene foam, or pop rice grain) coated with two layers: a subcoat of cellulose acetate phthalate, and an outer drug-containing coating of ethyl cellulose (EC)/hydroxypropyl methylcellulose (HPMC). This type of system is very useful for low-dose drugs but may not be suitable if larger amounts of drug are needed for an effective therapy.

6.2.2.4.1 Hollow Microspheres

Hollow microspheres are low-density systems that immediately float as soon as they come in contact with the gastric fluid, causing gastroretention and thereby improving drug bioavailability (Aloshi 2016). For instance, hollow microspheres (microballoons) consisting of Eudragit RS (an enteric polymer) containing the drug in the polymeric shell developed have been reported in the literature (Kawashima et al. 1989; Bansal et al. 2016a).

6.2.2.4.2 Floating Beads

Dosage forms containing spherical floating beads have been synthesized using lyophilized calcium alginate that can keep floating for 12 h. Floating beads have a prolonged gastroretention time of more than 5.5 h as compared to solid beads that show a shorter gastric retention of 1 h as diagrammatically represented in Figure 6.4. Both natural and synthetic polymeric systems have been used in the preparation of multiple-unit FDDS.

The floating properties of the devices strongly depended on the subsequent drying process. Oven dried beads did not float, whereas lyophilized beads remained floating for >12 h in hydrochloride buffer pH 1.5 due to the presence of air-filled hollow spaces within the system (Talukder and Fassihi 2004).

6.2.2.5 Intragastric Osmotically Controlled DDS

It consists of an osmotically driven DDS and an inflatable floating machinery in a bioerodible capsule, as shown in Figure 6.5. When the device reaches the stomach, bioerodible capsule quickly dissolves to release the DDS. These devices consist of two compartments, *viz.* the drug reservoir compartment and osmogen-containing compartment. The drug reservoir compartment is encapsulated in a pressure-responsive collapsible bag, which is impervious to vapors and liquids, and has a minute drug delivery orifice.

The second compartment contains an osmogen and is enclosed within a semipermeable sheath. In the stomach, the osmogen continuously absorbs water from the gastric fluid through the semipermeable membrane into the

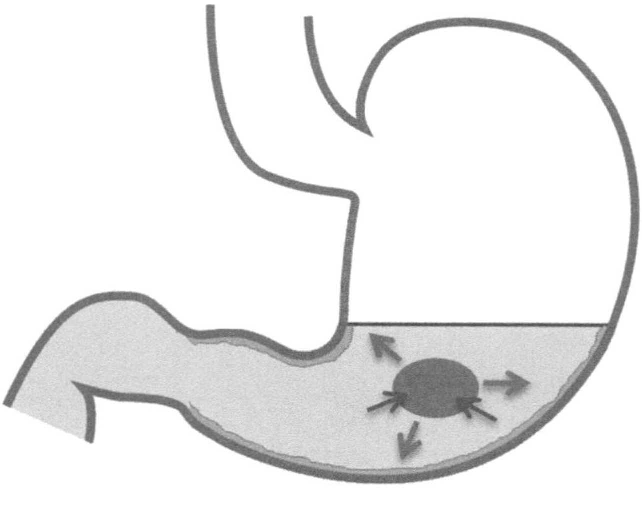

FIGURE 6.4
Floating bead system for gastroretention.

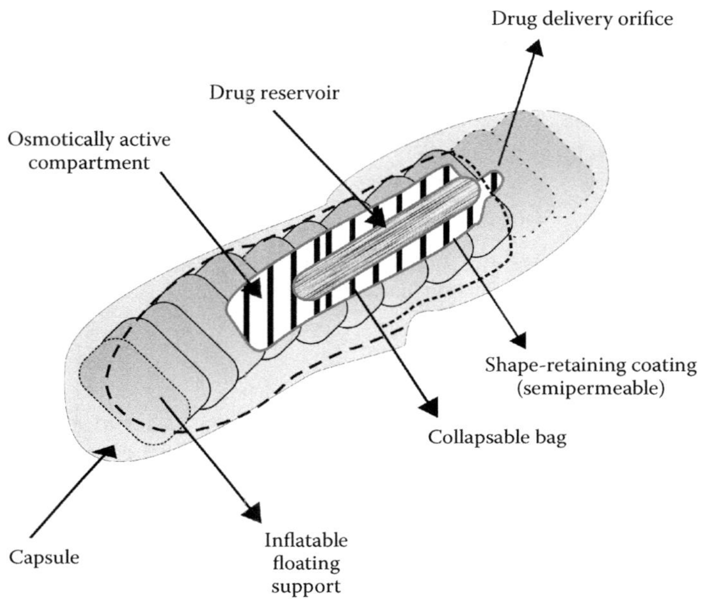

FIGURE 6.5
Osmotically controlled drug delivery system.

osmotically active compartment. This creates the necessary osmotic pressure which acts upon the collapsible bag and activates the release of the drug through the delivery orifice (Ali et al. 2018).

6.2.2.6 *Effervescent Systems*

6.2.2.6.1 Volatile Liquid Containing Systems

These systems incorporate an inflatable chamber containing a volatile liquid, such as ether or cyclopentane, which evaporates at body temperature leading eventually to inflation of the chamber in the stomach. These inflatable GI systems contain a hollow expandable and deformable unit that consists of two chambers separated by an impermeable, pressure-responsive, and movable bladder. The first chamber contains the drug and the second chamber contains the volatile liquid. In the stomach, the volatile liquid evaporates and inflates the device, leading to drug release from the reservoir into the gastric fluid, as shown in Figure 6.6a.

The device may also consist of a bioerodible plug made up of PVA, polyethylene, etc. that gradually dissolves causing the inflatable chamber to release gas and collapse after a predetermined time to permit spontaneous ejection of the inflatable system from the stomach (Rahim et al. 2015).

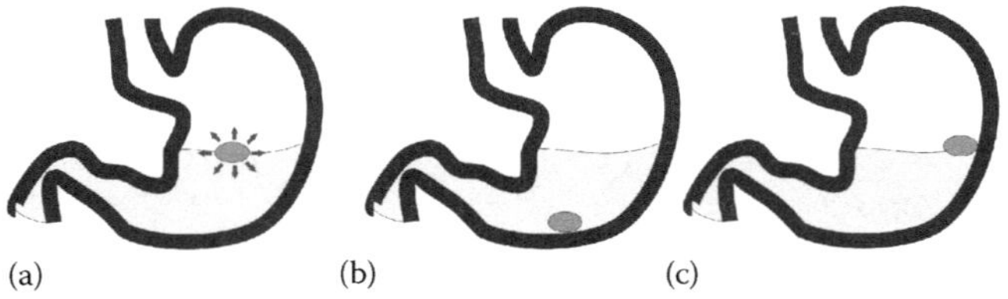

FIGURE 6.6
The three types of gastroretentive systems: (a) low density (effervescent/floating) system, (b) high-density system, and (c) floating bioadhesive system.

6.2.2.6.2 Gas Generating Systems

These systems incorporate, apart from the drug and the swelling polymers, such as chitosan and methylcellulose, some effervescent compounds, e.g., sodium bicarbonate ($NaHCO_3$), tartaric acid ($C_4H_6O_6$), and citric acid ($C_6H_8O_7$) that liberate CO_2 when they come in contact with acidic gastric contents. And, CO_2 in this case, gets entrapped in swollen hydrocolloids and provides buoyancy to the dosage forms (Mirani et al. 2016).

Generally, the effervescent systems suffer from a specific disadvantage that they do not float immediately after swallowing, as gas generation takes some time. Therefore, they could be cleared from the stomach before becoming effective.

6.2.2.7 Limitations of FDDS

A. The performance of low-density, floating DDS is strongly dependent on the fed/filling state of the stomach. Nevertheless, this approach can successfully prolong the gastric retention time and has already led to the production of pharmaceutical products, which are commercially available in the market (Talukder and Fassihi 2004).

B. An FDDS requires sufficiently high levels of fluid in the stomach to float and work efficiently. However, this can be overcome by administrating the dosage form with fluids (200–250 ml) and with frequent meals (Taranalli et al. 2015).

6.2.3 High-Density Systems

High-density devices employ weight as their retention mechanism. When the density of the system is significantly greater than that of the gastric content (approx. 1.004 g/cm^3), the device settles down to the bottom of the stomach. A density of 2.6–2.8 g/cm^3 acts as a threshold value, above which these systems remain located below the pylorus in the rugae of the stomach (Murphy et al. 2009).

High-density formulations employ the use of coated pellets whereby the drug is coated with heavy inert materials such as barium sulfate, zinc oxide, titanium dioxide, iron powder, etc. These materials increase density by up to 1.5–2.4 g/cm^3. Dense pellets (approximately 3 g/cm^3) trapped in rugae also tend to withstand the peristaltic movements of the stomach wall as shown in Figure 6.6b (Lopes et al. 2016). With pellets, the GI transit time can be extended from an average of 5.8–25 h, depending more on density than on diameter of the pellets, although many conflicting reports stating otherwise are also abundant in literature. These systems suffer from the drawback that it is technically difficult to manufacture them with a large amount of drug (>50%) and to achieve the required high density. No successful high-density system, therefore, has been marketed till date.

6.2.4 Bioadhesive Drug Delivery Systems

Bioadhesive or mucoadhesive DDS utilize the bioadhesive properties of certain water-soluble polymers that become adhesive on hydration and can adhere to the epithelial surface in the stomach. Hence, this can be used for targeting a drug to a particular region of the body. Some of the promising bioadhesives commonly used include polycarbophil, carbopol, chitosan, and carboxymethylcellulose. Gastric mucoadhesion alone does not tend to be strong enough to impart to dosage forms the ability to resist the strong propulsion forces of the stomach wall. The continuous production of mucous by the gastric mucosa to replace the mucous that is lost through peristaltic contractions and the dilution of the stomach content also seems to limit the potential of mucoadhesion as a GR force (Thombre and Gide 2016).

Mucoadhesives are synthetic or natural polymers, which are capable of interacting with biological materials and being retained on them for prolonged period of time (Roy et al. 2009). Polymers that can adhere to either hard or soft tissues are employed for the purpose of bioadhesion. They should have adhesiveness with the mucus layer to provide adequate contact and prolong the residence time of the drug at the site of administration. They should be able to swell and allow drug release. They should not interact with the active drug and be compatible with the mucosal surface. They should remain unaffected by hydrodynamic conditions, food, and pH as shown in Figure 6.6c. They should also be biodegradable, stable, and cost-effective (Thombre and Gide 2016).

6.2.4.1 Advantages of Bioadhesives

6.2.4.1.1 Stability

Bioadhesive dosage forms tend to increase stability of certain drugs by localizing the drug to an optimal site of its maximal stability. Also, these systems by disallowing a proper contact of drug with food components may protect the former from attack by the latter (Laffleur 2014).

6.2.4.1.2 *Improved Bioavailability*

Also, the bioadhesive systems have been successfully employed to improve the consistency of the drugs like atenolol by regulating their drug absorption and reducing fluctuation of their plasma levels (Singh et al. 2006).

6.2.4.1.3 *Peptide Delivery*

The susceptibility of the peptides to the diverse pH ranges is known to challenge their efficacy. The GR systems have been attempted in case of melatonin to effectively deliver them via an oral route (El-Gibaly 2002).

6.2.4.1.4 *Mucosal Protection*

Bioadhesive dosage forms could protect the GI mucosa from ulceration caused by NSAIDs (Gosswein 1989).

6.2.4.2 Disadvantages of Bioadhesives

6.2.4.2.1 *Ion-/pH- Sentivity*

Polyanionics as polyacrylic acid are highly sensitive to the ionic environment. Thus, the use of polyacrylates in an ion-rich environment may interfere with the adhesive properties of the polymer. Sufficient adhesiveness may be obtained at a specific pH range only. Rheological properties of Carbopol 934 samples were found to be substantially influenced by the environmental pH (Mehdizadeh and Yang 2013).

6.2.4.2.2 *High Viscosity*

Due to the high viscosity of the polymers, these systems could impede the delivery of the drug to the absorbing surface (Mehdizadeh and Yang 2013).

6.2.4.2.3 *Loss of Mucoadhesive Activity*

An increased wetting of the polymer may lead to the formation of nonadhesive, slippery mucilage that may cause loss of mucoadhesive activity. Due to this, bioadhesive system may move past the absorption site. Besides, mucoadhesion during the GI transit of the DDS will be limited by a relatively rapid mucus turnover. Thus, renewal of the mucus layer may be a limiting factor in bioadhesion. Rapid loss of the bioadhesive properties of DDS may occur by soluble mucins or its degradation products (Boddupalli et al. 2010).

6.2.4.2.4 *Gastrointestinal Irritation*

These systems are not suitable for drugs causing GI toxicity or irritation. The ingestion of such drugs with an insufficient amount of water may lead to the adherence of the dosage form to the esophageal mucosa, thus aggravating the local toxicity of such drugs.

6.2.5 Floating Bioadhesive Systems

Floating systems, however, possess multiple challenges like insufficient floatation when the fluid level is low in the stomach and chances of transit of the dosage form to the pylorus by forcible house-keeping waves are high, leading eventually to reduced buoyancy time and limited retention of the dosage form. Such limitations can largely be overcome by developing the floating systems, coupled with mucoadhesion characteristics, to adhere the dosage form to the mucous lining of stomach wall. The floating-bioadhesive systems, therefore, greatly improve the residence time of the drug resulting in effective extension in the absorption and oral bioavailability as shown in Figure 6.6c (Bansal et al. 2016b).

6.2.6 Size-Increasing (Expandable) Systems

Another approach to retain a dosage form in the stomach is by increasing its size above the diameter of the pyloric sphincter (>12 mm), which prevents its exit from the stomach. The formulation contains a polymer which, on coming in contact with gastric fluid, imbibes water and swells. As a result, the dosage form is retained in the stomach for a long period of time. These systems may be referred to as "plug type" systems, since they have a tendency to remain lodged at the pyloric sphincter. It is desirable to have an initial small size to facilitate swallowing, which increases significantly once the system is in the stomach. The system should expand quickly, in order to prevent premature emptying through the pylorus. These systems should also ensure their clearance from the gastric milieu after prescheduled time intervals, to circumvent dose dumping after multiple administrations (Sarojini and Manavalanb 2012). The expansion of the system size can be based upon a number of mechanisms, like unfolding in the stomach to complex geometric shapes or expanding due to swellable excipients.

6.2.6.1 Systems Unfolding in the Stomach

These systems are available in a number of geometric forms, like ring, tetrahedron, clover leaf, string, disk, and pellet/sphere, which can be fitted tightly into a gelatin capsule and unfold after dissolution of the capsule shell. Such dosage forms have three configurations, *viz.*: i) small collapsed configuration, enabling convenient oral intake; ii) swollen state achieved in the gastric environs, thus preventing the transit via pyloric sphincter; iii) another small form when the retention in stomach is no longer required following drug release, thereby enabling its evacuation (Sivaneswari et al. 2017). These systems consist of at least one erodible polymer (e.g., hydroxypropyl cellulose, Eudragit®), one nonerodible polymer (e.g., polyolefins, polyamides, polyurethanes), and a drug that is dispersed within the polymer matrix.

Erodible tetrahedron-shaped devices, consisting of rods ("arms"; made of poly [ortho ester]/polyethylene blends) and "corners" (based on a silastic elastomer) also showed prolonged gastric retention time (Chordiya et al. 2017).

The major disadvantages of most systems of this type are unfolding in the stomach in their complex shape and making them difficult to manufacture due to high cost on a larger scale. Moreover, their different geometrical shapes may also cause mucosal irritation.

6.2.6.2 Systems Expanding Due to Swellable Excipients

A significant increase in the size of such kind of DDS is usually accomplished using specific hydrogel formers, which imbibe water and swell following contact with aqueous media.

6.2.6.2.1 Superporous Hydrogel

Superporous hydrogels contain densely concentrated small pores of mean pore size greater than 100 μm. These swell dramatically and instantaneously due to rapid water uptake by capillary action through several interconnected open pores. These tend to swell to a remarkably large size with swelling ratio greater than 100 and intend to have adequate mechanical strength, to endure pressure generated from gastric contraction. Most commonly used excipients in such systems are polyacrylate polymers, HPMC, Carbopol, polyvinyl acetate, sodium alginate, agar, calcium chloride, polycarbonates, and polyethylene oxides (Beşkardeş et al. 2017).

Lately, superporous hydrogel hybrids, which are prepared by adding a water-soluble or water-dispersible polymer that can be crosslinked after the superporous hydrogel is formed (Vishal Gupta and Shivakumar 2010). Examples for hybrid agents are polysaccharides, including sodium alginate, pectin, chitosan or synthetic water-soluble hydrophilic polymers, such as poly(vinyl alcohol). Compared to conventional superporous hydrogels, superporous hydrogel hybrids are not easily breakable when stretched, because they possess highly elastic properties in the swollen state. This can be very useful for the development of GI devices.

A major merit of swellable systems is that their performance is independent of fed/fasting state of the stomach. On the other hand, they present the potential hazard of permanent retention and could lead to serious life-threatening effects after multiple dose administration. Consequently, these systems should consist of biodegradable materials or disintegrate after a desired time period (Nayak et al. 2013). Until this happens, the systems should have the sufficient mechanical strength so as to withstand the powerful housekeeping waves from the stomach. Such size-increasing systems has the propensity to cause polymeric obstruction too.

6.2.7 Miscellaneous

6.2.7.1 Incorporation of Passage-Delaying Agents

Food excipients, such as fatty acids like myristic acid salts, modify the pattern of stomach to a fed state. This causes considerable increase in gastric retention and permits marked prolongation of drug release. The delay in gastric emptying after diets rich in fats is mainly due to saturated fatty acids with chain length of C-10 to C-14 (Helin-Tanninen and Pinto 2015).

6.2.7.2 Magnetic Systems

This approach to enhance the GRT is based on the simple principle of placing a small internal magnet in the dosage form, while other being placed on the abdomen over the position of the stomach. In such a case, the external magnet needs to be adhered/positioned with a high degree of precision, thus challenging patient adherence (Mandal et al. 2016).

6.3 Patented Technologies for Gastroretentive Drug Delivery Systems

Using GR drug delivery technology, many novel drug delivery systems have been developed. Some of them have been patented. Table 6.4 summarizes the patented GR technologies.

6.4 Characterization of Gastroretentive Dosage Forms

6.4.1 Challenges Faced/Anticipated in Characterization of Gastroretentive Dosage Forms

The performance of the GR formulation is highly dependent on the physiological conditions of the stomach. A number of factors affect the gastric retention of the dosage form and hence the prediction of drug release profile is difficult. The high variability of gastric emptying time poses major challenge in determining the GR behavior of formulations. For instance, the presence of food that extends the GRT is represented by a higher gastric emptying time in the fed condition as compared to the fasting condition. The in-vitro–in-vivo correlation (IVIVC) often becomes difficult as the formulation

TABLE 6.4

Patented Technologies for Various Gastroretentive Drug Delivery Systems

Product	Drug	Clinical Indications	Company	Technology	Reference
Baclofen GRS®	Baclofen	Muscle spasticity	Sun Pharma Advanced Research Company Ltd.	High-density gastroretention	(Sun Pharma Advanced Research Company Ltd. 2017)
Cefaclor LP®	Cefaclor	Pneumonia and infections of lung, ear, skin, UTI and throat	Galenix, France	Floating system	(Dailymed. 2014)
Cifran OD®	Ciprofloxacin	Pneumonia and infections of the ear, lung, skin, throat, and urinary tract	Ranbaxy, India	Floating tablets	(Medline India 2017)
Conviron®	Ferrous sulphate	Iron-deficiency anemia	Ranbaxy, India	Colloidal gel forming systems	(CIMS 2017)
Cipro XR®	Ciprofloxacin HCL and betaine	Typhopid fever, gonorrhea, infections of skin, bone, joint and abdomen	Bayer, USA	Erodible matrix–based system	(U.S. National Library of Medicine 2016)
Coreg CR®	Carvedilol	Heart failure	GlaxoSmithKline	Gastroretention with osmotic system	(GlaxoSmithKline. 2017)
Cytotec®	Misoprostol	Antiulcer	Pharmacia/Pfizer Inc., USA	Bilayer floating capsules	(Medline Plus 2017)
Gaviscon®	Sodium bicarbonate and alginic acid	Gastric reflux	Reckett Benckiser Healthcare, UK	Effervescent floating	(Mandal et al. 2016)

(*Continued*)

TABLE 6.4 (CONTINUED)

Patented Technologies for Various Gastroretentive Drug Delivery Systems

Product	Drug	Clinical Indications	Company	Technology	Reference
Glumetza®	Metformin HCl	Diabetes	Depomed, Inc., USA	Polymer-based swelling technology	(Depomed Inc. 2009)
Gralise®	Gabapentin	Epilepsy	Depomed, Inc., USA	Polymer-based swelling technology	(Gralise 2017)
Inon Ace®	Simethicone	Gas such as uncomfortable or painful pressure, fullness, and bloating	Stao Pharma	Foam-based floating system	(Dailymed 2015)
Madopar®	Levodopa and benserazide	Parkinsonism	Roche, UK	Floating capsule	(Electronic Medicines Compendium (eMC). 2015)
Prazopress XL®	Prazosin hydrochloride	Hypertension	Sun Pharma, Japan	Effevescent and swelling based floating systems	(Singleton et al. 1989)
Proquin XR®	Ciprofloxacin	Pneumonia and infections of ear, lung, skin, throat, and urinary tract	Depomed, Inc., USA	Polymer-based swelling technology	(Depomed Inc. 2008)
Xifaxan®	Rifamixin	Traveler's diarrhea, encephalopathy	Lupin, India	Bioadhesive tablets	(U.S. National Library of Medicine 2016)

is designed in order to release the drug in the time interval during which it is expected to be in the stomach. However, due to high level of biological variation, the GRT may vary, ostensibly leading to undesired release profile (Singh et al. 2017).

6.4.2 In-Vitro Characterization and Gastroretention Study

Effective in-vitro characterization plays a crucial role in ensuring the quality and predicting the clinical utility of the developed formulations. In addition to routine evaluation parameters for final dosage form, like hardness, friability, general appearance, assay, uniformity of content, and weight variation for tablets, the following methods have also been reported to evaluate peculiar formulation characteristics accountable for gastric retention.

6.4.2.1 Floating Capacity (Buoyancy) Study

The study is performed in USP dissolution apparatus containing 900 mL of deionized water, 0.1N hydrochloric acid or more preferably simulated gastric fluid (SGF) as the dissolution medium at a temperature of 37 ± 0.5°C, with or without stirring. For tablet dosage forms, the time required by tablet to start floating (floating lag time) and the total duration for which the tablet remains floating (floating time) are often measured (Chen et al. 2013). For floating microparticulate drug delivery systems, the carrier is dispersed in continuously stirred testing medium for target duration. Subsequently, the floating as well as settled fraction of these carriers are separated and their dry weights (W_F and W_S, respectively) are measured to calculate percent buoyancy as a measure of gastric retention (Awasthi et al. 2012) using Equation 6.1.

$$\text{Percentage Buoyancy} = \left(\frac{W_F}{W_F + W_S}\right) \times 100 \tag{6.1}$$

6.4.2.2 Mucoadhesion Study

The mucoadhesive property of gastroretentive formulations is evaluated by measuring the strength, with which the formulation attaches to mucus lining of biological tissue samples and measuring the force required to detach the formulation as a measure of mucoadhesive strength. A universal tensile tester is often utilized to sensitively measure the detachment force for tablets while modified dynamic contact angle analyser or microtensiometer is utilized for individual microparticles (Santos 1999). Alternatively, an in-vitro wash-off test is also performed for multiparticulate systems, where the tissue is mounted on a glass slide, a predefined number of particles are allowed to attach to moistened tissue and their mucoadhesive strength is challenged by fastening the slide on one of the arm of USP tablet disintegration apparatus containing phosphate buffer at pH 6.8 at 37°C, giving regular up and down movement for target duration (Patel et al. 2016). Later, the percent mucoadhesion is obtained using Equation 6.2:

$$\text{Percent mucoadhesion} = \left(\frac{\text{Final number of particles remained adhered}}{\text{Initial number of particles taken}}\right) \times 100 \tag{6.2}$$

6.4.2.3 Swelling Index

The immersion method is used to study the swelling behavior of a GR systems. The method involves immersion of formulation in SGF at 37°C and the change in dimension (Garse et al. 2010), volume (Gromova et al. 2007), or weight (Arza et al. 2009) is measured at predetermined time points as a measure of swelling. The percent swelling is calculated by Equation 6.3, where M_0 and M_T are the measurements recorded initially and at time t, respectively:

$$\text{Percent swelling} = \left(\frac{M_T - M_0}{M_0}\right) \times 100 \tag{6.3}$$

6.4.2.4 Density of the Dosage Form

The density of formulation is calculated as mass to volume ratio. For instance, the density of a capsule shaped tablet is calculated by putting weight (M), radius (r) and side length (a) of the tablet, Equation 6.4 (Desai and Purohit 2017).

$$\text{Density} = \frac{M}{(4/3)\pi r^2 (r + a)} \tag{6.4}$$

6.4.2.5 In-Vitro Drug Release

The paddle (Jagdale et al. 2009; Meka et al. 2009) or basket type (Liu et al. 2011) dissolution test apparatus is commonly utilized for in-vitro drug release study using 0.1N HCl, FaSGF (SGF fasted condition) or FeSGF (SGF fed condition) as a release medium at 37 ± 0.5°C. FaSGF contains pepsin, sodium taurocholate, and lecithin at pH around 1.5, while FeSGF contains milk or buffer at pH 5, in combination to other ingredients of FaSGF. Drugs present in samples collected at regular time intervals are quantified to calculate the cumulative percentage of drug release. This drug release data can further be fitted in several mathematical models to establish the kinetics as well as mechanism involved in drug release from developed systems. Commonly employed drug release kinetic models include zero-order ($M_t = M_0 + K_0t$); first-order ($\ln M_t = \ln M_0 + K_1t$); Higuchi's diffusion ($M_t = M_0 + K_Ht^{0.5}$); Hixon-Crowell's dissolution $\left(M_t^{1/3} = M_0^{1/3} + K_{HC}t\right)$; and the Korsmeyer-Peppas model ($M_t/M_0 = K_{KP}t^n$), where M_t represents the drug amount released in time t, M_0 is the initial amount of drug and K's are respective release constants. The n is termed as release exponent whose magnitude reflects the releases mechanism based on the geometry of the formulation (Meka et al. 2009; Singh and Kapil 2010). A lower n value (n = 0.45 for cylindrical system) indicates Fickian diffusion, while higher value (n = 0.89) indicates a relaxational (case II) release. An intermediate value of n indicates anomalous non-Fickian diffusion reflecting a combination of diffusion through hydrated matrix and polymer relaxation. A still higher value of n (> 0.89) reflects rapid relaxation-controlled (super case II) release (Singh and Singh 1998).

6.4.3 Bioavailability or Bioequivalence Studies

The European agency provides biowaivers for the immediate release formulation of highly water-soluble drug releasing more than 85% drug within first 15 min, where gastroretentive dosage forms (GRDFs) don't fit due to controlled or sustained drug release. USFDA maintains the database of the biowaiver reports on approved federally compliant products that are used to prepare bioequivalence procedures as biowaivers; generating the relationship between the in-vitro and in-vivo data as IVIVC. This might aid in surmounting the problems associated with the biowaiver principles. In particular, a level A correlation suits the best for extended release systems like GRDF, as it involves point to point comparison of dissolution data directly with plasma drug concentration–time profile to provide better prediction of in-vivo performance.

6.4.3.1 In-Vivo Pharmacokinetics

The GI tract demonstrates varying absorption characteristics based on its region-specific differences in physiology and anatomy. Thus, drug delivery to different regions of the GI tract significantly impacts the corresponding pharmacokinetic profiles (Kagan and Hoffman 2008). As discussed earlier, the GRDF approach aims at providing continuous delivery of drugs to the upper part of the GI tract via overcoming the natural gastric activities responsible for evacuation of its content into the intestine (Singh et al. 2012; Misra and Bhardwaj 2016). The major approaches utilized so far for GRDF includes low-density systems that float in gastric fluid, high-density systems that sink to location anatomically lower than the pyloric sphincter, mucoadhesive systems that bind to gastric mucosa and swellable systems that swell to size much larger than pyloric sphincter. An in-vivo pharmacokinetic study plays a critical role in validating the rationale involved behind the selection of gastroretentive dosage form. Most often, bigger animals (like dogs) or human subjects are selected, due to handling and administration issues with smaller animals like mice, rats, guinea pigs, or rabbits, especially for large-size formulations like tablets, capsules, etc. Different modes of drug administration, like bolus oral solution and constant rate intraduodenal infusion, have been utilized as a crossover design to better demonstrate the altered pharmacokinetic as well as pharmacodynamic profile obtained with GRDF.

6.4.3.2 Pharmacokinetic Modeling and Simulation

Kagan and Hoffman (2008) generated a two-compartment (central and peripheral)-based pharmacokinetic model, where they demonstrated that dividing the GI tract into four regions *viz.*, stomach, upper small intestine, lower small intestine, and large intestine (Figure 6.7) resulted in a reasonably good fitting of experimental results, improving the predictive power of the model (Meka et al. 2009).

6.4.4 In-Vivo Visualization/Assessment of Gastroretention

A number of in-vivo techniques have been offered for the real-time visualization or assessment of an orally administered GRDF and provide the proof of concept about the unique formulation characteristics involved in enhancement of their GRT. The most commonly utilized techniques are as follows:

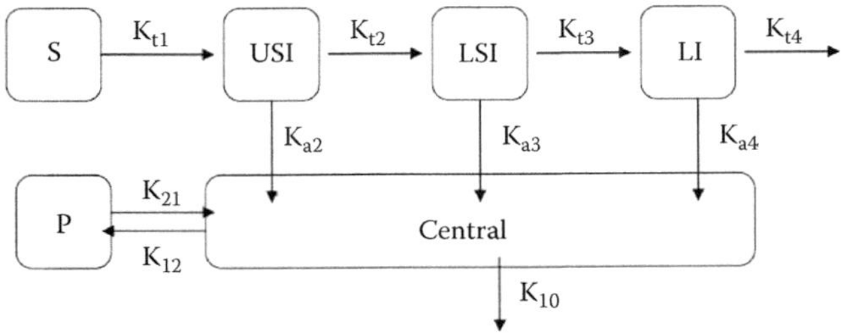

FIGURE 6.7
Scheme of the model describing the pharmacokinetics of Atenolol GRDF, where P stands for peripheral compartment and K_t, K_a, and K_{10} represent transit, absorption, and elimination rate constants, respectively. K_{12} and K_{21} are distribution rate constants representing drug distribution from the central to the peripheral compartment and the peripheral to the central compartment, respectively. (Adapted from Kagan, L. and A. Hoffman. 2008. Selection of drug candidates for gastroretentive dosage forms: Pharmacokinetics following continuous intragastric mode of administration in a rat model." *Eur. J. Pharm. Biopharm.* 69(1): 238–246.)

6.4.4.1 Gamma Scintigraphy

This technique involves administration of γ-emitting radioisotopes with short half-life (*viz.*, ^{99m}Tc (Technetium), ^{111}In (Indium), and ^{152}Sm (Samarium)) which are incorporated within the GR formulation, capture the emitted radiation by a gamma camera, and generate two-dimensional and three-dimensional images using a sophisticated computer software. However, the need of expensive instrument and radioactive isotopes hinder the utilization of this technique at frequent intervals, and also render the technique contra-indicated for children and pregnant women. Singh et al. (2012) optimized a GR polymeric blend to prepare floating bioadhesive tablets of lamivudine for enhanced gastric retention. Figure 6.8 depicts scintigraphic images of the stomachs of human volunteers following oral administration of ^{99m}Tc-tagged control drug formulation. Increased gastroretention was observed, along with improved drug bioavailability.

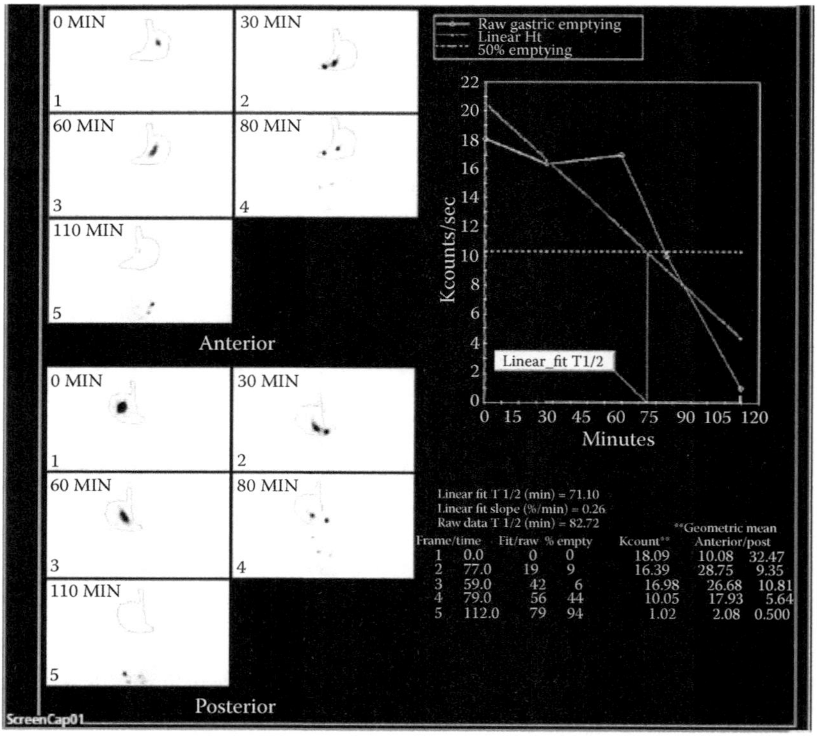

(a)

FIGURE 6.8
(a) Scintigraphic images of stomach of a human volunteer (anterior and posterior), following an oral administration of 99mTc-labelled a control drug formulation. Linear-fit time activity curve is showing GRT of radiolabeled drug from the stomach. *(Continued)*

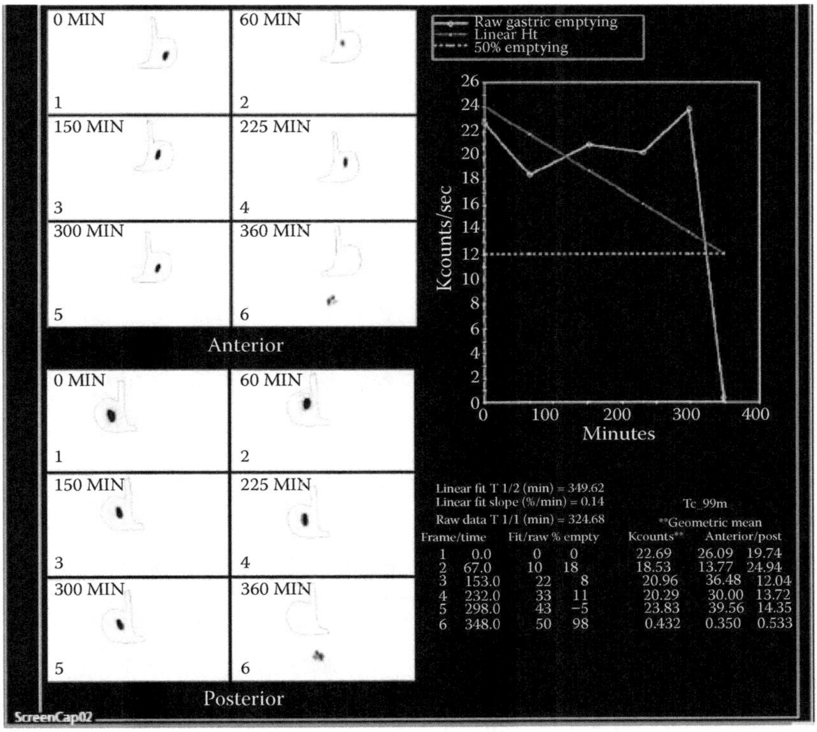

(b)

FIGURE 6.8 (CONTINUED)
(b) Anterior and posterior static scintigraphic images of stomach in a human volunteer following oral administration of 99mTc-labelled optimized drug formulation. Also shown is the linear fit time activity curve depicting gastric emptying time of the radiolabeled drug from the stomach.

6.4.4.2 Roentgenography

This is a cost-effective method that involves X-ray-based imaging of the digestive tract. Barium sulphate is most commonly used as a radiocontrast medium owing to its subsequent elimination through the feces. A fraction of drug (around 20% w/w) is replaced with barium sulphate in final formulations to impart adequate contrast for better visibility of the formulations by X-ray without affecting their gastroretentive properties. These labeled formulations are administered and the gastric radiography is performed at predetermined time points to measure the GRT of formulations (Lingam et al., 2008). Similar to gamma scintigraphy, this radiation may also pose a health risk, but owing to low radiation exposure from an X-ray, the benefits far overshadow the risks. Bansal et al. (2016a) systematically developed the itopride hydrochloride (ITH)–loaded multiunit GR microballoons, employing quality by design (QbD)–based approach. X-ray images of the rabbits' GI tracts successfully demonstrated the gastroretention phenomenon, as shown in Figure 6.9.

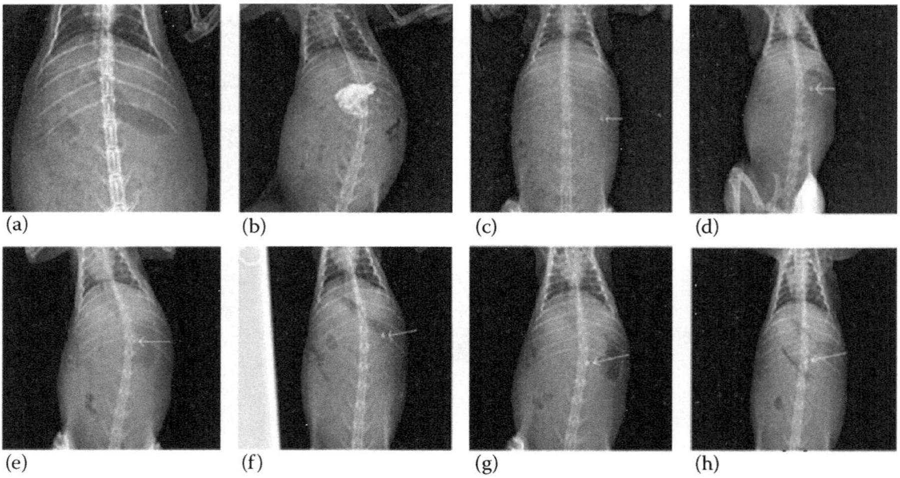

FIGURE 6.9
X-ray images of the rabbit stomach, (a) without microballoons, (b) with barium meal, (c) after 30 min administration of GR microballoons, (d) after 2 h administration of GR microballoons, (e) after at 3 h administration of GR microballoons, (f) after 4 h administration of GR microballoons, (g) after 6 h administration of GR microballoons, (h) after at 8 h administration of GR microballoons.

6.4.4.3 Magnetic Marker Monitoring

These techniques involve incorporation of magnetite (ferric oxide) in the GRDF as a super-paramagnetic contrast marker and perform MRI investigations at predetermined time intervals to estimate the mean gastric retention period (Tadros and Fahmy 2014). The low toxicity allowed daily intake of 0.5 mg/kg as additive in food processing (Steingoetter et al. 2003), together with the low cost, favors the use of magnetite, which affects the protons in its vicinity (Faas et al. 2001), leading to a local increase in signal intensity compared to the reference scan taken in its absence (at zero time). Hence, the gastric retention of developed formulation represents intensity greater than the highest intensity in the reference scan. As the method does not involve any radiation, it is completely safe.

6.4.4.4 Gastroscopy

Gastroscopy or gastric endoscopy is a very safe and effective tool that offers the visual examination of gastric region with the help of a slender, flexible, telescopic camera known as a gastroscope. The gastroscope is passed down through the mouth into the stomach via the oesophagus and displays the video on connected screen for variety of diagnostic and therapeutic purposes.

Klausner et al. (2003a) utilized gastroscopy to visualize GRDF administered with a glass of sugar water following an overnight fasting.

6.4.4.5 Ultrasonography

The imaging of some abdominal organs by ultrasonography relies on the reflection of ultrasonic waves at substantially different acoustic impedances across an interface. This technique is utilized for hydrogel-type GR formulations, where the solvent penetration within hydrogel takes place and the swollen hydrogels appear as sonolucent objects permitting passage of ultrasound waves without giving off echoes (Shalaby et al. 1992). However, a vast majority of oral formulations lack sharp acoustic mismatches at the gastric interface and hence are not suitable for visualization through ultrasonography. In addition, the presence of water is critical in ultrasonic imaging, which limits its use when water gets emptied from the stomach or during fasting conditions.

6.4.4.6 ^{13}C Octanoic Acid Breath Test (^{13}C-OBT)

This test was first introduced by Ghoos et al. (1993) to demonstrate the gastric emptying of solids. The test utilizes a medium chain fatty acid (^{13}C octanoic acid) which, as soon as it clears the stomach and reaches the duodenum, gets rapidly absorbed through intestinal mucosa, oxidizes in the liver, and the resulting $^{13}CO_2$ gets excreted into the breath almost instantaneously. The appearance of $^{13}CO_2$ in breath is estimated by continuous-flow type Isotope-Ratio Mass Spectrometry (CF-IRMS), and is utilized as a direct measure of gastric emptying (Perri et al. 2005).

6.5 Conclusion

GR dosage forms are finding increasing attention from researchers due to various applications such as prolonged drug action, enhanced oral bioavailability, reduced side effects, site specific retention of the drug, and the avoidance of premature metabolism. Various approaches have been used for prolonging the gastric residence time, each having its specific advantages and limitations. Taking this into consideration, the characteristics of the drug and the dosage form requirement, judicious selection of the approach should be done. There are a number of techniques available for characterization of GR dosage forms. Scientific advances in imaging techniques have enabled researchers to visualize the passage of drug within the GI tract and thereby have a clear picture of the residence of the dosage form in the gastric region.

With the advancement of such imaging techniques and continued need for gastroretention, the domain of GR systems is likely to be boosted in future.

References

Ali, J., S. Arora, A. Ahuja, A. K. Babbar, R. K. Sharma, R. K. Khar et al. (2007). "Formulation and development of hydrodynamically balanced system for metformin: In vitro and in vivo evaluation." *Eur. J. Pharm. Biopharm.* 67(1): 196–201.

Ali, R., M. Walther and R. Bodmeier (2018). "Cellulose acetate butyrate: Ammonio methacrylate copolymer blends as a novel coating in osmotic tablets." *AAPS PharmSciTech* 19(1): 148–154.

Aloshi, S. L. (2016). Formulation and evaluation of Gastroretentive microballoons (hollow Microspheres) of olmesartan medoxomil, KLE University, Belagavi, Karnataka.

Arza, R. A., C. S. Gonugunta and P. R. Veerareddy (2009). "Formulation and evaluation of swellable and floating gastroretentive ciprofloxacin hydrochloride tablets." *AAPS PharmSciTech* 10(1): 220–226.

Awasthi, R., G. T. Kulkarni, V. K. Pawar and G. Garg (2012). "Optimization studies on gastroretentive floating system using response surface methodology." *AAPS PharmSciTech* 13(1): 85–93.

Bansal, S., S. Beg, A. Asthana, B. Garg, G. S. Asthana, R. Kapil et al. (2016a). "QbD-enabled systematic development of gastroretentive multiple-unit microballoons of itopride hydrochloride." *Drug Deliv.* 23(2): 437–451.

Bansal, S., S. Beg, B. Garg, A. Asthana, G. S. Asthana and B. Singh (2016b). "QbD-oriented development and characterization of effervescent floating-bioadhesive tablets of cefuroxime axetil." *AAPS PharmSciTech* 17(5): 1086–1099.

Beşkardeş, I. G., R. S. Hayden, D. L. Glettig, D. L. Kaplan and M. Gümüşderelioğlu (2017). "Bone tissue engineering with scaffold-supported perfusion co-cultures of human stem cell-derived osteoblasts and cell line-derived osteoclasts." *Process Biochem.* 59(Part B): 303–311.

Boddupalli, B. M., Z. N. K. Mohammed, R. A. Nath and D. Banji (2010). "Mucoadhesive drug delivery system: An overview." *J Adv. Pharm. Technol. Res.* 1(4): 381–387.

Boldhane, S. P. and B. S. Kuchekar (2010). "Development and optimization of metoprolol succinate gastroretentive drug delivery system." *Acta Pharm.* 60(4): 415–425.

Chen, Y.-C., H.-O. Ho, D.-Z. Liu, W.-S. Siow and M.-T. Sheu (2015). "Swelling/floating capability and drug release characterizations of gastroretentive drug delivery system based on a combination of hydroxyethyl cellulose and sodium carboxymethyl cellulose." *PLoS ONE* 10(1): e0116914.

Chen, Y. C., H. O. Ho, T. Y. Lee and M. T. Sheu (2013). "Physical characterizations and sustained release profiling of gastroretentive drug delivery systems with improved floating and swelling capabilities." *Int. J. Pharm.* 441(1–2): 162–169.

Chordiya, M., H. Gangurde, V. Borkar and N. Chandwad (2017). "Technologies, optimization and analytical parameters in gastroretentive drug delivery systems." *Curr. Sci.* 112(5): 946–953.

CIMS. (2017). "CONVIRON-TR." Retrieved Sept, 20, 2017, from http://www.mims.com/india/drug/info/conviron-tr/conviron-tr%20cap.

Dailymed. (2014). "CEFACLOR-cefaclor tablet, film coated, extended release." Retrieved Sept, 20, 2017, from https://dailymed.nlm.nih.gov/dailymed/drugInfo.cfm?setid=0878bdc2-0410-4938-9890-96523aa81c2f.

Dailymed. (2015). "INON ACE- magnesium aluminosilicate, magnesium hydroxide solution." Retrieved Sept, 19, 2017, from https://dailymed.nlm.nih.gov/dailymed/drugInfo.cfm?setid=a6e4475c-dfac-493c-b98a-21473e3b59d5.

Darandale, S. S. and P. R. Vavia (2012). "Design of a gastroretentive mucoadhesive dosage form of furosemide for controlled release." *Acta Pharm. Sin. B.* 2(5): 509–517.

Depomed Inc. (2008). "Proquin® XR." Retrieved Sept, 22, 2017, from https://www.accessdata.fda.gov/drugsatfda_docs/label/2008/021744s008lbl.pdf.

Depomed Inc. (2009). "Glumetza®." Retrieved Sept, 20, 2017, from https://www.accessdata.fda.gov/drugsatfda_docs/label/2009/021748s005lbl.pdf.

Desai, N. and R. Purohit (2017). "Development of novel high density gastroretentive multiparticulate pulsatile tablet of clopidogrel bisulfate using quality by design approach." *AAPS PharmSciTech* 1–11.

Eisenächer, F., G. Garbacz and K. Mäder (2014). "Physiological relevant in vitro evaluation of polymer coats for gastroretentive floating tablets." *Eur. J Pharm. Biopharm.* 88(3): 778–786.

Electronic Medicines Compendium (eMC). (2015). "Madopar CR 100 mg/25 mg Prolonged Release Hard Capsules." Retrieved Sept, 19, 2017, from https://www.medicines.org.uk/emc/medicine/3211.

El-Gibaly I. (2002). "Development and in vitro evaluation of novel floating chitosan microcapsules for oral use: Comparison with non-floating chitosan microspheres". *Int. J. Pharm.* 249(1-2): 7–21.

Faas, H., W. Schwizer, C. Feinle, H. Lengsfeld, C. de Smidt, P. Boesiger et al. (2001). "Monitoring the intragastric distribution of a colloidal drug carrier model by magnetic resonance imaging460." *Pharm. Res.* 18(4): 460–466.

Garse, H., M. Vij, M. Yamgar, V. Kadam and R. Hirlekar (2010). "Formulation and evaluation of a gastroretentive dosage form of labetalol hydrochloride." *Arch. Pharm. Res.* 33(3): 405–410.

Ghoos, Y. F., B. D. Maes, B. J. Geypens, G. Mys, M. I. Hiele, P. J. Rutgeerts et al. (1993). "Measurement of gastric emptying rate of solids by means of a carbon-labeled octanoic acid breath test." *Gastroenterology* 104(6): 1640–1647.

GlaxoSmithKline. (2017). "COREG CR." Retrieved Sept, 22, 2017, from https://www.gsksource.com/pharma/content/dam/GlaxoSmithKline/US/en/Prescribing_Information/Coreg_CR/pdf/COREG-CR-PI-PIL.PDF.

Gosswein, C. (1989). Gastric mucosa protective agents. Germany. U. S. Patent. USA, Google Patents.

Gralise. (2017). "Gralise." Retrieved Sept, 20, 2017, from https://www.gralise.com/.

Gromova, L. I., D. Hoichman and J. Sela (2007). "Gastroretentive sustained release acyclovir tablets based on synergistically interacting polysaccharides." *Pharm. Chem. J.* 41(12): 656–658.

Gröning, R., M. Berntgen and M. Georgarakis (1998). "Acyclovir serum concentrations following peroral administration of magnetic depot tablets and the influence of extracorporal magnets to control gastrointestinal transit." *Eur. J. Pharm. Biopharm.* 46(3): 285–291.

Gupta, A. and S. Li (2013). "Safety and efficacy of once-daily gastroretentive gabapentin in patients with postherpetic neuralgia aged 75 years and over." *Drugs Aging* 30(12): 999–1008.

Hasler, W. L. and C. Owyang (2015). Approach to the patients with gastrointestinal diseases. *Harrison's Principles of Internal Medicine.* D. L. Kasper, A. S. Fauci, S. L. Hauser et al. (eds.) US, Mc Graw Hill: 1875–1879.

Helin-Tanninen, M. and J. Pinto (2015). *Oral solids. Practical Pharmaceutics: An International Guideline for the Preparation, Care and Use of Medicinal Products.* Y. Bouwman, V. I. Fenton-May and P. Le Brun. (eds.) US, Springer International Publishing: 51–76.

Ishak, R. A., G. A. Awad, N. D. Mortada and S. A. Nour (2007). "Preparation, in vitro and in vivo evaluation of stomach-specific metronidazole-loaded alginate beads as local anti-Helicobacter pylori therapy." *J. Control. Release* 119(2): 207–214.

Jagdale, S. C., A. J. Agavekar, S. V. Pandya, B. S. Kuchekar and A. R. Chabukswar (2009). "Formulation and evaluation of gastroretentive drug delivery system of propranolol hydrochloride." *AAPS PharmSciTech* 10(3): 1071–1079.

Johns Hopkins University. (2017). "Medicines and the digestive system." Retrieved Oct, 6, 2017, from http://www.hopkinsmedicine.org/healthlibrary/conditions/digestive_disorders/medications_and_the_digestive_system_85,P00389.

Kagan, L. and A. Hoffman (2008). "Selection of drug candidates for gastroretentive dosage forms: Pharmacokinetics following continuous intragastric mode of administration in a rat model." *Eur. J. Pharm. Biopharm.* 69(1): 238–246.

Kapil, R., S. Dhawan, B. Singh and B. Garg (2012). "Systematic formulation development of once-a-day gastroretentive controlled release tablets of rivastigmine using optimized polymer blends." *J. Drug. Deliv. Sci. Technol.* 22(6): 511–521.

Kawashima, Y., T. Niwa, T. Handa, H. Takeuchi, T. Iwamoto and Y. Itoh (1989). "Preparation of prolonged-release spherical micro-matrix of ibuprofen with acrylic polymer by the emulsion-solvent diffusion method for improving bioavailability." *Chem. Pharm. Bull.* (Tokyo) 37(2): 425–429.

Klausner, E. A., S. Eyal, E. Lavy, M. Friedman and A. Hoffman (2003b). "Novel levodopa gastroretentive dosage form: In-vivo evaluation in dogs." *J. Control. Release* 88(1): 117–126.

Klausner, E. A., E. Lavy, M. Barta, E. Cserepes, M. Friedman and A. Hoffman (2003a). "Novel gastroretentive dosage forms: Evaluation of gastroretentivity and its effect on levodopa absorption in humans." *Pharm. Res.* 20(9): 1466–1473.

Kong, F. and R. P. Singh (2008). "Disintegration of solid foods in human stomach." *J. Food Sci.* 73(5): R67–R80.

Laffleur, F. (2014). "Mucoadhesive polymers for buccal drug delivery." *Drug Dev. Ind. Pharm.* 40(5): 591–598.

Lingam, M., T. Ashok, V. Venkateswarlu and Y. Madhusudan Rao (2008). "Design and evaluation of a novel matrix type multiple units as biphasic gastroretentive drug delivery systems." *AAPS PharmSciTech* 9(4): 1253–1261.

Liu, Y., J. Zhang, Y. Gao and J. Zhu (2011). "Preparation and evaluation of glyceryl monooleate-coated hollow-bioadhesive microspheres for gastroretentive drug delivery." *Int. J. Pharm.* 413(1): 103–109.

Lopes, C. M., C. Bettencourt, A. Rossi, F. Buttini and P. Barata (2016). "Overview on gastroretentive drug delivery systems for improving drug bioavailability." *Int. J. Pharm.* 510(1): 144–158.

Mandal, U. K., B. Chatterjee and F. G. Senjoti (2016). "Gastro-retentive drug delivery systems and their in vivo success: A recent update." *Asian J. Pharm.* 11(5): 575–584.

Matta, E. J., J. F. Platt, Y. M. Elguindy and K. M. Elsayes (2015). *The Stomach. Cross-Sectional Imaging of the Abdomen and Pelvis: A Practical Algorithmic Approach.* K. M. Elsayes. (ed.) New York, NY, Springer New York: 263–306.

Medline India. (2017). "Ciprofloxacin." Retrieved Sept, 22, 2017, from http://www.medlineindia.com/antibiotic/ciprofloxacin.htm.

Medline Plus. (2017, Jan, 9, 2010). "Misoprostol." Retrieved Sept, 22, 2017, from https://medlineplus.gov/druginfo/meds/a689009.html.

Mehdizadeh, M. and J. Yang (2013). "Design strategies and applications of tissue bioadhesives." *Macromol. Biosci.* 13(3): 271–288.

Meka, L., B. Kesavan, K. M. Chinnala, V. Vobalaboina and M. R. Yamsani (2008). "Preparation of a matrix type multiple-unit gastro retentive floating drug delivery system for captopril based on gas formation technique: In vitro evaluation." *AAPS PharmSciTech* 9(2): 612.

Meka, L., B. Kesavan, V. N. Kalamata, C. M. Eaga, S. Bandari, V. Vobalaboina et al. (2009). "Design and evaluation of polymeric coated minitablets as multiple unit gastroretentive floating drug delivery systems for furosemide." *J. Pharm. Sci.* 98(6): 2122–2132.

Mirani, A. G., S. P. Patankar and V. J. Kadam (2016). "Risk-based approach for systematic development of gastroretentive drug delivery system." *Drug Deliv. Transl. Res.* 6(5): 579–596.

Misra, R. and P. Bhardwaj (2016). "Development and characterization of novel floating-mucoadhesive tablets bearing venlafaxine hydrochloride." *Scientifica* (Cairo) 2016: 4282986.

Murphy, C. S., V. Pillay, Y. E. Choonara and L. C. du Toit (2009). "Gastroretentive drug delivery systems: Current developments in novel system design and evaluation." *Curr. Drug Deliv.* 6(5): 451–460.

Nama, M., C. S. R. Gonugunta and P. Reddy Veerareddy (2008). "Formulation and evaluation of gastroretentive dosage forms of clarithromycin." *AAPS PharmSciTech* 9(1): 231.

Nayak, A. K., B. Das and R. Maji (2013). "Gastroretentive hydrodynamically balanced systems of ofloxacin: In vitro evaluation." *Saudi Pharm. J.* 21(1): 113–117.

OpenStax. (2016). *Digestive System Processes and Regulation. Anatomy and Physiology.* Rice University, USA, BC open textbooks, OpenStax.

Patel, N., J. Desai, P. Kumar and H. P. Thakkar (2016). "Development and in vitro characterization of capecitabine-loaded alginate–pectinate–chitosan beads for colon targeting." *J. Macromol. Sci. B* 55(1): 33–54.

Patil, G. B., B. B. Singh, K. P. Ramani, V. K. Chatap and P. K. Deshmukh (2013). "Design and development of novel dual-compartment capsule for improved gastroretention." *ISRN Pharm.* 2013: p. 7.

Paul, Y., M. Kumar and B. Singh (2011). "Formulation and in vitro evaluation of gastroretentive, drug delivery systems of cefixime trihydrate." *Int. J. Drug Dev. Res.* 3(4): 148–161.

Perri, F., M. R. Pastore and V. Annese (2005). "13C-octanoic acid breath test for measuring gastric emptying of solids." *Eur. Rev. Med. Pharmacol. Sci.* 9(5 Suppl 1): 3–8.

PubMed Health. (2017). "Gastrointestinal Tract (GI Tract)." Retrieved Oct, 2, 2017, from https://www.ncbi.nlm.nih.gov/pubmedhealth/PMHT0022855/.

Rahim, S. A., P. A. Carter and A. A. Elkordy (2015). "Design and evaluation of effervescent floating tablets based on hydroxyethyl cellulose and sodium alginate using pentoxifylline as a model drug." *Drug Des. Devel. Ther.* 9: 1843–1857.

Rajinikanth, P. S., J. Balasubramaniam and B. Mishra (2007). "Development and evaluation of a novel floating in situ gelling system of amoxicillin for eradication of Helicobacter pylori." *Int. J. Pharm.* 335(1): 114–122.

Rohith, G., B. K. Sridhar and A. Srinatha (2009). "Floating drug delivery of a locally acting H2-antagonist: An approach using an in situ gelling liquid formulation." *Acta Pharm.* 59(3): 345–354.

Roy, S., K. Pal, A. Anis, K. Pramanik and B. Prabhakar (2009). "Polymers in mucoadhesive drug-delivery systems: A brief note." *Des. Monomers Polym.* 12(6): 483–495.

Santos, C. (1999). Adaptation of a Microbalance to Measure Bioadhesive Properties of Microspheres. *Bioadhesive Drug Delivery Systems.* CRC Press: 131–146.

Sarojini, S. and R. Manavalanb (2012). "An overview on various approaches to gastroretentive dosage forms." *Int. J. Drug Dev. & Res.* 4(1): 1–13.

Sawicki, W. (2002). "Pharmacokinetics of verapamil and norverapamil from controlled release floating pellets in humans." *Eur. J. Pharm. Biopharm.* 53(1): 29–35.

Shah, A. K. and S. A. Agnihotri (2011). "Recent advances and novel strategies in preclinical formulation development: An overview." *J. Control. Release* 156(3): 281–296.

Shalaby, W. S., W. E. Blevins and K. Park (1992). "Use of ultrasound imaging and fluoroscopic imaging to study gastric retention of enzyme-digestible hydrogels." *Biomaterials* 13(5): 289–296.

Shishu., N. Gupta and N. Aggarwal (2007). "Stomach-specific drug delivery of 5-fluorouracil using floating alginate beads." *AAPS PharmSciTech* 8(2): E143–E149.

Singh, B., A. Kaur, S. Dhiman, B. Garg, R. K. Khurana and S. Beg (2016). "QbD-enabled development of novel stimuli-responsive gastroretentive systems of acyclovir for improved patient compliance and biopharmaceutical performance." *AAPS PharmSciTech* 17(2): 454–465.

Singh, B., A. Rani, N. Ahuja and R. Kapil (2010)."Formulation optimization of hydrodynamically balanced oral controlled release bioadhesive tablets of tramadol hydrochloride." *Sci. Pharm.* 78(2): 303–324.

Singh, B., B. Garg, R. Bhatowa, R. Kapil, S. Saini and S. Beg (2017). "Systematic development of a gastroretentive fixed dose combination of lamivudine and zidovudine for increased patient compliance." *J. Drug. Deliv. Sci. Technol.* 37: 204–215.

Singh, B., B. Garg, S. C. Chaturvedi, S. Arora, R. Mandsaurwale, R. Kapil, et al. (2012). "Formulation development of gastroretentive tablets of lamivudine using the floating-bioadhesive potential of optimized polymer blends." *J. Pharm. Pharmacol.* 64(5): 654–669.

Singh, B. and R. Kapil (2010). Drug release kinetics from extended release oral drug delivery systems: Mathematical modelling using various approaches. *Mathematical Modelling, Clustering Algorithms and Applications.* C. L. Wilson. (ed.) New York, Nova Science Publishers, Inc: 1–48.

Singh, B., S. K. Chakkal and N. Ahuja (2006). "Formulation and optimization of controlled release mucoadhesive tablets of atenolol using response surface methodology." *AAPS PharmSciTech* 7(1): E19–E28.

Singh, B., S. Pahuja, R. Kapil and N. Ahuja (2009)."Formulation development of oral controlled release tablets of hydralazine: Optimization of drug release and bioadhesive characteristics." *Acta Pharm.* 59(1): 1–13.

Singh, B. and S. Singh (1998). "A comprehensive computer program for the study of drug release kinetics from compressed matrices." *Indian J. Pharm. Sci.* 60(6): 358–360.

Singh, Y., V. K. Pawar, M. Chaurasia and M. K. Chourasia (2017). Gastroretentive Delivery. *Drug Delivery: An Integrated Clinical and Engineering Approach*. Y. Rosen, P. Gurman and N. Elman. (eds.) CRC Press: 84–111.

Singleton, W., R. K. Dix, L. Monsen, D. Moisey, M. Levenstein, D. F. Bottiglieri et al. (1989). "Efficacy and safety of Minipress XL, a new once-a-day formulation of prazosin." *Am. J. Med.* 87(2A): 45S–52S.

Sivaneswari, S., E. Karthikeyan and P. J. Chandana (2017). "Novel expandable gastro retentive system by unfolding mechanism of levetiracetam using simple lattice design–Formulation optimization and in vitro evaluation." *Bull. Fac. Pharm.* (Cairo Univ.) 55(1): 63–72.

Steingoetter, A., D. Weishaupt, P. Kunz, K. Mader, H. Lengsfeld, M. Thumshirn et al. (2003). "Magnetic resonance imaging for the in vivo evaluation of gastric-retentive tablets." *Pharm. Res.* 20(12): 2001–2007.

Sun Pharma Advanced Research Company Ltd. (2017). "Baclofen GRS." Retrieved Sept, 20, 2017, from http://www.sunpharma.in/BaclofenGRS.htm.

Tadros, M. I. and R. H. Fahmy (2014). "Controlled-release triple anti-inflammatory therapy based on novel gastroretentive sponges: Characterization and magnetic resonance imaging in healthy volunteers." *Int. J. Pharm.* 472(1–2): 27–39.

Takahashi, T. (2012). "Mechanism of interdigestive migrating motor complex." *Neurogastroenterol. Motil.* 18(3): 246–257.

Talukder, R. and R. Fassihi (2004). "Gastroretentive delivery systems: A mini review." *Drug Dev. Ind. Pharm.* 30(10): 1019–1028.

Tamizharasi, S., V. Rathi and J. Rathi (2011). "Floating drug delivery system." *Sys. Rev. Pharm.* 2(1): 333–337.

Taranalli, S. S., P. M. Dandagi and V. S. Mastiholimath (2015). "Development of hollow/porous floating beads of metoprolol for pulsatile drug delivery." *Eur. J. Drug Metab. Pharmacokinet.* 40(2): 225–233.

Thombre, N. A. and P. S. Gide (2016). "Floating-bioadhesive gastroretentive Caesalpinia pulcherrima-based beads of amoxicillin trihydrate for Helicobacter pylori eradication." *Drug Deliv.* 23(2): 405–419.

U.S. National Library of Medicine. (2016, Sept, 7). "Ciprofloxacin." Retrieved Sept, 22, 2017, from https://medlineplus.gov/druginfo/meds/a688016.html.

U.S. National Library of Medicine. (2016, May, 5). "Rifaximin." Retrieved Sept, 22, 2017, from https://medlineplus.gov/druginfo/meds/a604027.html.

Vishal Gupta, N. and H. G. Shivakumar (2010). "Preparation and characterization of superporous hydrogels as gastroretentive drug delivery system for rosiglitazone maleate." *DARU J. Pharm. Sci.* 18(3): 200–210.

Watanabe, S., M. Kayano, Y. Ishino and K. Miyao (1976). Solid therapeutic preparation remaining in stomach, Google Patents.

7

Colon Targeted Drug Delivery Systems

Naazneen Surti

CONTENTS

7.1 Introduction

The delivery of drugs to a specific target organ has many advantages. A smaller dose is required, which inevitably results in reduced incidence of undesirable systemic adverse reactions. Moreover, the drug is maintained in its intact form as close to the target as possible and can be made available only when required (Kinget et al. 1998). The targeting of drugs to the colon is desirable for the topical treatment of diseases of the colon such as Crohn's disease, ulcerative colitis, spastic colon, irritable bowel syndrome, and colorectal cancer. This approach has been found to provide safe and effective therapy with proven bioavailability enhancement.

Colonic drug delivery would additionally be of value in the treatment of diseases which have peak symptoms in the early morning, such as nocturnal asthma, angina or arthritis, when a delay in absorption is desired from a therapeutical point of view. In addition to this, with the rapid advancement of biotechnology and genetic engineering resulting into availability of peptides and proteins at reasonable costs, there has been an increased interest in utilizing the colon as site for drug absorption. The potential candidates in this respect include analgesic peptides, contraceptive peptides, oral vaccines, growth hormone, insulin, interferons, erythropoietin, and interleukins (Saffran et al. 1988) (Mackay and Tomlinson 1993).

7.2 Limitations and Challenges of Colonic Delivery

Due to the absence of well-defined villi in the colon, as found in small intestine, the absorptive surface area decreases drastically. Moreover, the high viscosity of colonic contents makes it even more difficult for a drug to diffuse from the lumen to the site of absorption. Further, there is reduced permeability to polar compounds due to tight junctions between cells (Madara and Dharmsathaphorn 1985). Many factors that impede colonic drug absorption have been reviewed (Mrsny 1992). A drug may bind specifically or non-specifically with dietary components or products released from bacteria in the lumen. This in turn, might result in facilitated enzymatic or environmental degradation by increasing the time that the drug remains in the colon. There remains the possibility of interaction between the negatively charged mucus layer and drug molecules which can result in either drug-mucus binding or drug-mucus repulsion. Examples of drugs that bind to the mucus are penicillins, cephalosporins, and aminoglycosides (Niibuchi, Aramaki, and Tsuchiya 1986). Additionally at the epithelium, the lipid bilayer of the individual colonocytes, and the occluding junctional complex between these cells, provide a physical barrier to the absorption of drugs.

There are enzymatic activities associated with the colonocytes, although significantly less than observed in the small intestine. Since the carrier-mediated uptake of drugs across the epithelial cell barrier is not extensive, lipid solubility, the ability of a drug molecule to partition between an aqueous and a lipid environment, the degree of ionization, and the pH at the site of absorption are very important factors. The unstirred water layer between mucus layer and epithelial surface area can also produce a diffusional barrier for the absorption of drugs, especially the very lipophilic ones (Rahman, Barrowman, and Rahimtula 1986).

7.3 Strategies for Targeting the Colon

An ideal colon-specific drug delivery system should prevent drug release in the upper gastrointestinal (GI) tract, and effect an abrupt onset of drug release upon entry into the colon. To achieve this, a triggering mechanism needs to be built in the delivery system which is responsive to the physiological changes particular to the colon. The physiological similarity between the distal small intestine and the proximal colon presents very limited options in selecting an appropriate drug release triggering mechanism, however, factors such as GI transit time, differential pH conditions, dissolution of the drug at the site of action, the intestinal fluid volume, and the propensity of the formulation or drug to be metabolized in the GI tract through enzymatic or microbial degradation can be exploited. Hua et al. exploited the physiological and microbial changes in the gastrointestinal tract (GIT) in inflammatory bowel disease, as depicted in Figure 7.1. (Hua et al. 2015)

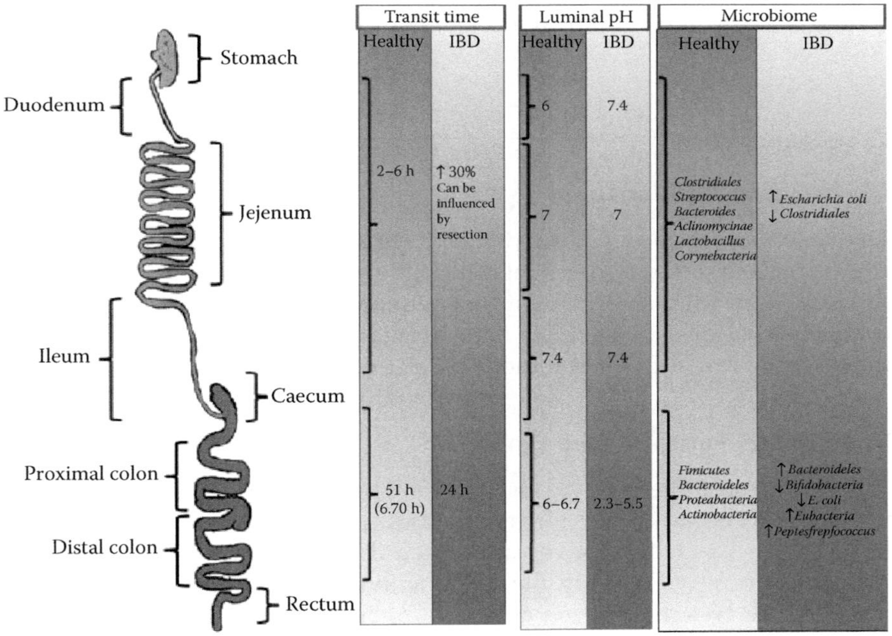

FIGURE 7.1
Physiological and microbial changes in the GIT in colonic diseases. (Reprinted from Hua, Susan, Ellen Marks, Jennifer J Schneider, and Simon Keely. 2015. Advances in oral nano-delivery systems for colon targeted drug delivery in inflammatory bowel disease: Selective targeting to diseased versus healthy tissue. *Nanomedicine: Nanotechnology, Biology and Medicine* no. 11 (5):1117–1132, with kind permission from the publisher.)

Recent advances in controlled-release techniques have allowed the delivery of drugs to the colon, following oral administration. Although the colon cannot match the small intestine in many of its morphological and functional aspects related to drug absorption, it notably has several carrier-mediated transport systems that might be used as drug targets for colonic absorption, and it is as permeable as the small intestine to some lipophilic drugs. (Ehrhardt and Kim 2007).

Colon-specific drug delivery exploits the differences in anatomical and physiological features of the upper and lower segments of the gut. Research has shown that the colon is more responsive to absorption enhancers, protease inhibitors, and bioadhesive and biodegradable polymers compared with other regions of the gut (Sinha et al. 2007). The challenge is to ensure that the drugs are intact when they eventually reach the colon, which is often a problem with traditional oral dosage forms.

Several approaches are being used to target drug release in the colon (Table 7.1), such as covalently linking drug to a carrier (prodrug approach), coating with or embedding in pH-sensitive or biodegradable polymers, the formulation of timed release systems, use of carriers that are specifically degraded by colonic bacteria, pressure sensitive systems, osmotically controlled systems, etc. (Basit 2005) (Rubinstein 1995) (Yang, Chu, and Fix 2002).

7.4 In-Vitro Evaluation

An efficient colon-specific drug delivery system is expected to remain intact in the physiological environment of upper gastrointestinal tract, but release the drug in the colon. No standard technique of evaluation of a colon-specific system has been established, but depending upon the method of preparation and dosage form, different methods have been proposed.

7.4.1 Dosage Form–Related Evaluation

Depending upon the type of dosage form prepared for colon specific delivery, it has to be evaluated for its standard parameters. For if it is a solid dosage form like tablets or capsules, then drug content, weight variation/content uniformity, friability, the swelling index, etc. have to be determined. If particulate system has been designed for colonic delivery, then they have to be characterized for parameters like micromeritic properties, percentage entrapment, particle size, etc.

7.4.2 Drug Release Studies In-Vitro Dissolution Test

An in-vitro model should ideally possess the in-vivo conditions of GIT such as pH, volume, stirring, bacteria, enzymes, enzyme activity, and other

TABLE 7.1

Approaches to Target Drugs to the Colon

Sr. No.	Approach	Basic Feature	References
1	Prodrug approach	Covalent linkage of the drug with carrier	
1.1	Azo conjugates	Drug is linked via azo bond	(Kinget et al. 1998) (Saffran et al. 1988)
1.2	Glucuronate conjugates	Drug is linked with glucuronate	(Simpkins et al. 1988) (Haeberlin et al. 1993)
1.3	Glycoside conjugates	Drug is linked with glycoside	(Friend and Chang 1985)
1.4	Dextran conjugates	Drug is linked with dextran	(McLeod, Friend, and Tozer 1993)
1.5	Cyclodextrin conjugates	Drug is linked with cyclodextrin	(Minami, Hirayama, and Uekama 1998)
1.6	Polypeptide conjugates	Drug is linked with poly(aspartic acid)	(Nakamura et al. 1992)
1.7	Polymeric prodrugs	Drug is linked with polymer	(Leopold and Friend 1995)
2	Delivery of intact drug molecule to colon		
2.1	Coating with polymers		
	Coating with pH-sensitive polymers	System coated with enteric polymers releases drug as it moves towards the alkaline pH	(Mardini et al. 1987)
	Coating with biodegradable polymers	Drug is released due to degradation of polymeric coating by colonic bacteria/enzyme	(Lamprecht et al. 2000)
2.2	Embedding in matrices		
	Embedding in pH-sensitive matrices	Dissolution of pH-sensitive matrices releases the drug	(Krogars et al. 2000)
	Embedding in biodegradable matrices	Drug embedded in polysaccharide matrices is released by swelling of the polymer and degradation by enzymes present in the colon	(Rubinstein et al. 1993)
3	Timed release systems	Lag time of 3–5 hrs is observed before drug is released	
4	Bioadhesive systems	System coated with bioadhesive polymer that selectively adheres to colonic mucosa	(Kopečková et al. 1994)
5	Osmotically active systems	After passing through the stomach intact, drug is released through the semipermeable membrane due to osmotic pressure	(Swanson et al. 1987)
6	Pressure-sensitive systems	Polymeric coating breakdown at pressure as encountered in the colon	(Shibata et al. 2001)

components of food. These conditions are generally influenced by the diet, physical stress, a diseased state, etc. and these factors make it difficult to design a standard in-vitro model. There are primarily two approaches utilized to evaluate drug release from a colon-specific delivery system in-vitro: (1) conventional USP dissolution methods and (2) in-vitro fermentation studies using rat caecal contents, human fecal slurries, multi-stage culture systems (or reactor), and isolated bacterial cultures (pure enzymes specific for the polymer used), with the latter four trying to mimic the colonic environment.

7.4.2.1 Conventional Dissolution Methods

A rationally developed dissolution method can help to assess drug release kinetics, the implications of formulation and changes in manufacturing process, the impact of pH, and hydrodynamic conditions on drug release characteristics, to elucidate the drug-release controlling mechanism, to ensure batch-to-batch consistency during manufacturing, and possibly to act as an in-vivo surrogate. This necessitates that the dissolution method be discriminative, reproducible, scientifically justifiable, and biorelevant. When developing dissolution methods for novel/special dosage forms, it is recommended that the compendial apparatus and methods be used first (Siewert et al. 2003). According to this recommendation, drug release from colon-specific drug delivery systems has been evaluated in a USP dissolution apparatus I, II, and III with slight modifications. To better represent the physiological conditions in the GI tract, three media are commonly used: simulated gastric, intestinal, and colonic fluids. The simulated colonic fluid usually contains the enzyme that degrades specifically the polysaccharides used in the delivery system (Wong et al. 1997) (Hovgaard and Brøndsted 1995) (Ji, Xu, and Wu 2007) (Fetzner et al. 2004) (Pitarresi et al. 2007) (Ugurlu et al. 2007). For instance, galactomannanase is usually included in the medium for the dissolution of guar gum–based system (Wong et al. 1997).

The duration of testing in each medium is chosen to simulate the transit times in the stomach and small intestine, approximately 2 h and 4 h. The drug release studies are performed, in-vitro, under different conditions to simulate the pH and times likely to be encountered during intestinal transit to the colon. The media chosen, for example, are pH 1.2 to simulate gastric fluid (0.1 M HCl with 0.32% w/v pepsin), pH 6.8 to simulate the jejunal region of the small intestine, and pH 7.2 (0.2 M phosphate buffer containing 1% w/w pancreatin) to simulate the ileum segment (Milojevic et al. 1996). Pancreatin is a mixture of digestive enzymes, containing lipase, amylase, and protease, obtained from porcine pancreas.

A multiparticulate system of chitosan hydrogel beads was investigated for colon-specific delivery of macromolecules using fluorescein isothiocyanate–labeled bovine serum albumin as a model protein. In the release studies, pH 7.0 was used to mimic the pH of the colon and the pH of small intestinal fluid was 7.5. A duration of 6 h and 14 h was used to simulate transit times in the small intestine and in the colon, respectively. One of those media was simulated intestinal fluid (SIF), which was 0.05 M phosphate buffer, pH 7.5, with 1% (w/v) pancreatin, as described in the United States Pharmacopeia (1995). Another medium was SIF without pancreatin, 0.05 M phosphate buffer, pH 7.5, and was used to compare the drug release behavior with that in SIF. Almond emulsin beta-glucosidase preparation, 0.5% (w/v), prepared in 0.05 M, pH 7.0 phosphate buffer was another release medium used for the study (Zhang, Alsarra, and Neau 2002).

A USP dissolution apparatus III (reciprocating cylinder) (Figure 7.2) has been employed to assess the performance of guar-based colonic formulations, containing dexamethasone or budesonide as the model drug, in-vitro (Wong et al. 1997). Because of the unique setup of the dissolution apparatus III (i.e., the dissolution tubes can be programmed to move along successive rows of vessels), drug release can be evaluated in different medium with ease successively.

Despite the simplicity and convenience, certain constraints associated with USP dissolution methods were recognized in the dissolution evaluation of complex drug delivery systems, such as: 1) USP dissolution testing primarily generates essential information on the processing of a colon-specific delivery system rather than being predictive of its in-vivo performance. 2) The selection of dissolution media is dependent on the design rationale of the delivery system. 3) The volume and composition of dissolution media and the mixing intensity are not representative of the conditions present in the colon. Because of the involvement of colon microflora as the mechanism in triggering drug release, the dissolution testing should be performed under a condition that the dynamic and ecologically diverse features of the colon can be incorporated into. It appears that this can hardly be accomplished with

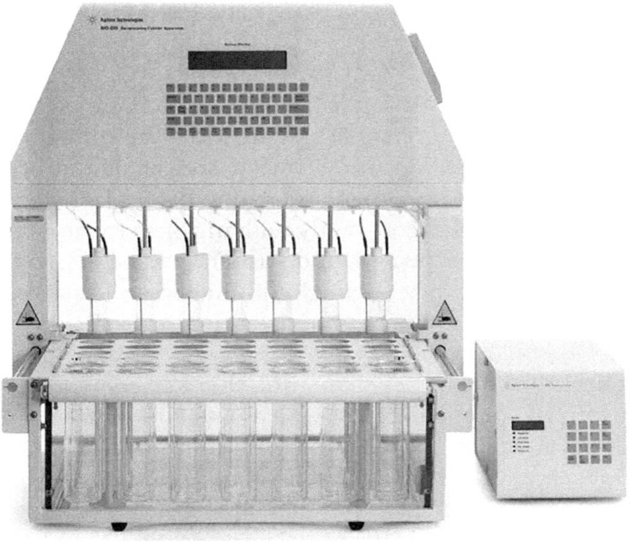

FIGURE 7.2
USP Dissolution Apparatus III.

the existing USP dissolution methods. Consequently, alternative approaches have been utilized and described in the literature. Hence, some modification of USP dissolution methods is deemed necessary (Pillay and Fassihi 1999).

7.4.2.2 In-Vitro Fermentation Studies

For the formulations containing polymers which are specially degraded by the enzymes and bacteria present in colon, it is difficult to obtain information about polymer's digestibility or permeability within the colonic environment by the general dissolution testing methodology. Instead, developing in-vitro methods which mimic the in-vivo system in terms of method are required, as well as the rate of breakdown of delivery agent using rat caecal matter or slurries of human fecal or multi-stage culture systems (Yang et al. 2003) (Rubinstein et al. 1993).

Following are methods commonly used:

7.4.2.2.1 Methods Using Rat Caecal Contents

To overcome the limitation of conventional dissolution testing, rat caecal contents have been widely utilized as alternative dissolution medium because of the similarity of human and rodent colonic microflora. For example, the average log 10 viable count of bacteroides and bifidobacteria, two numerically predominant polysaccharide-degrading bacteria, is 8.0 and 7.0, and 8.0 and 8.2 in the human large intestine and rat caecum, respectively (Hawksworth, Drasar, and Hili 1971). Another advantage of utilizing rat caecal contents is the relatively easy availability of rats. Rat caecal contents are prepared immediately prior to the initiation of drug release study due to the anaerobic nature of the caecum.

For preparation of rat caecal contents, Wistar rats weighing between 200 g and 250 g are maintained on a normal diet. One hour prior to the drug release studies, sufficient numbers of rats are killed by spinal traction. Their abdomens are opened and caecum is traced and ligated at both the ends. Thereafter, the caecum is dissected and immediately transferred to phosphate buffer (pH 6.5) bubbled constantly with nitrogen to maintain anaerobic condition. The contents are pooled, and suspended in the phosphate buffer (pH 6.5) which is continuously bubbled with nitrogen, These are eventually added to the dissolution media to obtain a final caecal dilution 4% (w/v) (Sinha et al. 2004).

The susceptibility of the locust bean gum, chitosan coats, and pectin to the enzymatic action of colonic bacteria was assessed by continuing the drug release studies in 100 ml of pH 6.8 phosphate-buffered saline (PBS) containing 4% w/v rat caecal contents (Raghavan et al. 2002) (Dev, Bali, and Pathak 2011).

The release of indomethacin from calcium pectinate tablets in 100 mL pH 7.0 PBS with/without 1.25% w/v rat caecal contents was studied by Rubinstein et al. (1993). Pectin, a non-starch, linear polysaccharide extracted from the plant cell walls, consisting of α-(1-4)-linked D-galacturonic acid residues interrupted by 1,2- linked L-rhamnose residues, is not digestible in the upper GI tract but is degraded by colonic bacteria. By itself, pectin is a poor carrier for colonic drug delivery due to its high solubility, therefore, the calcium salt of pectin was used to reduce solubility. The experiments were conducted in 100 mL PBS pH 7 under a CO_2 atmosphere at 37°C over a period of 25 h with or without the addition of 1.25% w/v rat caecal contents in sealed glass vials, shaken at 80 rpm during the experiments. Results indicated that indomethacin exhibited greater release in the presence of rat caecal contents when compared to the control. Because of the insolubility of indomethacin was at pH 7.0, it was able to discriminate the indomethacin release by matrix degradation from the contribution of diffusion. In the presence of rat caecal contents, 60.8 ± 15.7% of indomethacin was released within 24 h, in contrast to 4.9 ± 1.1% released in the control medium. This demonstrates that calcium pectinate can be degraded by bacterial enzymes and hence, exhibited the potential as a colon-specific drug delivery carrier.

The dissolution of CODES™ tablets was also investigated using acetaminophen as a model drug in 20 mL pH 6.8 buffer with or without 10% w/w rat caecal contents (Yang 2008). In the presence of rat caecal contents, acetaminophen release was essentially completed within 4 h while no drug release was observed in the medium containing no rat caecal contents within the same time period. This substantiated the design rationale that drug release from CODES™ system was triggered by the decrease in the microenvironment pH due to the degradation of lactulose by colonic bacteria.

7.4.2.2.2 Methods Using Human Fecal Contents

Freshly prepared human fecal slurries have been commonly used to investigate the fermentation of non-starch polysaccharides in which the production of SCFA, acetate, propionate, and butyrate was monitored as a function of fermentation time, because approximately 55% of fecal solids consist of bacteria (Stephen and Cummings 1980a) (Stephen and Cummings 1980b).

Milojevic et al. studied the release of 5-aminosalicylic acid (5-ASA) from pellets coated with amylose/ethylcellulose of different ratios using a batch culture fermenter inoculated with 5% w/v human fecal slurries (Milojevic et al. 1996). The slurries were prepared by homogenizing fresh feces obtained from healthy human volunteers in anaerobic 0.1 M sodium phosphate buffer (pH 7.0). Gas head space and liquid samples were collected at predetermined time intervals from the fermentor for 48 h period. Since gases (H2 and CO2) and volatile fatty acids (acetate, propionate, and butyrate) are the principal fermentation products of polysaccharides, they were quantified to assess the degree of the film fermentation process. It was observed that at the ratio of 1:4 for amylose/ethylcellulose, 5-ASA was completely released within 6 h in the fermenter after 2 h lag time, but the release of 5-ASA was negligible for 12 h of dissolution testing in 900 mL pH 1.2 and 7.2 buffers. It was also found that the concentration of volatile fatty acids increased as a function of the fermentation time, indicating the degradation of amylose in the coating film. Therefore, it can be concluded that the degradation of amylose rather than the physical breakage of the coating is responsible for the release of 5-ASA from amylose/ethylcellulose coated pellets.

McConnell et al. studied release of theophylline from amylose/ethylcellulose coated pellets under colonic conditions using a batch fermentation system. The pellets, containing 150 mg of theophylline, were introduced into 100 ml fermenters (37°C, under positive nitrogen pressure) inoculated with human faecal material (freshly obtained and pooled from three volunteers 10% w/v in phosphate buffer pH 6.8). Control experiments were carried out under the same conditions, using pH 6.8 buffer in the absence of faecal material. Each experiment was carried out six times. At 0, 1, 2, 4, 6, and 12 h samples (2 ml) were removed, centrifuged at 13,000 rpm for 5 min, filtered (0.2 mm) and analyzed. The release of theohylline was faster in the presence of faecal material than that which occurred under control conditions (McConnell, Short, and Basit 2008).

7.4.2.2.3 Multi-Stage Culture Systems

The colon continuously receives the indigestible residues from ileum, which are then absorbed, fermented, and condensed during transit through the colon and ultimately discharged in the form of feces. In this regard, the colon, especially the caecum and the ascending colon, is similar to a continuous culture system. Because of the physical inaccessibility of the ascending colon, Macfarlane et al. designed a three-stage compound continuous culture system as shown in Figure 7.3 to reproduce the spatial, temporal, nutritional, and physicochemical characteristics of the microbiota in different regions of the colon (Gibson, Cummings, and Macfarlane 1988) (Macfarlane, Macfarlane, and Gibson 1998). This culture system consists of three glass fermentation vessels arranged in series with working volumes of 200 mL (V1, proximal colon), 200 mL (V2, transverse colon), and 280 mL (V3, distal colon), respectively. Each vessel is magnetically stirred and kept at 37°C under an atmosphere of CO2. The pH is maintained at 5.5, 6.2, and 6.8 for vessels 1 to 3, respectively to represent the pH in the proximal, transverse, and distal colon. Each fermentation vessel was inoculated with 100 mL of freshly prepared 20% (w/v) fecal slurries from healthy non-methanogenic donors. Upon reaching equilibrium, the growth medium, (containing different sources of polymerized carbon and organic nitrogen such as pectin, guar gum, inulin, xylan, and arabinogalactan to enhance the species diversity of the bacteria) was continuously sparged with O2 free N2 and fed to V1 by a peristaltic pump, which was sequentially supplied to V2 and V3. Thereafter, samples of spent culture medium were removed at regular time intervals and processed for analysis, such as enzyme activity, bacteria composition, or substrate degradation. This system has been validated on the basis of chemical and microbiological measurements on intestinal contents obtained from human sudden death victims. Correlations between in-vivo chemical and bacteriological measurements and data obtained in-vitro

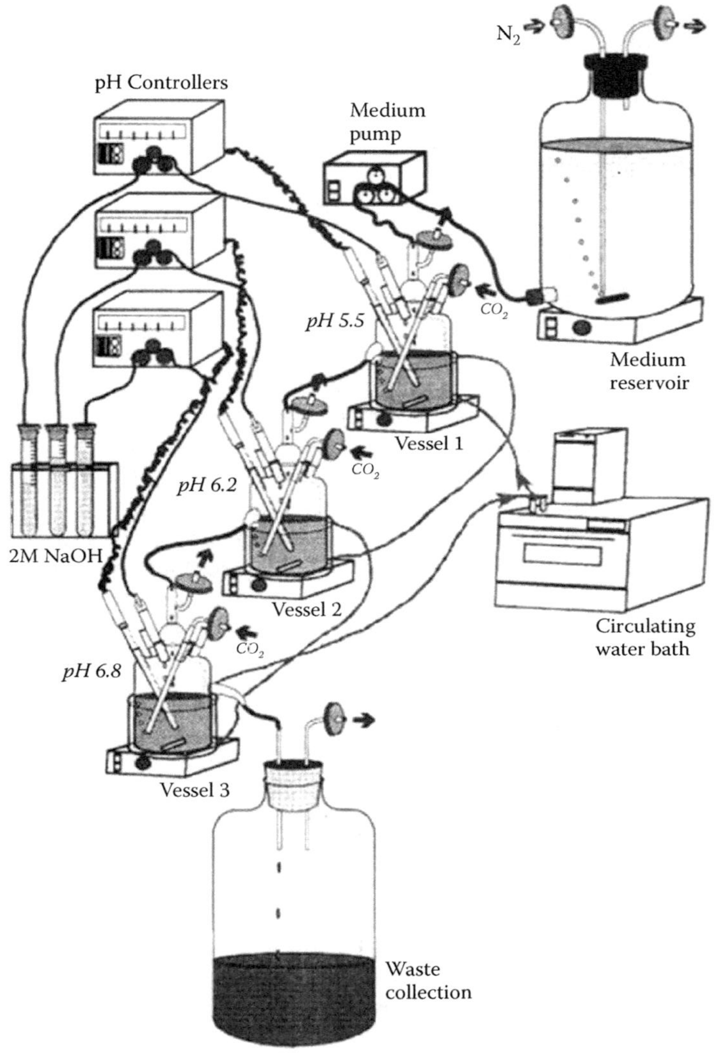

FIGURE 7.3
Three-stage compound continuous culture system. (Reprinted from Macfarlane, GT, S Macfarlane, and GR Gibson. 1998. Validation of a three-stage compound continuous culture system for investigating the effect of retention time on the ecology and metabolism of bacteria in the human colon. *Microbial Ecology* no. 35 (2):180–187, with kind permission from the publisher.)

demonstrate that the three-stage fermentation system provide a useful model for studying the physiology and ecology of large intestinal microorganisms under different nutritional and environmental conditions.

In a similar study, Molly et al. (Molly, Woestyne, and Verstraete 1993) developed a five-step multi-chamber reactor (SHIME) to simulate both the small and large intestinal microbial ecosystem. Three polymeric prodrugs of 5-ASA, poly(1-vinyl-2-pyrrolidone-co-maleic anhydride) (PVP-MA), poly(N-(2-hydroxyethyl)-DL-aspartamide) (PHEA), and dextran, were evaluated in the SHIME reactor. Little or no hydrolysis of the three prodrugs was observed in the reactors representing stomach and small intestine. The release of 5-ASA was most pronounced in the reactor simulating caecum and proximal colon. The results also indicated that the extent of drug release in the reactor simulating caecum and proximal colon depends on the nature of polymeric carriers (Schacht et al. 1996).

7.4.2.3 Isolated Bacterial Cultures

Pure cultures of colonic bacteria can be used to assess the drug release from CDDS. The delivery system is incubated in a fermenter containing suitable medium for bacteria (*Streptococcus faecium* and *Bacteroides ovatus*). The amount of drug released at different time intervals is determined, which is directly proportional to the rate of degradation of polymer carrier.

The ability of pure cultures of colonic bacteria to utilize amylose as a growth substrate was determined in batch culture. Bacterial cultures were grown overnight and then were used to inoculate serum bottles containing basal growth medium with amylose. Bacteroids and clostridia bacteria were found to utilize amylose to a higher degree than other genera tested (Milojevic et al. 1996).

Wakerly employed a method using both pure enzymes and bacterial cultures to evaluate a pectin-based colonic delivery system. Bacteroides ovatus was selected because of its known pectinolytic activity. The dissolution apparatus was designed to withstand heat sterilization to allow use of pure enzymes and pure bacterial cultures. The nutritional medium composed of all the necessary ingredients to support growth of bacteria excluding the carbon source, which was to be derived from the dosage form. Sterile nitrogen was bubbled into the dissolution vessels to maintain anaerobic conditions. Compression-coated pectin tablets were introduced into the basket assemblies and samples were withdrawn periodically to determine the amount of dye released. Similarity found in the timing and profile of release of dye from the enzyme and bacterial experiments indicated similarity in mode of breakdown of pectin (Wakerly et al. 1996).

7.4.3 Tests for Bioadhesion

Bioadhesion has been proposed as a means of improving the performance and extending the mean residence time of colonic drug delivery systems (Duchene and Ponchel 1993) (Chickering and Mathiowitz 1995). Despite the fact that goblet cells increase in numbers along the GIT (Lenaerts and Gurny 1989), the turnover rate of colonic mucus is relatively low. As a result of this, adherence of acrylic polymers to colonic mucosa is much stronger than the mucosal tissue of the stomach and the small intestine (Rubinstein and Tirosh 1994).

It has also been observed that some bacteria, e.g., *Shigellu jlexneri* adhere to the colonic cells of guinea pigs (Izhar, Nuchamowitz, and Mirelman 1982). This adherence was found to reside in the host cells and is fucose and glucose-specific (Duchene and Ponchel 1993). This suggests that there is a potential for the synthesis of water soluble copolymers which would show a two-fold specificity in bioadhesion and drug release.

A new monomer of 5-aminosalicylic acid-containing was synthesized and incorporated into N- (2-Hydroxypropyl) methacrylamide (HPMA) copolymer together with the fucosylamine (bioadhesive moiety)-containing a comonomer using radical copolymerization. The design of this polymer was based on the concept of site-specific binding of carbohydrate moieties complementary to colonic mucosal lectins and site-specific drug (5aminosalicylic acid) release by the microbial azoreductase activity present in the colon. For bioadhesion experiments (13,14), radiolabeled copolymers were incubated with everted sacs isolated from guinea pig small intestine and colon in preoxidized MEM (minimum essential medium) containing 5% FCS (fetal calf serum) for 30 min. at 37°C. Each segment (one fourth of small intestine or one third of colon) was incubated in 10 ml of media containing 2 µCi of ^{125}I labeled HPMA copolymer. After washing, the radioactivity of the segments was determined using a Packard gamma counter. HPMA copolymer-containing side chains terminating in fucosylamine showed a higher adherence to guinea pig colon when compared to HPMA copolymer without fucosylamine moieties, both in-vitro and in-vivo. The incorporation of 5-ASA-containing aromatic side-chains into HPMA copolymers further increased their adherence, which may be due to a combination of nonspecific hydrophobic binding with specific recognition (Kopečková et al. 1994).

One study reports the performance of a novel polymeric material that is capable of providing site specificity in delivery of active agent and the development of mucoadhesive interactions. Azo-networks, based on an acrylic backbone crosslinked with 4,49-divinylazobenzene, were subjected to in-vitro degradation and mucoadhesion studies (before and after degradation), in order to predict their performance in the gastrointestinal tract. For bioadhesion studies, the GIT of male Wistar rats (200–300 g) was used as the biological substrate. The GIT was isolated and divided into four regions: stomach, proximal small intestine, distal small intestine, and colon. The sections were opened laterally to expose the mucosal surface and washed gently with physiological (pH 7). The tissue under test was fixed onto a tissue clamp so as to expose a 3.8-cm circular area. The clamped tissue was then immersed in 75 ml of an isotonic solution, buffered at a predetermined pH (depending on the intestinal section tested) and kept at 37°C, where it was allowed to stand for 5 min before each measurement. The water bath was located on a vertically moving platform which was positioned beneath the sensor arm of a balance (GEC-Avery, UK) as shown in Figure 7.4. The polymeric materials under test

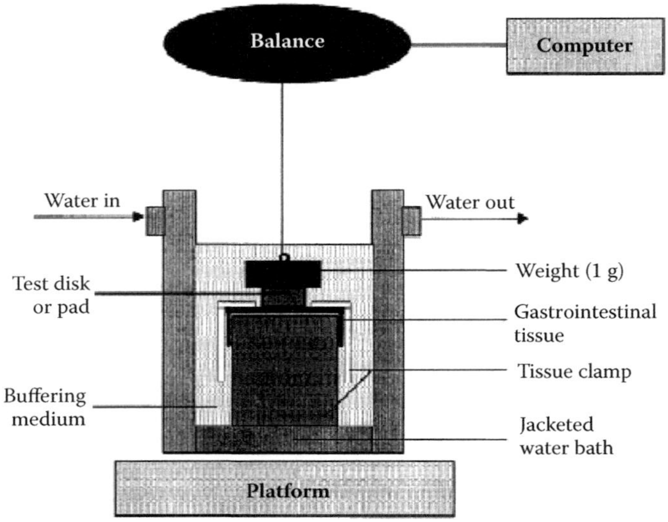

FIGURE 7.4
Apparatus for assessing mucoadhesion. (Reprinted from Kakoulides, Elias P, John D Smart, and John Tsibouklis. 1998. Azocrosslinked poly (acrylic acid) for colonic delivery and adhesion specificity: In vitro degradation and preliminary ex vivo bioadhesion studies. *Journal of Controlled Release* no. 54 (1):95–109, with kind permission from the publisher.)

were milled using a pestle and mortar. The powder fraction that passed through a 55-μm sieve was collected and used for the adhesion studies. Circles, 6 mm in diameter, were cut from double-sided adhesive film and one side was attached to a 1.4-g weight. The protective film was peeled off the free side and the weight was repeatedly placed against a sample of test powder until its surface was completely covered with particles. The surface was cleared of loosely held particles, by exposing it to compressed air for 2 min, and then attached to the sensor arm which was positioned directly above the mucosal surface. The platform was raised at maximum speed until the disk and the weight were completely immersed in the buffer. The balance was tared and the platform was raised slowly until the negative reading of the balance remained constant for 5 sec. The surfaces were allowed to interact for 2 min and then the platform was lowered at a rate of 1 mm min^{-1} until adhesive joint failure occurred. The maximum detachment force and the work of adhesion were calculated as the mean of six sets of measurements. The tissue was also examined in order to identify the position at which the fracture of the adhesive joint had occurred. The observations of these studies indicate that there is an optimum crosslinking density to allow nonadhesive particles to reach the colon. The azo network degrades within the colonic environment to produce a structure capable of developing mucoadhesive interactions with the colonic mucosa (Kakoulides, Smart, and Tsibouklis 1998).

7.4.4 Cell Line Studies: In-Vitro Permeability, Toxicity Evaluation and Adhesion Studies

During the last few years the use of intestinal cell lines has been increased to study the intestinal permeability and for toxicity evaluations. The human epithelial cell line Caco-2 has been widely used as a model of the intestinal epithelial barrier. Caco-2 (Cancer coli-2) was established from a human colorectal adenocarcinoma by Jorgen Fogh at the Sloan-Kettering Cancer Research Institute. Caco-2 cells develop morphologic characteristics of normal enterocytes when grown on plastic dishes or nitrocellulose filters. The monolayers of Caco-2 mimic intestinal absorptive epithelium and represent a very useful tool for studying transepithelial transport. Another cell line of interest is the human colon adenocarcinoma cell line HT29. It is receiving special interest in studies focused on bioavailability, due to the ability to express characteristics of mature intestinal cells (Lea 2015).

Beloqui et al. prepared PLGA/Eudragit S100 pH-sensitive polymeric nanoparticles for selective colonic curcumin release in IBD treatment and studied their transport behaviour across Caco-2 cell monolayers. The cytotoxicity of curcumin nanoparticles (CC-NPs) was evaluated on J774 cells and Caco-2 cells using the MTT method. For this, 20,000 cells/well were seeded in 96-well plates (Nunc, Roskilde, DK) and maintained for 24 h at 37°C. Cells were then exposed to 100 mL of PLGA/ES100-CC NPs dispersed in culture medium for 4 h (J774 cells) or 2 h (Caco-2 cells) at different concentrations. Cells were then washed 3 times with HBSS and incubated for other 3 h with a 0.5 mg/mL MTT solution in RPMI. The medium was then removed, and the purple formazan crystals were dissolved in 100 mL of DMSO. The absorbance was measured at 560 nm using a MultiSkan EX plate reader. Cells with Triton-X 100 and cells with culture medium were considered as positive and negative controls, respectively. The IC50s for the different formulations were calculated (Beloqui et al. 2014). In the case of Caco-2 cells, the IC50 was found to be around 3.0 mg/mL. In J774 cytotoxic studies, IC50 was around 0.5 mg/mL for NPs.

pH-sensitive CC-NPs have been studied for their cell association behavior across Caco-2 cell monolayers. The localization of CC-NPs in Caco-2 cell monolayers was studied qualitatively by confocal laser scanning microscopy. Neither CC-suspension nor CC-NPs were found to accumulate within the Caco-2 monolayers (Beloqui et al. 2014).

HT-29 human colon cancer cell lines have also been used to assess the cytotoxic potential of a colon specific formulation of 5-Fluorouracil. HT-29 cell cultures were grown in Earl's minimal essential medium supplemented with 2mMl-glutamine, 10% fetal bovine serum, penicillin (100 µg/ml), streptomycin (100 µg/ml), and amphoterecin B (5 µg/ml). The cells were maintained at 37°C in a humidified atmosphere with 5% CO2. SRB (sulphorhodamine B) is a dark pink amino xanthene dye with sulfonic groups. Under mild conditions, SRB binds to protein basic amino acid residues of protein in trichloroacetic acid (TCA)–fixed cells to provide a sensitive index of cellular protein content that is linear over a cell density range of at least two orders of magnitude. Color development in SRB assay is rapid, stable, and visible. The developed color can be measured over a broad range of visible wavelength in either a spectrophotometer or a 96-well plate reader. When TCA-fixed, SRB stained samples are air-dried, they can be stored indefinitely without deterioration. The monolayer cell culture was trypsinized and the cell count was adjusted to 1.0 × 105 cells/ml using minimal essential medium containing 10% new born calf serum. To each well of the 96-well microliter plate, 0.1 ml of diluted cell suspension (approximately 10,000 cells) was added. After 24 h, when a partial monolayer was formed, the supernatant was flicked off and the monolayer was washed once. 100 µl of the medium and different drug concentrations were added to the culture in microliter plates. The plates were then incubated at 37°C for 3 days in 5% CO_2 atmosphere. Microscopic examination was carried out and observations were recorded every 24 h. After 72 h, 25 µl of 50% TCA was added to the wells gently such that it formed a thin layer over the drug dilution to form an overall concentration of 10%. The plates were incubated at 4°C for 1 h. The culture plates were flicked and washed five times with water to remove traces of medium, drug and serum, and were then air-dried. The air-dried plates were stained with SRB for 30 min. The unbound dye was then removed by rapidly washing four times with 1% acetic acid. The plates were then air dried. 100 µl of 10 mM tris base was then added to the wells to solubilize the dye. The plates were shaken vigorously for 5 min. The absorbance was measured using micro plate reader at a wavelength of 540 nm. The percentage growth inhibition was calculated using the Equation 7.1.

$$\begin{aligned}&\%\text{Growth inhibition} =\\&100 - (\text{Mean OD of individual test group}/\text{Mean OD of control group}) \times 100\end{aligned} \tag{7.1}$$

The results were reported in terms of CTC50 (cytotoxic concentration for 50% cells) (Dev, Bali, and Pathak 2011).

Human mucus–secreting and non-mucus-secreting intestinal cell monolayers, E12 and HT29 can be used to assess polymer mucoadhesion. The results using these cell lines are comparable to data obtained from isolated rat intestinal sacs. Polymers based on poly(2-(dimethylamino-ethyl) methacrylate (pDMAEMA) have been studied for adhesion using these cell lines. The adherence of pDMAEMA polymers was compared to that obtained with the mucoadhesive, *N*-trimethylated chitosan. pDMAEMA displayed similar levels of mucoadhesion and lower levels of bioadhesion than chitosan derivative and was not cytotoxic (Keely et al. 2005).

7.4.5 Miscellaneous Studies

Light microscopy has been used to determine the degradation of polymeric films after exposure to colonic conditions. A combination of amylose and Ethocel was used to coat pellets of 5-ASA for delivery to the colon. The films, after being stained with iodine solution, were viewed microscopically before and after treatment with faecal inoculum. The images of the film, after 48-h treatment with faecal inoculum, showed no blue staining, indicating that all the amylose had degraded (Milojevic et al. 1996).

7.5 In-Vivo Assessment Techniques

As in other controlled release delivery systems, the successful development of a colon-specific drug delivery system is ultimately determined by its ability to achieve colon-specific drug release and thus exert the intended therapeutic effect. After the successful optimization of the colonic formulation and its in-vitro characterization, in-vivo studies are usually undertaken to evaluate the site specificity of drug release and to obtain relevant pharmacokinetics information of the delivery system. Though animal models have obvious advantages in assessing colon-specific drug delivery systems, human subjects are increasingly utilized for evaluation of this type of delivery systems with visualization techniques, such as γ-scintigraphy imaging (Yang, Chu, and Fix 2002).

7.5.1 Animal Models

Different animals have been used to evaluate the performance of colon-specific drug delivery systems, such as rats (Van den Mooter, Samyn, and Kinget 1995) (Jung, Lee, and Kim 2000), (Tozaki et al. 2001), pigs (Friend et al. 1991) (Gardner et al. 1996), and dogs (Saffran et al. 1991) (Takaya et al. 1995) (Shibata et al. 2001) (Yang et al. 2001). To closely simulate the human physiological environment of the colon, the selection of an appropriate animal model for evaluating a colon-specific delivery system depends on its triggering mechanism and system design. For instance, guinea pigs have comparable glycosidase and glucuronidase activities in the colon and similar digestive anatomy and physiology to that of humans (Hawksworth, Drasar, and Hili 1971) (Kararli 1995), so they are more suitable in evaluating glucoside and glucuronate conjugated prodrugs intended for colon delivery. Additionally, the carrageenan-induced IBD model in guinea pig is available (Watt and Marcus 1971). Friend et al. evaluated the therapeutic efficacy of dexamethasone-_-D-glucoside with dexamethasone in guinea pigs with experimentally induced IBD (Friend et al. 1991). Comparable therapeutic outcome was observed with both prodrugs and drugs, as indicated by the reduction on ulcer number. However, a half-dose of the prodrug was needed to achieve the same effect. Even though guinea pig is the preferred animal model to investigate the in-vivo performance of certain colon specific delivery systems, it is difficult to administer the delivery system orally. More often, gastric intubation has to be utilized.

Rats also have been used to evaluate colon-specific drug delivery systems based on azo-polymers or prodrugs containing azo bonds because the distribution of azoreductase activity in the GI tract is similar between rats and human subjects (Renwick 2013). Owing to their small size, oral administration of large solid dosage forms is difficult. Therefore, capsules were surgically inserted directly to the region of interest in rats (Van den Mooter, Samyn, and Kinget 1995). Higher plasma concentrations of theophylline were observed when the capsule was inserted in the cecum as compared with the small intestine.

Another animal commonly used to evaluate the oral controlled–release delivery systems is the dog (Renwick 2013). Though data obtained from dogs does not extrapolate well to human due to the difference of intestinal anatomy and physiology, dogs are increasingly used to evaluate the colon-specific delivery systems (Saffran et al. 1991) (Takaya et al. 1995) (Shibata et al. 2001) (Yang et al. 2001). In these cases, the performance of the delivery system was evaluated indirectly by measuring the plasma concentration profiles of a model drug delivered by the system.

However, the location of the delivery system in the GI tract at the onset of drug release can only be estimated by comparing the plasma concentration profiles of drug released from a colon-specific delivery system and the reference dosage-enteric coated tablet. The in-vivo performance of CODES™ was evaluated in beagle dogs using acetaminophen as a model drug and lactulose as the matrix-forming excipient in the core tablet (Yang et al. 2001). Compared with enteric-coated core tablet, the onset of acetaminophen release from CODES™ was delayed more than 3 h. Since the first appearance of acetaminophen from the enteric-coated core tablet was 0.5 h later following oral administration, this suggested that the stomach residence time was very short in this case. The transit time of tablet in the small intestine of Beagle dogs has been characterized to be about 2 h (Davis, Wilding, and Wilding 1993). Therefore, it can be inferred that the onset of drug release from CODES™ took place in the proximal colon of the beagle dogs.

The ability of intestinal pressure controlled colon delivery capsules (PCDCs) to obtain colon-specific delivery has also been investigated in Beagle dogs. Glycyrrhizin, a model drug, did not appear in the systemic circulation until 3.33 ± 1.76 h following oral administration of the system (Shibata et al. 2001). This was consistent with the colon arrival time of 3.5 ± 0.3 h determined with a sulfasalazine study of PCDCs (Hu et al. 1999).

For induction of colitis the most commonly used chemical agents used are trinitrobenzene sulfonic acid (TNBS), dextran sulphate sodium (DSS), or oxazolone. For evaluation of therapeutic efficiency of budesonide and budesonide nanoaprticles, all these three models were used in BALB/c mice. The purpose was to test the novel nanoformulation on both, acute and chronic colitis animal models, because the pathophysiological changes, such as clinical symptoms and immunological responses, may differ from model to model (Gottfries, Melgar, and Michaëlsson 2012). Also, it has been noted that the TNBS induced colitis has similar histologic features to Crohn's disease, while, the DSS induced colitis resembles to ulcerative colitis (Alex et al. 2009). Therefore, such a comparative study was done to compare results from endoscopy, histology, and cytokine profiles from these different animal models. It was showed that nanoparticle delivery may improve the anti-inflammatory efficacy of budesonide in terms of endoscopical, histological, and biochemical parameters in comparison to the free drug. Moreover, such nanoparticle delivery via oral administration can be further improved by implementing pH-sensitive release characteristics, for example by using the appropriate coating (Ali et al. 2014).

There exist significant differences between human subjects and commonly used laboratory animals in GI tract anatomy and physiology, including GI transit time, pH, distribution of enzyme activity, population of bacteria, etc. Hence, the data obtained from animal models should be interpreted with caution. In the case of evaluating colon-specific drug delivery systems, the success of a colon-specific delivery system will be primarily decided by the accomplishment of in-vivo drug release in the desired location (i.e., colon). Therefore, this event can only be ascertained through visualization.

7.5.2 γ-Scintigraphy

With growing complexity in the design of novel drug delivery systems, including colon-specific delivery systems, and associated developmental processes, it is critical to understand their in-vivo performance and prove that the system functions in-vivo in accordance with the proposed rationale. Conventional pharmacokinetic evaluation may not generate sufficient data to justify the intended rationale of system design, in majority of cases. γ-Scintigraphy is a non-invasive imaging modality through which the in-vivo performance of drug delivery systems can be visualized under normal physiological conditions. It has been employed to investigate the functionality of tablets and capsules in-vivo since more than two decades (Alpsten, Ekenved, and Soelvell 1975) (Casey et al. 1976), and has become an established technique and extensively used to monitor the performance of novel drug delivery systems within human GI tract. The underlying principles of γ-scintigraphy and its applications in pharmaceutical research and development have been reported (Digenis and Sandefer 1991) (Wilding, Coupe, and Davis 1991) (Newman and Wilding 1999). γ-Scintigraphy imaging gives information regarding the performance of a colon-specific delivery system within the GI tract in terms of the location as a function of time, the time and location of both initial and complete disintegration of the system, the extent of dispersion, stomach residence time, small intestine transit times, and the colon arrival time.

γ-Scintigraphy study of placebo CODES™ was conducted in eight male healthy volunteers to ascertain the time and location of tablet disintegration in the GI tract (Takemura et al. 2000). Radiolabelled resin (1 MBq of 111In) was incorporated in the core tablet of CODES™, which was then coated with the pH-sensitive polymer coating. Gamma camera images were recorded throughout a period of 24 h. The average in-vivo small intestine transit time was 5.2 h after the system was emptied from the stomach, which is well consistent with the established value of 4 ± 1 h (Davis, Hardy, and Fara 1986). The difference between the initial disintegration and colon arrival times is considered as the induction period for acid generation and dissolution of the acid-soluble polymer coating. It was also observed that the system disintegration in the colon was completed within 60 min. Results further indicated that fed conditions did not adversely affect the CODES™ disintegration profile in-vivo even though the gastric residence time was increased.

The in-vivo performance of the colonic delivery system based on pectin and galactomannan coating was also evaluated in healthy human subjects with γ-scintigraphy together with conventional pharmacokinetic analysis using nifedipine as a model drug (Pai et al. 2000). Overall, γ-scintigraphic results demonstrated that it took 5.44 ± 1.77 h for the tablets to reach the ascending colon in 92% of 12 subjects. Upon arrival in the ascending colon, approximately additional 1 h was required to initiate the tablet disintegration.

For more than 5 h post-dose, the mean plasma concentration of nifedipine was negligible and then increased rapidly. The pharmacokinetic profile exhibited a good correlation with the scintigraphic results. It should be noted that the appearance of nifedipine in the systemic circulation before the average colon arrival time could be primarily attributed to the variation of system colon arrival time between individual subjects.

In a nutshell, γ-scintigraphic evaluation of a colon-specific drug delivery system provides "proof of concept," i.e., visualization of system disintegration event and confirmation of disintegration location in the GI tract. Mechanistically, in-vivo functioning of colon-specific drug delivery systems involves the interaction between the system and the gut physiology. Thus, it appears that the precise mechanism responsible for the disintegration of a colon-specific drug delivery system cannot be determined with γ-scintigraphy imaging.

Given the complexity in the functioning of a colon-specific drug delivery system within the GI tract, it is likely that more than one mechanism is involved in the disintegration of the system. For the systems with a microflora degradable film-coating, time-dependent erosion and mechanical failure may play a synergistic role in the disintegration of film-coating.

7.6 *In Silico* Models

The effectiveness of most of the triggering mechanisms for colon targeted systems, are limited by the large variations in the GIT (intra/intersubject), especially transit time and physiological conditions. Because the pH gradient along the GIT forms the basis of several targeted colon delivery systems, understanding how this gradient varies in health and disease states is important. Application of mechanistic computational models can be instrumental in investigating the effects of GIT environmental variability to design alternative and more efficient delivery systems. A true mechanistic model is one that: (i) considers the underlying physical principles to describe the observed phenomena, (ii) considers the stochastic nature of the involved processes, (iii) addresses the empiricism that is often assumed in estimating the model parameters, and (iv) incorporates the true geometry of the domain under consideration.

By means of a mathematical model that considers the pertinent physicochemical processes and biological variability of the GIT environment, *in silico* experiments that accurately replicate in-vivo clinical experiments can be performed. Moreover, such a model can be used for design and optimization of broad range of new and existing targeted and controlled release formulations to the distal intestine and colon. The investigation of the effect of transit time and interluminal pH in the GIT as confounding factors on the performance of the different delivery systems can be carried out, precisely and practically, using computational approaches. These methods, in concert with the in-vivo and clinical experimentation, could result in better treatment strategies and provide greater understanding and improvement in drug design. Haddish-Berhane et al. have developed and validated a multiscale model which considers biological variability and demonstrated that the model can be a practical tool for assessing the bioavailability of the therapeutic agent for treatment of IBD (Haddish-Berhane et al. 2006). These scientists assessed (*in silico*) the effect of variability in GI transit time and pH profile on the drug pharmacokinetics and targeting for one of the commercially available GIT targeted delivery formulations (ASACOL™) using a multi-scale stochastic simulation model. The effect of regional variability of the GIT on the dissolution of the delayed release coating and, in turn, on drug release was addressed for healthy and diseased (UC-ulcerative colitis) subjects. From the *in silico* analysis, it was possible to construct performance curves of the delivery system with an average drug release of 44 ± 19%

and 48 ± 21% for healthy and diseased subjects, respectively. Favorable agreement between clinical literature results and predicted values were obtained for both healthy and UC patients for total urinary and fecal recovery of the active drug (5-ASA) and its metabolite (+N-Acetyl-5-ASA). With the help of the *in silico* analysis, characteristic curves of the delivery systems can be obtained. Physicians can use the characteristic curves of different delivery systems to select a delivery system that would work best for their patients based upon the patient's pH and transit time profiles. Moreover, these curves can be used for design purposes to improve the targeting efficacy of the existing dosage forms and to advance new targeted delivery systems. Although similar evaluation results for a range of oral targeted delivery systems exist from in-vivo evaluations, the present *in silico* evaluation provides valuable information about the performance of the targeted delivery system in the presence of biological variability thereby reducing the number of clinical trials to be conducted saving resources and lives. In the future, drug absorption model can be incorporated with the multiscale model to predict local drug pharmacokinetics and drug efficacy (Haddish-Berhane et al. 2007).

7.7 Future Perspectives

In the past two decades, a great deal of research has been done in developing delivery systems for targeting drug release in the colon. This is a region of the gastrointestinal tract which does not have the hostile environment that is responsible for drug degradation in the stomach and the small intestine. Local therapy of pathologies of the colon and reduced drug availability attributable to degradation of the active ingredients by digestive or mucosal enzymes can benefit from colonic delivery.

Colon-specific drug delivery systems are becoming increasingly sophisticated but the current dissolution methods are still uncertain in establishing possible in-vitro–in-vivo correlation. Therefore it becomes a challenge for the pharmaceutical scientists to develop and validate a dissolution method that incorporates the physiological features of the colon and yet can be used routinely in an industry setting for the evaluation of colon-specific drug delivery systems. A major problem in comparing different delivery systems to the colon is that the degradation studies are carried out in different experimental conditions. Moreover, despite promising results in animal studies, none of the polymeric systems have yet been tested clinically. Much work remains to be done to satisfactorily answer the concerns that have surfaced about colon-specific drug delivery.

References

Alex, Philip, Nicholas C Zachos, Thuan Nguyen, Liberty Gonzales, Tian-E Chen, Laurie S Conklin, Michael Centola, and Xuhang Li. 2009. "Distinct cytokine patterns identified from multiplex profiles of murine DSS and TNBS-induced colitis." *Inflammatory Bowel Diseases* no. 15 (3):341–352.

Ali, H, B Weigmann, MF Neurath, EM Collnot, M Windbergs, and C-M Lehr. 2014. "Budesonide loaded nanoparticles with pH-sensitive coating for improved mucosal targeting in mouse models of inflammatory bowel diseases." *Journal of Controlled Release* no. 183:167–177.

Alpsten, M, G Ekenved, and L Soelvell. 1975. "A profile scanning method of studying the release properties of different types of tablets in man." *Acta Pharmaceutica Suecica* no. 13 (2):107–122.

Basit, Abdul W. 2005. "Advances in colonic drug delivery." *Drugs* no. 65 (14):1991–2007.

Beloqui, Ana, Régis Coco, Patrick B Memvanga, Bernard Ucakar, Anne des Rieux, and Véronique Préat. 2014. "pH-sensitive nanoparticles for colonic delivery of curcumin in inflammatory bowel disease." *International Journal of Pharmaceutics* no. 473 (1):203–212.

Casey, Dennis L, Robert M Beihn, George A Digenis, and Manvendra B Shambhu. 1976. "Method for monitoring hard gelatin capsule disintegration times in humans using external scintigraphy." *Journal of Pharmaceutical Sciences* no. 65 (9):1412–1413.

Chickering, DE, and E Mathiowitz. 1995. "Bioadhesive microspheres: I. A novel electrobalance-based method to study adhesive interactions between individual microspheres and intestinal mucosa." *Journal of Controlled Release* no. 34 (3):251–262.

Davis, SS, JG Hardy, and JW Fara. 1986. "Transit of pharmaceutical dosage forms through the small intestine." *Gut* no. 27 (8):886–892.

Davis, SS, EA Wilding, and IR Wilding. 1993. "Gastrointestinal transit of a matrix tablet formulation: Comparison of canine and human data." *International Journal of Pharmaceutics* no. 94 (1–3):235–238.

Dev, Rakesh Kumar, Vikas Bali, and Kamla Pathak. 2011. "Novel microbially triggered colon specific delivery system of 5-Fluorouracil: Statistical optimization, in vitro, in vivo, cytotoxic and stability assessment." *International Journal of Pharmaceutics* no. 411 (1):142–151.

Digenis, GA, and E Sandefer. 1991. "Gamma scintigraphy and neutron activation techniques in the in vivo assessment of orally administered dosage forms." *Critical Reviews in Therapeutic Drug Carrier Systems* no. 7 (4):309–345.

Duchene, D, and G Ponchel. 1993. "Colonic administration, development of drug delivery systems, contribution of bioadhesion." *STP Pharma Sciences* no. 3 (4):277–285.

Ehrhardt, Carsten, and Kwang-Jin Kim. 2007. *Drug Absorption Studies: In Situ, In Vitro and In Silico Models*. Springer Science & Business Media.

Fetzner, Axel, Stefan Böhm, Sven Schreder, and Rolf Schubert. 2004. "Degradation of raw or film-incorporated β-cyclodextrin by enzymes and colonic bacteria." *European Journal of Pharmaceutics and Biopharmaceutics* no. 58 (1):91–97.

Friend, David R, and George W Chang. 1985. "Drug glycosides: Potential prodrugs for colon-specific drug delivery." *Journal of Medicinal Chemistry* no. 28 (1):51–57.

Friend, DR, S Phillips, A McLeod, and TN Tozer. 1991. "Relative anti-inflammatory effect of oral dexamethasone-β-D-glucoside and dexamethasone in experimental inflammatory bowel disease in guinea-pigs." *Journal of Pharmacy and Pharmacology* no. 43 (5):353–355.

Gardner, Naomi, Will Haresign, Robin Spiller, NIDAL FARRAJ, Julian Wiseman, Heather Norbury, and Lisbeth Illum. 1996. "Development and Validation of a Pig Model for Colon-specific Drug Delivery." *Journal of Pharmacy and Pharmacology* no. 48 (7):689–693.

Gibson, Glenn R, John H Cummings, and George T Macfarlane. 1988. "Use of a three-stage continuous culture system to study the effect of mucin on dissimilatory sulfate reduction and methanogenesis by mixed populations of human gut bacteria." *Applied and Environmental Microbiology* no. 54 (11):2750–2755.

Gottfries, Johan, Silvia Melgar, and Erik Michaëlsson. 2012. "Modelling of mouse experimental colitis by global property screens: A holistic approach to assess drug effects in inflammatory bowel disease." *PloS one* no. 7 (1):e30005.

Haddish-Berhane, Nahor, Ashkan Farhadi, Chell Nyquist, Kamyar Haghighi, and Ali Keshavarzian. 2007. "Biological variability and targeted delivery of therapeutics for inflammatory bowel diseases: An in silico approach." *Inflammation & Allergy-Drug Targets (Formerly Current Drug Targets-Inflammation & Allergy)* no. 6 (1):47–55.

Haddish-Berhane, Nahor, Chell Nyquist, Kamyar Haghighi, Carlos Corvalan, Ali Keshavarzian, Osvaldo Campanella, Jenna Rickus, and Ashkan Farhadi. 2006. "A multi-scale stochastic drug release model for polymer-coated targeted drug delivery systems." *Journal of Controlled release* no. 110 (2):314–322.

Haeberlin, Barbara, Werner Rubas, Harold W Nolen III, and David R Friend. 1993. "In vitro evaluation of dexamethasone-β-d-glucuronide for colon-specific drug delivery." *Pharmaceutical Research* no. 10 (11):1553–1562.

Hawksworth, Gabrielle, BS Drasar, and MJ Hili. 1971. "Intestinal bacteria and the hydrolysis of glycosidic bonds." *Journal of Medical Microbiology* no. 4 (4):451–459.

Hovgaard, Lars, and Helle Brøndsted. 1995. "Dextran hydrogels for colon-specific drug delivery." *Journal of Controlled Release* no. 36 (1–2):159–166.

Hu, Zhaopeng, Tatsuharu Shimokawa, Tomoya Ohno, Go Kimura, Shun-Suke Mawatari, Megumi Kamitsuna, Yukako Yoshikawa, Shigeki Masuda, and Kanji Takada. 1999. "Characterization of norfloxacine release from tablet coated with a new pH-sensitive polymer, P-4135F." *Journal of Drug Targeting* no. 7 (3):223–232.

Hua, Susan, Ellen Marks, Jennifer J Schneider, and Simon Keely. 2015. "Advances in oral nano-delivery systems for colon targeted drug delivery in inflammatory bowel disease: Selective targeting to diseased versus healthy tissue." *Nanomedicine: Nanotechnology, Biology and Medicine* no. 11 (5):1117–1132.

Izhar, M, Y Nuchamowitz, and D Mirelman. 1982. "Adherence of Shigella flexneri to guinea pig intestinal cells is mediated by a mucosal adhesion." *Infection and Immunity* no. 35 (3):1110–1118.

Ji, Chongmin, Huinan Xu, and Wei Wu. 2007. "In vitro evaluation and pharmacokinetics in dogs of guar gum and Eudragit FS30D-coated colon-targeted pellets of indomethacin." *Journal of Drug Targeting* no. 15 (2):123–131.

Jung, Yun Jin, Jeoung Soo Lee, and Young Mi Kim. 2000. "Synthesis and in vitro/in vivo evaluation of 5-aminosalicyl-glycine as a colon-specific prodrug of 5-aminosalicylic acid." *Journal of Pharmaceutical Sciences* no. 89 (5):594–602.

Kakoulides, Elias P, John D Smart, and John Tsibouklis. 1998. "Azocrosslinked poly (acrylic acid) for colonic delivery and adhesion specificity: In vitro degradation and preliminary ex vivo bioadhesion studies." *Journal of Controlled Release* no. 54 (1):95–109.

Kararli, Tugrul T. 1995. "Comparison of the gastrointestinal anatomy, physiology, and biochemistry of humans and commonly used laboratory animals." *Biopharmaceutics & Drug Disposition* no. 16 (5):351–380.

Keely, Simon, Atvinder Rullay, Carolyn Wilson, Adrian Carmichael, Steve Carrington, Anthony Corfield, David M Haddleton, and David J Brayden. 2005. "In vitro and ex vivo intestinal tissue models to measure mucoadhesion of poly (methacrylate) and N-trimethylated chitosan polymers." *Pharmaceutical Research* no. 22 (1):38–49.

Kinget, Renaat, Willbrord Kalala, Liesbeth Vervoort, and Guy Van den Mooter. 1998. "Colonic drug targeting." *Journal of Drug Targeting* no. 6 (2):129–149.

Kopečková, P, R Rathi, S Takada, B Říhová, MM Berenson, and J Kopeček. 1994. "Bioadhesive N-(2-hydroxypropyl) methacrylamide copolymers for colon-specific drug delivery." *Journal of Controlled Release* no. 28 (1–3):211–222.

Krogars, Karin, Jyrki Heinämäki, Johanna Vesalahti, Martti Marvola, Osmo Antikainen, and Jouko Yliruusi. 2000. "Extrusion–spheronization of pH-sensitive polymeric matrix pellets for possible colonic drug delivery." *International Journal of Pharmaceutics* no. 199 (2):187–194.

Lamprecht, Alf, Helena Rodero Torres, Ulrich Schäfer, and Claus-Michael Lehr. 2000. "Biodegradable microparticles as a two-drug controlled release formulation: A potential treatment of inflammatory bowel disease." *Journal of Controlled Release* no. 69 (3):445–454.

Lea, Tor. 2015. "Caco-2 cell line." In *The Impact of Food Bioactives on Health*, 103–111. Springer.

Lenaerts, Vincent M, and Robert Gurny. 1989. *Bioadhesive Drug Delivery Systems*. CRC Press.

Leopold, Claudia S, and David R Friend. 1995. "In vivo pharmacokinetic study for the assessment of poly (L-aspartic acid) as a drug carrier for colon-specific drug delivery." *Journal of Pharmacokinetics and Pharmacodynamics* no. 23 (4):397–406.

Macfarlane, GT, S Macfarlane, and GR Gibson. 1998. "Validation of a three-stage compound continuous culture system for investigating the effect of retention time on the ecology and metabolism of bacteria in the human colon." *Microbial Ecology* no. 35 (2):180–187.

Mackay, M, and E Tomlinson. 1993. "Colonic delivery of therapeutic peptides and proteins." *Drugs and the Pharmaceutical Sciences* no. 60:159–176.

Madara, James L, and Kiertisin Dharmsathaphorn. 1985. "Occluding junction structure-function relationships in a cultured epithelial monolayer." *Journal of Cell Biology* no. 101 (6):2124–2133.

Mardini, H AL, DC Lindsay, CM Deighton, and CO Record. 1987. "Effect of polymer coating on faecal recovery of ingested 5-amino salicylic acid in patients with ulcerative colitis." *Gut* no. 28 (9):1084–1089.

McConnell, Emma L, Michael D Short, and Abdul W Basit. 2008. "An in vivo comparison of intestinal pH and bacteria as physiological trigger mechanisms for colonic targeting in man." *Journal of Controlled Release* no. 130 (2):154–160.

McLeod, Andrew D, David R Friend, and Thomas N Tozer. 1993. "Synthesis and chemical stability of glucocorticoid-dextran esters: Potential prodrugs for colon-specific delivery." *International Journal of Pharmaceutics* no. 92 (1–3):105–114.

Milojevic, Snezana, John Michael Newton, John H Cummings, Glenn R Gibson, R Louise Botham, Stephen G Ring, Mike Stockham, and Mike C Allwood. 1996. "Amylose as a coating for drug delivery to the colon: Preparation and in vitro evaluation using 5-aminosalicylic acid pellets." *Journal of Controlled Release* no. 38 (1):75–84.

Minami, Kunihiro, Fumitoshi Hirayama, and Kaneto Uekama. 1998. "Colon-specific drug delivery based on a cyclodextrin prodrug: Release behavior of biphenylylacetic acid from its cyclodextrin conjugates in rat intestinal tracts after oral administration." *Journal of Pharmaceutical Sciences* no. 87 (6):715–720.

Molly, Koen, M Vande Woestyne, and Willy Verstraete. 1993. "Development of a 5-step multi-chamber reactor as a simulation of the human intestinal microbial ecosystem." *Applied Microbiology and Biotechnology* no. 39 (2):254–258.

Mrsny, Randall J. 1992. "The colon as a site for drug delivery." *Journal of Controlled Release* no. 22 (1):15–34.

Nakamura, Junzo, Mitsuhiko Kido, Koyo Nishida, and Hitoshi Sasaki. 1992. "Hydrolysis of salicylic acid-tyrosine and salicylic acid-methionine prodrugs in the rabbit." *International Journal of Pharmaceutics* no. 87 (1–3):59–66.

Newman, Stephen P, and Ian R Wilding. 1999. "Imaging techniques for assessing drug delivery in man." *Pharmaceutical Science & Technology Today* no. 2 (5):181–189.

Niibuchi, Jun-Ji, Yukihiko Aramaki, and Seishi Tsuchiya. 1986. "Binding of antibiotics to rat intestinal mucin." *International Journal of Pharmaceutics* no. 30 (2–3):181–187.

Pai, CM, CB Lim, SJ Lee, I Park, HN Park, G Seomoon, AL Connor, and IR Wilding. 2000. Pharmacoscintigraphic and pharmacokinetic evaluation of colon specific delivery system in healthy volunteers. Paper read at Proceedings of the International Symposium on Controlled Release Bioactive Materials.

Pillay, Viness, and Reza Fassihi. 1999. "Unconventional dissolution methodologies." *Journal of Pharmaceutical Sciences* no. 88 (9):843–851.

Pitarresi, Giovanna, Maria Antonietta Casadei, Delia Mandracchia, Patrizia Paolicelli, Fabio Salvatore Palumbo, and Gaetano Giammona. 2007. "Photocrosslinking of dextran and polyaspartamide derivatives: A combination suitable for colon-specific drug delivery." *Journal of Controlled Release* no. 119 (3):328–338.

Raghavan, Chellan Vijaya, Chithambaram Muthulingam, Joseph Amaladoss Josephine Leno Jenita, and Thengungal Kochupapy Ravi. 2002. "An in vitro and in vivo investigation into the suitability of bacterially triggered delivery system for colon targeting." *Chemical and Pharmaceutical Bulletin* no. 50 (7):892–895.

Rahman, Anisur, JA Barrowman, and A Rahimtula. 1986. "The influence of bile on the bioavailability of polynuclear aromatic hydrocarbons from the rat intestine." *Canadian Journal of Physiology and Pharmacology* no. 64 (9):1214–1218.

Renwick, AG. 2013. "First-pass metabolism within the lumen of the gastrointestinal tract." *Presystemic Drug Elimination: Butterworths International Medical Reviews: Clinical Pharmacology and Therapeutics* no. 1:1.

Rubinstein, Abraham. 1995. "Approaches and opportunities in colon-specific drug delivery." *Critical Reviews™ in Therapeutic Drug Carrier Systems* no. 12 (2–3).

Rubinstein, Abraham, and Boaz Tirosh. 1994. "Mucus gel thickness and turnover in the gastrointestinal tract of the rat: Response to cholinergic stimulus and implication for mucoadhesion." *Pharmaceutical Research* no. 11 (6):794–799.

Rubinstein, Abraham, Raphael Radai, Miriam Ezra, Seema Pathak, and J Stefan Rokem. 1993. "In vitro evaluation of calcium pectinate: A potential colon-specific drug delivery carrier." *Pharmaceutical Research* no. 10 (2):258–263.

Saffran, M, JB Field, J Pena, RH Jones, and Y Okuda. 1991. "Oral insulin in diabetic dogs." *Journal of Endocrinology* no. 131 (2):267–278.

Saffran, Murray, Craig Bedra, G Sudesh Kumar, and Douglas C Neckers. 1988. "Vasopressin: A model for the study of effects of additives on the oral and rectal administration of peptide drugs." *Journal of Pharmaceutical Sciences* no. 77 (1):33–38.

Schacht, Etienne, An Gevaert, Koen Molly, Willy Verstraete, Peter Adriaensens, Robert Carleer, and Jan Gelan. 1996. "Polymers for colon specific drug delivery." *Journal of Controlled Release* no. 39 (2–3):327–338.

Shibata, Nobuhito, Tomoya Ohno, Tatsuharu Shimokawa, Zhaopeng Hu, Yukako Yoshikawa, Kenjiro Koga, Masahiro Murakami, and Kanji Takada. 2001. "Application of pressure-controlled colon delivery capsule to oral administration of glycyrrhizin in dogs." *Journal of Pharmacy and Pharmacology* no. 53 (4):441–447.

Siewert, Martin, Jennifer Dressman, Cynthia K Brown, Vinod P Shah, Jean-Marc Aiache, Nobuo Aoyagi, Dennis Bashaw, Cynthia Brown, William Brown, and Diane Burgess. 2003. "FIP/AAPS guidelines to dissolution/in vitro release testing of novel/special dosage forms." *AAPS PharmSciTech* no. 4 (1):43–52.

Simpkins, James W, Maciej Smulkowski, Ross Dixon, and Ronald Tuttle. 1988. "Evidence for the delivery of narcotic antagonists to the colon as their glucuronide conjugates." *Journal of Pharmacology and Experimental Therapeutics* no. 244 (1):195–205.

Sinha, VR, BR Mittal, KK Bhutani, and Rachna Kumria. 2004. "Colonic drug delivery of 5-fluorouracil: An in vitro evaluation." *International Journal of Pharmaceutics* no. 269 (1):101–108.

Sinha, Vivek Ranjan, Asmita Singh, Ruchita V Kumar, Sanjay Singh, Rachana Kumria, and JR Bhinge. 2007. "Oral colon-specific drug delivery of protein and peptide drugs." *Critical Reviews™ in Therapeutic Drug Carrier Systems* no. 24 (1).

Stephen, Alison M, and JH Cummings. 1980a. "The microbial contribution to human faecal mass." *Journal of Medical Microbiology* no. 13 (1):45–56.

Stephen, Alison M, and John H Cummings. 1980b. "Mechanism of action of dietary fibre in the human colon." *Nature* no. 284 (5753):283–284.

Swanson, David R, Brian L Barclay, Patrick SL Wong, and Felix Theeuwes. 1987. "Nifedipine gastrointestinal therapeutic system." *American Journal of Medicine* no. 83 (6):3–9.

Takaya, Tomohiro, Chikako Ikeda, Naoya Imagawa, Kiyoshi Niwa, and Kanji Takada. 1995. "Development of a colon delivery capsule and the pharmacological activity of recombinant human granulocyte colony-stimulating factor (rhG-CSF) in beagle dogs." *Journal of Pharmacy and Pharmacology* no. 47 (6):474–478.

Takemura, S, S Watanabe, M Katsuma, and M Fukui. 2000. "Human gastrointestinal treatment study of a novel colon delivery system (CODES) using scintography." *Pro Int Sym Control Rel Bioact Mat* no. 27.

Tozaki, Hideyuki, Junko Nishioka, Junta Komoike, Naoki Okada, Takuya Fujita, Shozo Muranishi, Sooh-Ih Kim, Hiroshi Terashima, and Akira Yamamoto. 2001. "Enhanced absorption of insulin and (Asu1, 7) eel-calcitonin using novel azopolymer-coated pellets for colon-specific drug delivery." *Journal of Pharmaceutical Sciences* no. 90 (1):89–97.

Ugurlu, Timucin, Murat Turkoglu, Umran Soyogul Gurer, and Burcak Gurbuz Akarsu. 2007. "Colonic delivery of compression coated nisin tablets using pectin/HPMC polymer mixture." *European Journal of Pharmaceutics and Biopharmaceutics* no. 67 (1):202–210.

Van den Mooter, Guy, Celest Samyn, and Renaat Kinget. 1995. "In vivo evaluation of a colon-specific drug delivery system: An absorption study of theophylline from capsules coated with azo polymers in rats." *Pharmaceutical Research* no. 12 (2):244–247.

Wakerly, Zoë, John T Fell, David Attwood, and David A Parkins. 1996. "In vitro evaluation of pectin-based colonic drug delivery systems." *International Journal of Pharmaceutics* no. 129 (1–2):73–77.

Watt, Jo, and R Marcus. 1971. "Carrageenan-induced ulceration of the large intestine in the guinea pig." *Gut* no. 12 (2):164–171.

Wilding, IR, AJ Coupe, and SS Davis. 1991. "The role of γ-scintigraphy in oral drug delivery." *Advanced Drug Delivery Reviews* no. 7 (1):87–117.

Wong, D, S Larrabee, K Clifford, J Tremblay, and DR Friend. 1997. "USP dissolution apparatus III (reciprocating cylinder) for screening of guar-based colonic delivery formulations." *Journal of Controlled Release* no. 47 (2):173–179.

Yang, LB, S Watanabe, Y Shi, F Liu, JS Chu, J Li, R Gu, and JA Fix. 2001. Design and in vitro/in vivo evaluation of a novel colon-specific drug delivery system (CODES™). Paper read at AAPS. Annual. Meeting. Poster. Presentation.

Yang, Libo. 2008. "Biorelevant dissolution testing of colon-specific delivery systems activated by colonic microflora." *Journal of Controlled Release* no. 125 (2):77–86.

Yang, Libo, James S Chu, and Joseph A Fix. 2002. "Colon-specific drug delivery: New approaches and in vitro/in vivo evaluation." *International Journal of Pharmaceutics* no. 235 (1):1–15.

Yang, Libo, Shunsuke Watanabe, Jinhe Li, James S Chu, Masataka Katsuma, Shigeharu Yokohama, and Joseph A Fix. 2003. "Effect of colonic lactulose availability on the timing of drug release onset in vivo from a unique colon-specific drug delivery system (CODES™)." *Pharmaceutical Research* no. 20 (3):429–434.

Zhang, Hua, Ibrahim A Alsarra, and Steven H Neau. 2002. "An in vitro evaluation of a chitosan-containing multiparticulate system for macromolecule delivery to the colon." *International Journal of Pharmaceutics* no. 239 (1):197–205.

8

Brain Targeted Drug Delivery Systems

Manisha Lalan, Rohan Lalani, Vivek Patel, and Ambikanandan Misra

CONTENTS

8.1 Introduction

Unlike other parts of the body where there is free and easy movement of ions, electrolytes, and blood components across the various cells and tissues, the scenario is quite different when the regulation of movement of such blood components and accessory nutritional counterparts are concerned. The blood vessels that supply vital nutrients and oxygen to the brain are highly specialized and tightly regulate the transport process. Thus, they exhibit a sort of barrier function, and their location is defined as the blood–brain barrier (BBB). The need for such barrier function lies in the fact that the brain is the most important organ in the body for the regulation of each and every metabolic activity that involves a range of voluntary and involuntary functions. Thus, it was of utmost importance that a special defense mechanism be established. This mechanism bars off and prevents access of harmful chemicals, toxic metabolites, and infectious organisms to gain easy access to the central nervous system thus compromising homeostasis. The structural and functional unit of brain are neurons and the signal transduction mechanisms exist for processing the signals in the brain. Cell types in the brain includes neural, vascular, and immune ones that regulate the various functions. Further, the walls of the blood vessels are line with endothelial cells that coordinate various metabolic, barrier, and transport properties. Taking into consideration the basic aspects of the structural organization and transport mechanisms, Chapter 8 provides a brief discussion of the blood–brain barrier along with nature of the barrier and various strategies to overcome the BBB which is essential in the delivery of therapeutics to the brain in several disorders.

8.2 Blood–Brain Barrier

The exclusive nature of the transport of molecules to the brain against the rest of the body is depicted by the term BBB due to the unique vascular modelling of the cells of the CNS. The microvasculature of the CNS consists of vessels that are continuous and are nonfenestrated ones, however, this does not imply that movement is restricted altogether (Daneman 2012). Transport of essential molecules across the barrier and among the cells across the CNS occurs across the cells and also involves specific transport mechanisms for some essential molecules (Zloković 2008). However, challenge to the integrity of the barrier and loss in barrier function occurs during various diseases such as neurodegenerative ones, stroke, trauma, etc. that may lead to the dysfunction of neurons and subsequently their degeneration. In the delivery of drug therapeutics too, this barrier imposes a restrictive nature and thus various methods to bypass it or modifying it have been investigated in brain drug delivery research (Larsen, Martin, and Byrne 2014).

The concept of the BBB marked its inception in the observations made by various scientist during the end of 19th century and in the beginning of 20th century and the term was introduced by Stern and Gautier in 1921 (Rascher and Wolburg 2002). These were the observations using several dyes that were injected systemically that either nonselectively stained all organs except for choroid plexus or those injected into cerebrospinal fluid that selectively stained CNS (Sage 1982; King 1968; Alavijeh et al. 2005). These observations led to the findings that there exists a permeability barrier across the CNS. This was due to the compositional and structural differences between the endothelial cells that lines the capillaries in the case of different organs where fenestrations exist between adjacent cells and that in case of CNS, where there are no fenestrations and there exist tight junctions interconnecting endothelial cells due to interaction of transmembrane proteins paracellularly (Ballabh, Braun, and Nedergaard 2004; Abbott 2005). Due to this, the transport or diffusion of polar solutes is prevented through the blood that flows through the paracellular pathway. However, specific transporters exist in the brain endothelial cells for polar solutes, lipophilic molecules, and molecules with low molecular weight to gain easy access across the endothelial tight junction, along with few essential nutrients and cofactors that are transported across by active transport mechanism (Zloković et al. 1985; Zloković and Apuzzo 1997; Zlokovic et al. 1990). The BBB is thus a dynamic interface and is involved in maintenance of extracellular homeostasis and protection against toxic compounds and changes in blood composition that may impact the normal functioning of the neurons (Erdő, Denes, and de Lange 2017). The organization of brain is quite complex and a lot of studies have been carried out to decode each and every element and pathways of the brain and their function.

8.2.1 Nature of Barrier

The neurovascular unit is comprised of the BBB, neurons, and the neuronal environment that intriguingly control the cerebral functions. Pericytes, endothelial cells (ECs), and astrocytes are the cellular elements of the BBB (Figure 8.1). The former two are interspersed in basement membrane that is abluminal one. As described earlier, the ECs in the brain do not have fenestrations and are closely associated due to presence of tight junctions in between adjacent cells. Over and above tight junctions, adherent junctions also exist for providing interconnections.

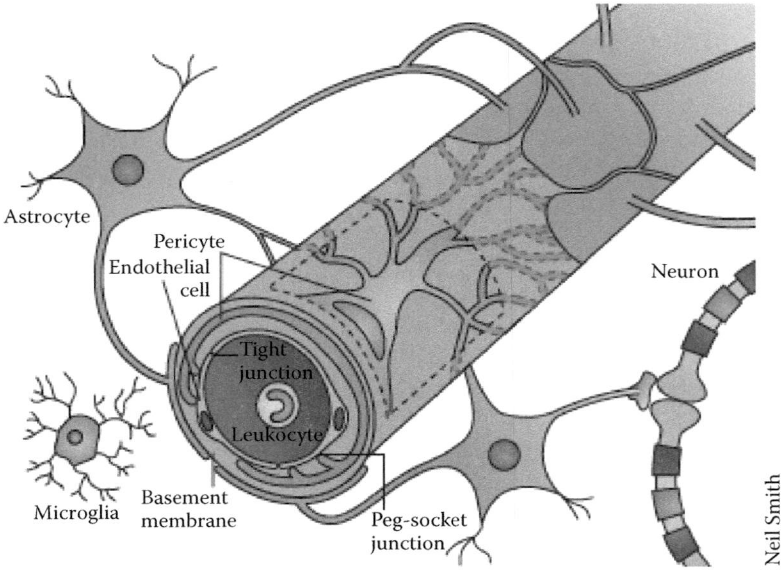

FIGURE 8.1
Cellular interplay at the neurovascular unit (capillary level). The BBB is part of the NVU, which represents an elaborate interplay of central and peripheral cells. Vascular endothelial cells sealed by tight junctions constitute the BBB. The endothelium's abluminal surface is covered by a basement membrane in which pericytes and their processes are embedded. Direct intercellular crosstalk between endothelial cells and pericytes are implemented by peg-socket junctions. Astrocytes extend foot processes that encircle the abluminal side of the vessel to an extent of nearly 100%. Although at the capillary level the basement membrane is regarded as a composite basement membrane, it is separated into endothelial and parenchymal basement membranes at the level of the postcapillary venule, delimiting the perivascular space (not shown). Neurons and microglia are considered members of the NVU as they interact with core elements of the BBB and influence barrier functions. Peripheral blood cells, including leukocytes, also participate in this cellular interplay as they modulate BBB functions under pathological conditions such as inflammation. (Adapted with permission from Macmillan Publishers Ltd: [Nature Medicine] Obermeier, Daneman, and Ransohoff 2013, copyright 2013.)

For maintaining the functional and structural integrity of the BBB, the primary structural protein in the BBB, actin, is linked by the zona occludens to transmembrane protein components, namely claudins, occludins that exhibit paracellular functions and junction adhesion molecules (Hawkins and Davis 2005). Accessory proteins, such as Z0-1 to 3 and cingulin, are cytoplasmic proteins that anchors cytoskeleton with occludin and claudins (Suzuki, Nagai, and Umemura 2016). Due to the above structural organization, vesicular transport that occurs in the form of pinocytosis is very sparse, and a high electrical resistance due to which only lipophiles can easily transport by diffusion mechanism or by dissolving across the junctions (Butt and Jones 1992). The function of pericytes involves maintenance of structural integrity, stabilization of endothelium, vascular differentiation, and the foundation

TABLE 8.1

Regulation of Formation of Tight Junctions

Effect	Effector	Contributing NVU Member
Upregulation of claudin-3	Wnt–β-catenin	Not described
Sealing of TJs by occludin and claudin-5	PDGF-B–PDGFR-β	Pericytes
Induction of claudin-5 expression	TGF-β–TGF-βR	Pericytes
Upregulation of occludin and claudin-5	SHH–PTC1	Astrocytes
Upregulation and subcellular distribution of TJ proteins	ANG-1–TIE-2	Astrocytes
Post-translational modification of occludin and subcellular distribution	ANG-II–AT1	Astrocytes
Post-translational modification of occluding	APOE–LRP-1	Astrocytes
Maintenance of claudin-5 expression and localization	β_1 integrin–ECM ligands	Basement membrane
Stabilization of TJs	Agrin	Basement membrane
Upregulation of claudin-3, claudin-5 and ZO-1	Shear stress	Blood flow

Source: Adapted by permission from Macmillan Publishers Ltd: [Nature Medicine] Obermeier, Daneman, and Ransohoff 2013, copyright 2013.

of the tight junction. Similarly, astrocytes are also involved in maintaining integrity of BBB by ensheathing pericytes and vessel wall of endothelial cells (Tolia et al. 2005). An overview of the various effector molecules functioning along with the above neurovascular unit in regulation of tight junction formation is provided in Table 8.1.

For absorption and transport of molecules, ions, and compounds across the barriers several mechanisms exist such as passive diffusion, facilitated transport, carrier-mediated transport, and transcytosis. These are selectively distributed at the endothelial abluminal and luminal surface (del Pino, Hawkins, and Peterson 1995). Whereas, P-glycoprotein, anion, and cation transporter receptors are efflux mediators from the endothelium of brain to the basolateral membrane along with several enzymes that are involved in degradation of drugs, and neurotransmitters (Smith 2003).

At several stages, however, the intactness of BBB gets compromised, which involves several neurodegenerative diseases. A list of diseases linked to BBB dysfunction are provided in Table 8.2.

8.2.2 Strategies to Overcome the BBB

The BBB serves as an immunological, physical, and metabolic barrier for permitting transport of drugs. Several strategies have been devised for facilitating the delivery of therapeutics across the barrier and can be classified as active and passive targeting systems (Pardridge 2007). Few of them are discussed here for quick reference.

TABLE 8.2

Disease Linked to BBB Dysfunction

Disease	Level of BBB Effect[a]	Comment
Stroke	Primary	Microvascular injury induced by oxidative stress during ischemia-reperfusion
Epilepsy	Primary	Systemic inflammation can disturb brain homeostasis by allowing entry of ions and epileptogenic substances across the BBB
	Secondary	Seizures reduce BBB integrity, which enables entry of plasma proteins into the brain that sustain the epileptogenic state
AD	Primary	BBB dysfunction, including defective amyloid-β clearance from brain and congophilic angiopathy
Familial ALS	Primary	Loss of BBB integrity at an ultrastructural level associated with expression of mutant SOD1 in brain capillary endothelial cells
PD	Secondary	Increased BBB permeability and decreased transport activity across the BBB, including inefficient efflux of toxic molecules via P-glycoprotein
MS	Secondary	Extravasation of autoreactive T cells and monocytes across a compromised BBB
Natalizumab-PML with IRIS	Secondary	Infiltration of T cells in perivascular space and parenchyma after discontinuation of natalizumab in context of PML
NMO	Primary	BBB breakdown including loss of AQP4 and of astrocytes caused by AQP4-specific IgG
Primary CNS vasculitis	Primary	Inflammation of cerebral vessels without systemic disorder
Secondary CNS vasculitis	Primary	Inflammation of cerebral vessels associated with systemic inflammatory illness
VZV vasculopathy	Primary	Viral infection (primary or upon reactivation) of cerebral arteries
Cerebral malaria	Primary	Sequestration of parasitized red blood cells in lumen of cerebral microvasculature
Primary CNS lymphoma	Secondary	Leaky angiogenic vessels in malignant tissue
Glioblastoma	Secondary	Leaky neoangiogenic vessels and loss of BBB integrity in preexisting vessels (by subcellular mislocalization of astroglial AQP4) in malignant tissue
PRES	Primary	Vascular injury by systemic influence, such as disorders of clotting or bleeding, and chemotherapy agents (particularly those which inhibit VEGFR kinase)
TBI	Secondary	Mechanical disruption of the BBB followed by post-traumatic BBB dysfunction

(Continued)

TABLE 8.2 (CONTINUED)

Disease Linked to BBB Dysfunction

Disease	Level of BBB Effect[a]	Comment
Migraine	Secondary	Cortical spreading depression with subsequent vascular reaction
Diabetes	Secondary	Increased BBB permeability, possibly leading to cognitive impairment

Source: Adapted by permission from Macmillan Publishers Ltd: (Nature Medicine) Obermeier, Daneman, and Ransohoff 2013, copyright 2013.

[a] Primary level of BBB effect indicates that the cerebrovasculature is probably compromised upstream from CNS pathogenesis, whereas secondary level of BBB effect is interpreted as happening downstream from the initial insult and aggravating disease. AD, Alzheimer's disease; PD, Parkinson's disease; PML, progressive multifocal leukoencephalopathy; IRIS, immune reconstitution inflammatory syndrome; VZV, varicella zoster virus; PRES, posterior reversible encephalopathy syndrome; TBI, traumatic brain injury.

8.2.2.1 Increasing the Permeability or Influx of Drugs Across the BBB

a. Lipophilic approach

The permeability across the cerebrovascular region is strongly correlated with lipophilicity of the drug. An optimal partition coefficient is required with a log P value around 1.5–2.5. One of the strategy for imparting lipophilic character is by modification with fatty acids for protein molecules (Kabanov, Levashov, and Alakhov 1989; Kabanov, Levashov, and Martinek 1987). As protein and polypeptides are hydrophilic molecules, their transport across the BBB is a challenge. With the increasing number of novel therapeutics being investigated, this class is an emerging one and for taking benefit of their therapeutic potential for management of brain disorders, strategies for their transport across the barrier needs to be investigated. One such strategy is by use of surfactants. Herein, reverse micellar systems are created where water-soluble protein and polypeptide molecules become entrapped in the inner layer or the aqueous compartment of the micelles and surfactant chains lines on the outside. Such modifications have been investigated for improving the rates of internalization along with enhancing cell binding characteristics (Slepnev et al. 1995; Ekrami, Kennedy, and Shen 1995; Chopineau et al. 1998; Chekhonin et al. 1991). Another, strategy is to synthesize the lipophilic precursors of the drugs with a hydrophilic character. However, the strategy has had limited success. This is due to the fact that increasing the lipophilicity leads to significant increase in binding with plasma proteins, which decreases the amount of drug available for partitioning in to the CNS. Thus, it was implicated that a careful balance between the plasma solubility and cerebral permeability is required for efficient delivery of lipophilic-modified drugs. However, there are several limitations of using this approach such as lessening of pharmacological activity, altered bio-distribution, and chances of rebound drug efflux mechanism from the BBB.

b. Prodrug approach

The prodrug strategy involves coupling of drug to a functional group via covalent bond formation that will impart desirable physicochemical property (of improved lipid solubility or membrane permeability) to the parent drug molecule. It generally involves chemical modifications and the end product is a pharmacologically inactive drug molecule (Varsha et al. 2014). However, after reaching to the target site, the functional moieties get detached either due to enzymatic action or hydrolysis to release the free active drug that is available to act on the receptors or exhibit therapeutic effect (Oldendorf 1971; Bodor and Buchwald 2003). Thus, it can be stated as a one-way transfer of the drug from circulation to the brain compartment; once it is inside the brain and metabolized, the drug cannot diffuse or transport back to the vascular site due to its conversion to a hydrophilic form. Functional modification of drugs containing amino-, hydroxy- and carboxylic acid group with amide or ester bonds is shown to enhance lipophilicity. Over and above this, phospholipids, glycerides, or fatty acids have also been used. This strategy faces the same limitations as reported in lipophilic approach. Further, the prodrug may nonselectively get transformed at other sites in the circulation or organs to get converted to other reactive metabolites that may pose toxicity issues (Rautio et al. 2008). Some of the examples where this strategy has provided promising results are for morphine, levodopa, valproic acid, vigabatrin, testosterone, etc.

8.2.2.2 Disruption of BBB

a. Chemical modification

Chemical modification is an invasive technique. This strategy involves prior administration of vasoactive compounds by intra-articular route to open up the tight junctions between the brain endothelial cells. It thus facilitates the entry of the subsequently administered therapeutic agent that now can gain easy access to the brain compartment. Few of these vasoactive substances includes bradykinin, leukotriene, interleukins, etc. One specific example of a bradykinin derivative is RMP-7 (Cereport) (Emerich et al. 1998; Borlongan and Emerich 2003). Herein, bradykinin B2 receptor are involved and are act through caveolin receptors (Liu, Xue, and Liu 2010). The disruption of BBB using such compounds are purely temporary and the intactness of the barrier is regained within minutes once the infusion of the agent is stopped (Sanovich et al. 1995; Fike et al. 1998). This has been used for increasing the transport of anticancer and diagnostic agents to CNS such as cyclosporine A, carboplatin, etc. (Elliott et al. 1996; Borlongan et al. 2002).

b. Osmolysis

This technique takes advantage of the manipulation of tight junction by use of hyperosmolar solutions. Mannitol, urea, saline, arabinose, etc. are the various agents used in this category (Bellavance, Blanchette, and Fortin 2008; Rapoport 2000). As a general mechanism, due to osmotic difference, there is shrinkage of endothelial cells at tight junctions paving the way for delivery of therapeutic agents (Kroll, Neuwelt, and Neuwelt 1998). This strategy has been used for increasing the concentration of anticancer agents in brain in the CNS lymphoma (Neuwelt et al. 1991). However, the strategy also impacts the normal endothelium with a resultant higher opening of these cells leading to the manifestation of toxic effects over and above increasing the permeability of therapeutic agents (Williams et al. 1995).

c. Ultrasound based

The use of ultrasound is quite safe and effective compared to the other techniques due to the advantage of non-invasiveness, easy repeatability, and ability of selectively target specific area (Hynynen et al. 2005). It has produced the only transient disruption of the brain with focal lesions, by either tight junction protein disruption and/or producing shear stress on the cells and regulating cell permeability via signaling pathway (Hynynen et al. 2003). This technique can be used in conjunction with other techniques, such as microbubbles, for minimizing the side effects of ultrasound (Jalali et al. 2010; van Wamel et al. 2006). This technique has been successfully evaluated for increasing the brain concentration of chemotherapeutic agents such as carmustine, trastuzumab, doxorubicin, etc. (Aryal et al. 2013; Liu et al. 2010).

d. Using Radiation

The use of high frequency radiation for inducing damage to DNA thereby leading to cell death is the recent treatment modality for CNS disorders particularly cancer along with disrupting the BBB (Cao et al. 2005). Radiation through lasers are extremely precise, and can be shaped as desired using beam shaping techniques (Chacko et al. 2013). Due to precision, the laser has a minimal effect on surrounding normal tissues and cells (Patel and Mehta 2007; Qin et al. 1990).

Other strategies involve bypassing the BBB using either an intranasal route, intracarotid injection, microdialysis, or a transmucosal route. Further, the use of nanocarriers, such as liposomes, nanoparticles, microspheres, or polymeric micelles, have also been investigated.

8.3 In-Vitro Studies

The treatment of various neurological disorders suffers a setback because of the insufficiency in delivering the drugs across the formidable impediment of the BBB. Over 98% of the therapeutic molecules are unable to breach the barrier and reach meaningful levels in CNS (Wilhelm and Krizbai 2014). This inability to successfully cross the BBB is a function of the physicochemical properties of the molecule and the nature of barrier. The BBB serves as an active and dynamic interface regulating the cellular and molecular trafficking between the systemic circulation and the CNS. The loss of functionality of BBB has been strongly indicted in the pathologies of various neurological and autoimmune disorders like Alzheimer's disease, Parkinsonism, meningitis, stroke, etc. (Wilhelm and Krizbai 2014; Cecchelli et al. 2007).

The availability of reliable and efficient in-vitro models to assess permeation across BBB is very crucial and significant in drug development cycle of neurotherapeutics. Although in-vivo techniques are the most accurate to predict the permeability, they are expensive, difficult, and not suitable for high throughput screening. In-vitro models have been developed to assist researchers by mimicking the in-vivo physiology. These models are very useful to facilitate and expedite cerebrovascular research. The ideal BBB model should be a fingerprint of BBB functionality, reflecting the composite cellular nature of the barrier; restrictive paracellular pathways; functional expression of transcellular membrane transporters; enzyme mileu; and for permitting the rapid, efficient, and realistic screening of permeation (Helms et al. 2016; Nielsen et al. 2011; Mensch et al. 2009).

No model has been able to completely replicate the physiological architecture and functionality of the BBB. However, validation markers of the in-vitro models are important parameters which can be useful tools in decision. The most significant parameter is functional tightness of the cellular architecture. This can be estimated by measurement of transendothelial electrical resistance (TEER) and apparent permeability of markers like lucifer yellow (444 Da), sodium fluorescein (376 Da), sucrose (342 Da), and mannitol (182 Da). The tight junction proteins like occludin, claudin, and ZO-I expression also help to estimate the models' utility. Transporters like ABC (ATP binding cassette transporter) family and SLC (solute carrier transporter) are implicated in transport of both endogenous and exogenous substances. The efflux ratio of different substances may be taken as the functional marker of transporters. The efflux ratio reflects the permeability in both directions, apical to basal and basal to apical. Efflux ratio of greater than 1.5–2.0 or less than 0.5 indicates drug efflux or uptake. Further, mRNA quantification of these transporters (P-gp, BCRP, Mrp, Glut–I, LAT–I, MCT–I) can also help to ascertain their expression levels in the model. Of the various receptors, transferrin is believed to be highly expressed in brain endothelial cells.

Hence, its expression may also be treated as a validation marker (Gumbleton and Audus 2001; Wilhelm, Fazakas, and Krizbai 2011).

8.3.1 Isolated Brain Capillaries

Isolated brain capillaries have been also used in assessment of permeation. Mechanical homogenization and sucrose gradient centrifugation helped to isolate intact capillaries devoid of contamination. The brain capillaries mimic the in-vivo conditions closely. Some of the studies attempted to optimize the method to yield valuable inputs on transporter activities or CNS toxicity. However, reaching the luminal side of the capillaries is difficult, and the model necessitates freshly isolated brain microvessels, which have very short useful life of about 6 hours. Hence, this is not very successful as in-vitro model of BBB transport in the pharmaceutical industry (Pardridge 1998; Cecchelli et al. 1999).

8.3.2 Static Models

These models do not mimic the shear stress produced by blood flow in-vivo. They may be further classified as monocultures and cocultures.

8.3.2.1 Monoculture Models

The use of cell culture techniques as a measure of BBB functionality can be traced to the early 1970s. Isolated brain capillaries were subjected to treatment with enzymes to remove basement membrane and pericytes. The monocultures of brain microvessel endothelial cells were grown on microporous membranes. Isolated BMVECs can be subcultured for up to 8 passages and stored. The monoculture model can be developed from various species including, rat, porcine, bovine, monkey, and human cerebral endothelial cells. A noteworthy advantage of rodent models is the availability of these experimental animals and the possibility to use transgenic animals in the case of mouse models. Rodent models are the best characterized ones, and antibodies and cloned genes for these models are also easily available which favors their use. However, their small size limits the amount of endothelial cells that can be obtained from them. Further, these models exhibit low TEER values because of the lack of stimulating factors derived from other cells like astrocytes and pericytes. As they are simple and less costly, they are apt for rapid screening. Hence, it may be used in studying permeability of potential drug candidates in early stages. Human brain primary cultures would be the ideal model for drug delivery studies. Such cultures have also been developed, but they are limited by the availability of human brain on regular basis. They have been useful for studying low-density lipoprotein transport and receptor-mediated endocytosis (Wilhelm and Krizbai 2014; Mensch et al. 2009; Aparicio-Blanco, Martín-Sabroso, and Torres-Suárez 2016).

8.3.2.2 Coculture Models

Astrocytes are not in direct contact with the endothelial cells, but are separated by the basal membrane. Still, they are known to exert a considerable influence on the evolvement of tight junction integrity and the regulation of transporter expression. A multitude of experimental setups can be designed to coculture astrocytes with endothelial cells. Astrocytes and brain endothelial cells can be cocultured on two different sides of a porous filter insert, astrocytes may be cultured at the bottom of the transwell or a medium conditioned with soluble growth factors from astrocytes may be used for culturing (Wilhelm and Krizbai 2014; Helms et al. 2016; Gaillard et al. 2001; Abbott et al. 2012; Bicker et al. 2014).

Rodent endothelial cell coculture with glial cells have been most extensively explored. They have been used for investigation on BBB pathologies also. A stable BBB model could be developed by using bovine brain endothelial cell cocultures along with rat glial cells. The glial cells are known to induce changes in the culture to resemble the BBB, by increasing the junctional tightness. The configurations explored include monocultures as wells as cocultures with rat astrocytes. The model exhibited marked tightness of the junctions with a TEER of 600 Ω. Modification of this method allowed for chronic testing as well, because the models could be maintained for 2 weeks. The induction of brain endothelial cells by glial cells can be replicated by the use of a "BBB–inducing medium" which is a coculture-conditioned medium. This allows the researchers the convenience of ready-to-use models (Wilhelm and Krizbai 2014; Abbott et al. 2012; Ramsohoye and Fritz 1998; Deracinois et al. 2013).

Bovine BBB models have exhibited substantial variations in transporter activity and junctional tightness. The assurance of reproducibility in these models is less. They have been mostly used for studying receptor-mediated transcytosis, paracellular permeability, and ABC-mediated efflux (Shawahna, Declèves, and Scherrmann 2013; Summerfield and Dong 2013; Franke, Galla, and Beuckmann 2000; Culot et al. 2008; Helms and Brodin 2014). Porcine brain endothelial cells were initially isolated by Mischeck et al. (1989). Quantitative proteomics have revealed that transporter expression in porcine models were closer to human in comparison to rodents. Porcine models (monoculture and coculture with rat astrocytes) have been useful tool to study macromolecule transport through the BBB, mainly the receptor-mediated transport (Mischeck, Meyer, and Galla 1989; Patabendige, Skinner, and Abbott 2013; Patabendige et al. 2013; Zhang et al. 2006).

Pericytes are the closest to endothelial cells in-vivo, but their influence is less characterized. Apart from their contractile, immune, phagocytic, and angiogenic functions, they regulate paracellular permeability also. The pericytes are better than astrocytes in inducing junctional tightness. They are also involved in induction of ABC transporters. They increase the MRP-6 expression in endothelial cells (Nakagawa et al. 2007; Berezowski et al. 2004). This recognition of the activity of pericytes drove the generation of triple culture models. Endothelial cells are cultured on the top of the porous support, while astrocytes and pericytes may be localized at the bottom of the transwell and on the other side of the porous support. Reverse configuration has also been explored where the pericytes and astrocytes are on the top. Triple cultures have exhibited good in-vivo co-relationships. These models also displayed age related variations in the barrier properties. Advancements in the model led to the generation of a triple culture model where endothelial cell were of primate origin and combined with rat astrocytes and pericytes. Further, neuronal cells have also been introduced in these systems. A coculture model of RBE4.B cells and cortical neurons was developed by Cestelli et al. (2001). They demonstrated that both mature and embryonic neural progenitor cells could bring about the induction of occludin expression. Coculture models, in general, face one major issue of obtaining high quality human neurovascular unit cells like astrocytes, neurons, and pericytes (Nakagawa et al. 2009; Cestelli et al. 2001; Shawahna, Declèves, and Scherrmann 2013).

8.3.2.3 Cell Lines

RBE4 is a well-characterized rat brain endothelial cell line obtained by the transfection of rat brain microvessel endothelial cells with a plasmid containing the E1A adenovirus gene. They exhibit many BBB characteristics, like high alkaline phosphatase and gamma-glutamyl transpeptidase activity, expression, and regulation of P-glycoprotein (Wilhelm, Fazakas, and Krizbai 2011; Roux et al. 1994). Another well-characterized rat brain endothelial cell line is GP8, generated by the immortalization of rat cerebral endothelial cells using SV40 large T antigen. GP8 cell line has been used with reasonable success in various signaling studies and the regulation of P-glycoprotein activity (Wilhelm, Fazakas, and Krizbai 2011; Omidi et al. 2003).

Mouse cerebral endothelial cell lines like bEND.3 and MBEC4 have been used as a model for studying permeability. They exhibit inferior barrier properties but tight junction proteins; claudin-5, occludin, and ZO-1 were expressed. Considerable efforts have been invested in development of immortalized endothelial cell lines. The murine cerebrovascular endothelial cell line (cEND) was developed. The cEND cell line has TEER varying from 300 to 800 Ω cm^2 and expression of occludin and claudin-5 in the tight junctions. Glucocorticoids have a strong inducing effect on cytoskeletal rearrangements, regulating tight junction proteins occludin and claudin-5, and TEER increases to 1000 Ω cm^2 (Wilhelm, Fazakas, and Krizbai 2011; Helms et al. 2016; Förster et al. 2006; Poller et al. 2008).

Human immortalized brain endothelial cell lines have been developed, and the hCMEC/D3 cell line has been maximally investigated in particular. It is derived from human temporal lobe microvessels immortalized by hTERT/SV40. Although, hCMEC/D3 cell monolayers express the characteristic tight junction proteins of the BBB and show a lower expression level of claudin-5. The culture protocols define the precise barrier properties of the model. The hCMEC/D3 cell line is an easy-to-use, well-established model of human origin, which can be used for the characterization of drug uptake, revealing the response of the brain endothelium to human pathogens and neuroinflammatory stimuli. However, low junctional tightness under routine culture conditions poses a challenge for it to become a successful in-vivo permeation prediction model (Poller et al. 2008; Helms et al. 2016). Some other immortalized human brain endothelial cell lines include BB 19 and hBMEC, which have been established and explored in permeability prediction. hBMEC was generated by transfection of early passage endothelial cells with a pBR322-based plasmid containing 40 large simian virus T antigens (Eigenmann et al. 2013). However, none of them comes very close to the actual in-vivo conditions. Cell lines are capable of retaining the characteristics of BBB phenotypes, but do not form tight monolayers like primary cultures. This leads to leakiness and low junctional tightness reflected by the low TEER values. Further, human origin cells are sensitive and fragile, and require careful control of culturing conditions (Shawahna, Declèves, and Scherrmann 2013; Bicker et al. 2014).

8.3.2.4 Human Stem Cell–Derived Models

Advances in the field have led to the development of cell lines from human stem cells. Human pluripotent stem cells and human cord blood-derived stem cells have been harnessed to yield brain endothelial cell lines. These present a possibility of BBB cell line resources which are renewable and scalable as well. Human pluripotent stem cells are initially cultured in unconditioned media to codifferentiate them into endothelial cells and neural progenitor cells. They mimic the embryonic brain compartment and are then subcultured to maintain them as pure monolayers on collagen/fibronectin-coated transwell filters or plates. They exhibit expression of claudin-5, occludin, and ZO-1 and junctional tightness. The cultured monolayers produce baseline TEER values of 250 Ω cm^2 but cocultures with rat astrocytes can improve the TEER to 1450 Ω cm^2.

Human stem cell models have been established on cord blood-derived stem cells also. The astrocytes or pericytes are cocultured to introduce differentiation. Both the cocultures gave endothelial phenotype to the model. While pericytes were found to be more effective in inducing junctional tightness when compared to the astrocytes. It was also reported that such cultures may be regulated by growing them in stimulating environs like in presence of Wnt3a or Wnt7a. In conclusion, stem cell–derived BBB models are useful and potential tools for both mechanistic studies of human brain endothelial cell biology and as well as screening tool for CNS-drug permeability studies. However, these models are yet to be extensively characterized (Eigenmann et al. 2013; Lippmann et al. 2012; Cecchelli et al. 2014).

8.3.3 Dynamic Models

Systemic blood flow induces shear stress of the order 5 dyn/cm^2 which has been shown to markedly affect the endothelial properties including barrier functions. There are three major types of dynamic BBB models: the cone and plate model, the dynamic in-vitro BBB model, and microfluidic in-vitro BBB models. The brain endothelial cells' monolayer is seeded on the bottom of the plate and the shear stress is generated by a rotating cone. This system *per se* is not designed to mimic the physiology of BBB but study the effects of laminar or turbulent fluid on cells and the contribution of fluid viscosity, time of exposure, and other physiological variables. It was the first attempt at development of dynamic model of BBB (He et al. 2014; Bussolari, Dewey Jr, and Gimbrone Jr 1982).

Dynamic in-vitro models try to recreate the in-vivo blood flow simulating conditions by culturing endothelial cells in hollow polypropylene fibers to mimic microvessels and circulating the culture media through it. Coculture configurations can also be applied in this model, where two types of the cells are grown on the opposite sides of the support. The human immortalized cell line hCMEC/D3 has been utilized in this way also. In one of the models, immortalized porcine brain endothelial cells PBMEC/C1-2 were seeded onto the inner surface of the hollow fibers (polypropylene capillaries coated with ProNectin F) and C6 glioma cells were seeded in the extracapillary space. Another advancement of the humanized dynamic in-vitro BBB model enabled the study of transendothelial trafficking of immune cells when hollow fibers with pores of 2–4 μm were used. These have been used to study the pathophysiology of various CNS diseases, including ischemia-reperfusion-induced injury and epilepsy. Ischemia-induced injury is accompanied by flow cessation and inflammation, which is replicated in the model by stopping the flow of culture media and reperfusion with media enriched with white blood cells. However, it suffers from the disadvantage that direct visualization of the endothelial morphology in the luminal compartment is not possible. Although the TEER values closely resemble the in-vivo conditions, they take a long time to reach steady state. This model is not suitable for rapid screening and requires technical expertise (Naik and Cucullo 2012; Neuhaus et al. 2006; Wilhelm and Krizbai 2014).

The microfluidic model is another newly introduced configuration used for studying BBB permeation. Neurovascular cells are grown on porous membranes; the culture media flows through two microchannels made up of PDMS, which are placed at the interface with the cell carrying porous support. hCMEC/D3 cells are cultured on a microfluidic device. bEND3 brain endothelial cells and C8-D1A astrocytes have been cocultured in this configuration. Even RBE4 endothelial cells are cultured in an astrocyte conditioned medium in microfluidic device. Another latest advancement is the brain-on-a-chip model, wherein the three-dimensional (3D) physiological architecture of BBB would be developed. It would comprise of two microfabricated compartments representing CSF and brain. The brain compartment will contain hollow fibers, wherein endothelial cells along with astrocytes and pericytes would be cultured. These all models are technically difficult to use and have yet to be standardized (Wilhelm and Krizbai 2014; Naik and Cucullo 2012; Griep et al. 2013; Yeon et al. 2012; Santaguida et al. 2006).

8.3.4 Epithelial Cell Lines

The culturing brain endothelial cell is a labor-intensive and expensive affair and often with poor in-vivo relevance. Attempts have been made to use modify epithelial cell line to simulate blood–brain barrier properties. Though, they are not actual BBB model, but they may be valuable tool for initial phases of drug development cycle. The Madin–Darby canine kidney (MDCK) cell line is the most widely used cell line model. The combination of MDCKII and MDCK-ABCB1 cell monolayers and mathematical modeling has helped to predict the in-vivo permeability characteristics. Porcine kidney epithelial cells LLCPK1 stably transfected with Abcb1a-enabled efflux ratio measurements for neurotherapeutics. Vinblastine-resistant Caco-2 and the MDCK-ABCB1 models helped to identify the efflux transporter substrates but did not closely simulate BBB. Epithelial cells differ fundamentally from endothelial cells in terms of the morphology of their intercellular junctions and the expression of profiles of transporters and enzymes (Wilhelm and Krizbai 2014; Hellinger et al. 2012; Mangas-Sanjuan et al. 2013).

8.3.5 Parallel Artificial Membrane Permeability Assay

The parallel artificial membrane permeability assay (PAMPA) method was developed as a model for gastrointestinal absorption. It is based on permeability of molecules across nonbiological artificial membrane made up of lipids. The transwell configuration is adopted wherein the artificial membrane made up of lipids separates the two compartments. The change in drug content in the two compartments serves to indicate the permeability of the molecule. The method may be adapted for studying BBB permeation by using porcine brain lipids. A classification range is established to differentiate the compounds as high (CNS+) and low (CNS–) BBB permeability potential. The model does not exactly mimic the physiology of the BBB, and does not account for the transport mechanisms operational there as enzymes and transporters which are absent in the PAMPA-BBB model. Different configurations with variable incubation time, lipid composition, stirring, etc. are used to assess the permeation characteristics. However, it gives a reliable estimate of passive transcellular diffusion (Aparicio-Blanco, Martín-Sabroso, and Torres-Suárez 2016; Mensch et al. 2010; Jhala, Chettiar, and Singh 2012).

8.3.6 Immobilized Artificial Membranes

An immobilized artificial membrane is a solid phase model of the biological membrane. The artificial membrane is used as a chromatographic interface, wherein the phosphatidylcholine residues are covalently bound to silica propylamine. This design will resemble a lipid bilayer. This model has also been used for evaluating BBB permeation, but is relevant only for molecules permeating by passive transport mechanisms. It reflects basically partitioning of drugs into membrane and bears poor corelation to brain uptake, as an artificial membrane is devoid of proteins and enzymes and it presents only a lipid monolayer not a bilayer (Stewart and Chan 1998).

8.3.7 Optimization of Cell-Based Models

The cell-based models are sensitive and culture protocols with control over the conditions will help to regulate their characteristics. Generally, the porous base used for culturing the cells is made up of polycarbonate or polyethylene filter inserts with pore size ranging from 0.4 to 8.0 μm. These supports are coated with type 4 collagen and fibronectin before seeding the cells for culturing. The culture media is very crucial in maintaining viability and helps in conditioning of the culture cells. The media is generally enriched with permeable cyclic adenosine monophosphate (cAMP) analogs and substances which inhibit phosphodiesterases. Hydrocortisone treatment is employed to improve barrier properties. Activation of Wnt (wingless type family of proteins) signaling and higher buffering activity of the media also aids in increasing junctional tightness. It is now agreed that fetal bovine serum in a concentration between 1% and 20% should be present in the media. The species choices vary according to the objective of the experiment. Porcine models are closest to humans and specific diseases may be studied in transgenic rodent models (Wilhelm and Krizbai 2014; Hoheisel et al. 1998; Paolinelli et al. 2013).

8.3.8 Neurotoxicity Assessment

Neurotoxicity events are the adverse effects on the chemistry, anatomy, and physiology of the nervous system (during development or at maturity) caused by chemical, biological, or physical means. Some of the drugs, metals, industrial chemicals, etc. are known neurotoxic agents. They may cause neuropathies or affect neurotransmission. The neurotoxicity assessment required by the Environmental Protection Agency and Organisation for Economic Co-operation and Development guidelines focus on behavioral and histopathology evaluations of the nervous system, which are expensive, time consuming, do not correlate well with human neurotoxicity, and are not apt for high throughput screening. In-vitro models have been generated to seek information on cell biology and neuronal functioning. A range of in-vitro

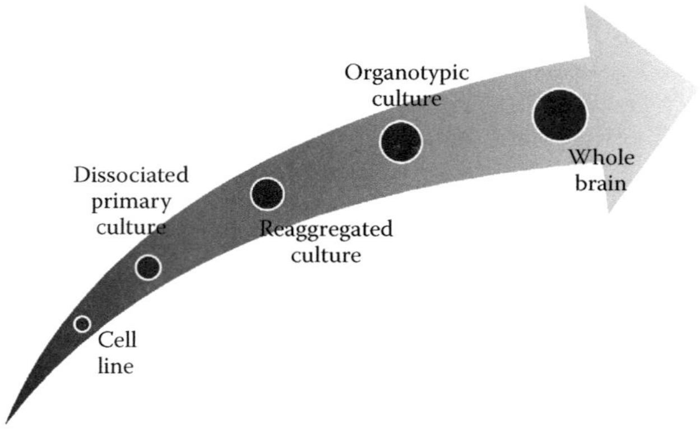

FIGURE 8.2
In-vivo–like physiological architecture and predictability to humans.

systems are available with ascending biological complexity to complement the in-vivo assessment as shown in Figure 8.2. Some of the key parameters guiding the success of such evaluations are maintaining the heterogeneity of nervous system in-vitro, lack of cellular hemostatic mechanisms, interspecies differences, exposure time, age of the neuronal units in model, etc. However, the major drawback of in-vitro systems is the inability to provide behavioral end points (Barbosa et al. 2015; Radio and Mundy 2008).

Neurotoxicity can be induced in the cultures by treating them with glutamate or other excitotoxic chemicals. Oxidative stress may be induced by hydrogen peroxide or UV irradiation. Culturing in serum deprived the media and indicated the presence of neurotoxic agents rotenone, 1-methyl-4-phenyl-1,2,3,6-tetrahydropyridine (MPTP), and 3-nitropropionic acid, which can help create a hostile milieu for the cultures. The most common method to assess viability of cells after the treatment of stress inducers is an MTT assay. The principle of these studies is based on the conversion of yellow colored MTT to blue formazan by active mitochondria. LDH assay is another biochemical approach for evaluating cell viability. The morphological assessment of the cells can also indicate the cell state. Dyes like DAPI, and Hoechst, may be used for staining the cells. Hoechst and propidium iodide may be used in conjugation also for costaining the cells. Apoptosis in the cells may be detected by immunocytochemical staining of the cleaved/activated caspase-3 and TUNEL staining. Reactive oxygen species and nitric oxide in the cultures may also be quantitated (Harry et al. 1998; Atterwill et al. 1994; Barbosa et al. 2015).

8.3.8.1 Subcellular Systems

Subcellular systems have been useful in study of signaling pathways and receptor-mediated signal transduction studies. Isolated mitochondria are separated from the whole brain and they may be subclassified as synaptosomal fractions, heavy and light mitochondria, and free mitochondria from nonsynaptic origin. They have been used extensively for study of the interference of several neurotoxicants, including drugs, on mitochondrial bioenergetics. A synaptosome model is another popular subcellular model derived from neurons by brain tissue homogenization and functions as small anucleated cells that retain neuronal vesicles, enzymes, and active ion transport systems. This simple preparation is a tool to study metabolic pathways, ion movements, neurotransmitters' synthesis, storage and release, and insults to mitochondria, as well as a way to replicate the mitochondrial deficits found in neurodegenerative diseases (Pereira et al. 2009; Barbosa et al. 2015).

8.3.8.2 Cellular Systems

A number of cell lines are at the disposal of the researcher for the evaluating the different aspects of neuronal functioning and toxicology. A diverse array of immortalized cell lines have been examined and employed for neurotoxicological studies. A SH-SY5Y neuroblastoma cell line was generated from the bone marrow of a neuroblastoma patient. They may be differentiated using specific treatment protocols and can be used extensively to establish the neurotoxicity of classical neurotoxicants, like MPP+, 6-hydroxydopamine, or organic pollutants. Further disruption in intracellular Ca^{2+} levels, mitochondrial dysfunction, and oxidative stress mechanisms can be studied in the cell line (Barbosa et al. 2015). Neuro-2a is a mouse-derived neuroblastoma cell line, ND7/23 is a mouse neurobalstoma and rat neuronal hybrid cell line, which is also used for screening (Barbosa et al. 2015). The rat-derived pheochromocytoma cell line PC12 has been extensively used in neurotoxicological research. Pheochromocytoma-derived cell lines present high exocytotic activity, and hence become a very suitable model for neurosecretory studies, as well as drugs that modulate the process of neurotransmitters' release (Barbosa et al. 2015, 58). Human U87-MG cell line and the rat C6 cell line are glioma cell lines and have proved useful in toxicity studies and study of basic cellular mechanisms. C6 glioma cells present cancer stem cell–like characteristics and provide inputs on regulation and modulation of myelin specific genes which is implicated in lead toxicity (Barbosa et al. 2015, 59).

The dissociated primary cultures are prepared from suspensions of individual cells, and are obtained by deaggregation of the brain tissue. The serum and trophic factors' presence, oxygen content, composition of substratum, and seeding density influence the viability and differentiation of the cells. Under a proper culture environment, the cells acquire properties of mature neurons like distinct axons and dendrites, synapses, receptors, and ion channels. Their main advantage lies in the accessibility to individual living cells. They are less suited for traditional biochemical approaches. Primary neuronal cultures may be sourced from different regions of the brain (the hippocampus, cortex, striatum, or cerebellum) or from the peripheral nervous system. Different species like rats, mice, or the fetal brains from humans and chickens have been used for establishing the cultures. The prime disadvantage of the system is the difficulty in obtaining pure cultures devoid of other cellular contaminants. The neurotoxicity studies in these cultures have assessed endpoints like cell viability, interference with neurite outgrowth, and impaired mitochondrial function. Advancements in the field have helped develop sandwich cocultures of neurons and glials cells to permit study of the influence of chemicals on their interdependence (Barbosa et al. 2015; Silva et al. 2006; Harry et al. 1998).

Most of the in-vitro models are two-dimensional (2D) models and hence, to replicate in-vivo conditions, 3D models were developed. Neural stem cells, neural progenitor cells, immature postmitotic neurons, and glial cells from the fetal origin are grown in chemically enriched media. Under controlled conditions, they form neurospheres, which are used as neural tissue models as they resembles the physiological architecture. They may be established in concave microwells or by seeding the cells on biomatrices so as to mimic an extracellular matrix. The system has the potential for screening teratogenicity and cell-specific toxicity. Their major advantages are high yield and reproducibility of the cultures and the opportunity to perform multiparametric endpoint analyses (Barbosa et al. 2015; Fritsche, Gassmann, and Schreiber 2011; Harry et al. 1998).

Organs/explant cultures and brain tissue slice cultures from early postnatal organs offers a method with high degree of cellular maturation and differentiation. It allows the study of a gamut of heterogenous cells in a similar way, as they are found in-vivo. These organotypic cultures permit the local distribution signaling molecules in a fashion similar to in-vivo conditions and not allowing free diffusion and dilution in the culture medium. They are important assets for studying nervous system development processes. However, the preparation techniques may lead to massive denervation and cell death, which may limit its resemblance to in-vivo conditions (Barbosa et al. 2015; Becker and Liu 2006; Pena 2010).

Differentiated human embryonic stem cells derived from embryos at the blastocyst stage of development are unique developmental models for understanding drug-induced adverse reactions. Along with human-induced pluripotent stem cells, they are used for screening studies. The technique is comparatively complex and requires technical skills (Barbosa et al. 2015; Wilson, Graham, and Ball 2014; Hou et al. 2013).

8.3.9 In Silico Studies

Methods based on computer simulation (computer-assisted prediction), also known as "in silico" modeling, have become very powerful tools which have become possible due to advancements in computer technology. The major attraction for these studies in medicine are their predictability of efficacy and the bioavailability of potential candidates. Using this technology, compounds are synthesized, prescreened, and virtually tested to predict their penetration across the BBB into the brain. This allows for the rapid and inexpensive screening of lead molecules and limits the need for labor-intensive laboratory experiments and expensive clinical trials. At this stage, the in silico models available for BBB permeation prediction are still in their infancy, however, they do have all the possibility of becoming a reliable surrogate. In silico models need to be supported by cell-based in-vitro techniques and in-vivo experimentation. Physicochemical parameters (e.g., solubility, lipophilicity, molecular size, hydrogen-bonding capacity, and charge) are used as molecular descriptors in generating the model. In silico models are available for prediction of multiple parameters like PS, logBB, $f_u$brain, and P-gp efflux (Fan et al. 2010; Di, Kerns, and Carter 2008; Chen et al. 2012; Shityakov et al. 2017; Chen et al. 2011).

The large virtual libraries of chemicals are screened using several models that have been developed based on the molecular descriptors or the fingerprints. The models have been designed based on certain general rules. The rule of thumb that has been evaluated and agreed upon by a number of researchers is that if the sum of nitrogen and oxygen (N + O) atoms in a molecule is five or less, the log P – (N + O) > 0, polar surface area (PSA) is below 90 Å^2, and the molecular weight is below 450, the compound is expected to show BBB permeability. High-throughput methods may be based on molecular docking or descriptor-based strategies. Molecular descriptor-based strategies have given less false negative outcomes, but more false positive outcomes, as these models don't account for the effect of transporters. The molecular docking technique determines the P-glycoprotein substrates or inhibitors dealing with the phenomenon of active multidrug efflux by P-gp in the brain (Mensch et al. 2009; Norinder and Haeberlein 2002; Kelder et al. 1999).

Apart from this, several quantitative structure–activity relationship (QSAR) regression models based on BBB partitioning values, such as log BB, are derived from experimental data sets for various drug-like molecules. The logBB parameter is defined as the logarithm value of a steady-state brain-to-blood (plasma) concentration ratio for a drug of interest according to following equation:

$$\text{logBB} = \log(c_{\text{brain}} / c_{\text{blood}}) \tag{8.1}$$

A large number of models are based on prediction of logBB value, which may also be predicted from the molecular descriptors. It is said that PSA, log P value, and H Bonding capacity influences the logBB value (Abbott 2004). Clark and Rishton equations (Clark 1999; Rishton et al. 2006) are used for predicting the log BB value:

$$\text{logBB} = 0.152\text{logP} - 0.0148\text{PSA} + 0.139 \tag{8.2}$$

TABLE 8.3

Major in Silico Models for BBB Permeation

Model by Researchers	Molecular Descriptors	Inference for Improved BBB Permeation
Young et al. 1988 Österberg and Norinder 2000 Feher, Sourial, and Schmidt 2000	Lipophillicity and hydrogen bonding	Increasing lipophillicity and decreasing hydrogen bonding
Subramanian and Kitchen 2003	Lipophillicity, hydrogen bonding, and electrotopological parameters	
Lombardo, Blake, and Curatolo 1996 Keserü and Molnár 2001	Solvation free energy in water	Lipophillic compounds and solvation energy less than 40 KJ/mol
Calder and Ganellin 1994	Size and hydrogen bonding	Increased size, hydrogen bonding, and hydrophillicity
Kaliszan and Markuszewski 1996	Size and lipophillicity	
Kelder et al. 1999	PSA	Low PSA (less than 60–70 Å^2), high lipophilicity, small size
Clark 1999	PSA and lipophillicity	
Fu et al. 2001	PSA and molecular volume	
Norinder, Sjöberg, and Österberg 1998	Polarity, Lewis base strength, and hydrogen bonding	Polarizable electrons, lipophilicity, low hydrogen bonding
Hou and Xu 2003	Lipophilicity, size, and hydrogen bonding	MW less than 360, high lipophilicity and reduced H bonding
Fu et al. 2004	Molecular volume, hydrogen bond donors, and acceptors	Molecular volume decreases logBB but increases lipophilicity. Balance required.
Hutter 2003	Molecular size, polarizability, hydrogen bonding, and other quantum chemically-derived parameters	PSA, ionization potential, and hydrogen bonding determine permeation
Fan et al. 2010	2D molecular descriptors	Polar surface area, number of aromatic rings, number of amines have a major influence on BBB permeation.
Lanevskij et al. 2011	2D molecular descriptors	Corrected logBB, log P and Pka determines permeation
Muehlbacher et al. 2011	2D molecular descriptors	Topological polar surface area, Sum of all partial charges on nitrogen atoms. (QSUMN), QSUMN/Weight, I3 descriptors give reliable prediction.

Note: PSA: Polar surface area.

$$\log BB = 0.155\log P - 0.01PSA + 0.164 \tag{8.3}$$

The logBB value is not able to guide the BBB permeation prediction because it is highly affected by the drug binding to plasma and brain tissue proteins and, thus, it does not reflect the blood–brain barrier permeability process, instead it provides a partition between these tissues. This is quantified from total brain tissue concentrations, which may not reflect a correlation with the unbound or free drug concentration in brain which is in the pharmacologically active concentrations in the brain. Further, the datasets often violate the conditions of steady state for logBB determination. Hence, more and more evidence discourage its use as a reflection of BBB permeation (Martin 2004; Pardridge 2004).

Some researchers advocate the use of BBB-PS (permeability surface area) as a better indicator of BBB permeability, as it predicts the level of free drug in the brain. This is because the level of free drug is determined by the total drug concentration in plasma, the PS product, and the fraction of drug in plasma that is available for transport into the brain. The models are essentially based on lipophillicity along with other physicochemical properties. However, a limited number of data sets are available for model generation. The lack of in-vivo blood–brain barrier permeability data (e.g., permeability–surface area product determined with in situ brain perfusion technique) is responsible for only few reports using the parameter (Nicolazzo, Charman, and Charman 2006; Klon 2009; Pardridge 2004).

Further, any model that is said to have widespread acceptability has to satisfy structure–activity relationships (SAR) of all BBB transporter interactions with drug molecules. The SAR requirements for substrate specificity for different transporters vary significantly (Goodwin and Clark 2005; Shityakov, Salvador, and Förster 2013). The major work in this area, along with its salient features, has been highlighted in Table 8.3.

The industry is demanding customizable and fully scalable models which will mimic the in-vivo BBB as closely as possible. Although in silico models are in their developmental stage, they hold immense potential in future breakthroughs. Current scenario is definitely encouraging however, as transporter effects needs to be accounted reliably in the forthcoming models.

8.4 In-Vivo Evaluation of Brain-Targeting Methodologies

8.4.1 Pharmacokinetic Studies

Various in silico, in-vitro, and in-vivo methods are utilized to assess the drug transportation across the blood–brain barrier. Unfortunately, there are very few in-vivo studies in the literature for measuring reliable brain uptake and there is not any high-throughput in-vitro or in silico model available in drug discovery. Moreover, the construction of in silico models are done by in-vivo model data (Liu, Chen, and Smith 2008; Cory Kalvass and Maurer 2002). Several methods have been developed to test brain uptake of drugs. Nanoparticles targeting the brain are used for the same. Since nanoparticles differ in size, composition, and uptake mechanisms, they lead to developments of brain uptake approaches which may be classified into the following noninvasive and invasive techniques:

Invasive Techniques

1. Brain/plasma ratio (logBB, Kp)
2. Brain uptake index (BUI)
3. Permeability surface area product (PS) and permeability coefficient (PC) through in situ brain perfusion
4. Brain efflux index (BEX)
5. IV injection PS technique
6. Intracerebral microdialysis
7. Quantitative autoradiography (QAR)

Noninvasive Techniques

1. Positron Emission Tomography (PET)
2. Magnetic Resonance Imaging (MRI)
3. Single-photon emission computed tomography (SPECT)

Mensch et al. described a schematic overview of experimental setup for different invasive in-vivo brain uptake methods (Abbott 2004; Mensch et al. 2009).

8.4.1.1 Invasive Techniques

1. Brain/plasma ratio (logBB, Kp)

   The ratio of concentrations in brain and blood at equilibrium distribution is termed as logBB; brain/plasma ratio(K_p) which is the logarithm of the ratio of the steady-state concentration of a drug in the brain to that in the blood [$logBB=log(C_{brain}/C_{plasma})$]. The binding affinity differences of drug between the plasma proteins and brain tissue, passive diffusion characteristics, and uptake and efflux transporters at the BBB junction and metabolism, are the few significant parameters of log BB (Mensch et al. 2009). There are lots of variations in dose size, route of administration, and sampling points in order to generate logBB data because of a gap in standard

TABLE 8.4

LogBB Value Inference

LogBB Value	Inference
0.3–0.5	Sufficient access to CNS
Greater than 1	Free to cross BBB
Smaller than 0.1	Unable to enter the CNS

protocol. The brain/plasma ratio is measured at multiple time-points after oral, intravenous (IV), or subcutaneous (SC) administration to reduce the time dependence. The brain/plasma ratio is calculated from the brain and plasma concentrations area under the curve. The following Table 8.4 gives indication for a drug to access CNS (Doran et al. 2004).

There are several disadvantages associated with measurement of logBB. It requires several animals; moreover, it is labor intensive and costly. It does not give any ideas about unbound free drug which are responsible for the pharmacological effect. It only gives the partitioning of drug in lipid layers so CNS activity of drug cannot be predicted from logBB. It suggests the need for the unified approach to improve prediction of the CNS penetration in which permeability, efflux/influx, plasma protein, and tissue binding were considered. Boström et al. clarified further the factors affecting CNS penetration by transformation of the logBB value (Cbrain/Cplasma) into unbound partition ratios with following equations (Boström, Simonsson, and Hammarlund-Udenaes 2006).

$$K_p = C_{brain}/C_{plasma} \text{ (with } C_{brain} = AUC_{brain} \text{ and } C_{plasma} = AUC_{plasma}) \quad (8.4)$$

$$K_{p,u} = C_{brain}/C_{u,plasma} \text{ (with } C_{u;plasma} = \text{the unbound concentration in the plasma)} \quad (8.5)$$

$$K_{p,uu} = C_{u,brain}/C_{u,plasma} \text{ (with } C_{u;brain} = \text{the unbound concentration in the brain interstitial fluid)} \quad (8.6)$$

where $K_{p,\,uu}$ abolishes the effect of plasma protein binding, while $K_{p,uu}$ factors out both the influence of plasma protein binding and the influence of brain tissue binding which is directly linked to the BBB's passive transport for drugs as well as the effects of efflux and influx (Mensch et al. 2009) (Boström, Simonsson, and Hammarlund-Udenaes 2006).

2. Brain uptake index (BUI)

   The brain uptake index is one of the oldest methods to evaluate the brain uptake of drugs. The fast carotid intra-arterial administration of a test compound (3H-labeled) and a reference compound (14C-labeled) was originally described by Oldendorf et al. in a single pass technique (Oldendorf 1970). The reference compound enters the brain easily, which is an internal standard to verify the amount of injected material that goes into the brain. The animal is executed after 15 s followed by single injection of the bolus which passes through the brain. Pardridge et al. (Pardridge 1995) derived an equation to calculate the BUI by the brain concentrations of test and reference compounds and relating it to the plasma concentrations.

$$\text{BUI} = ([\text{H}]\text{dpm}/[^{14}\text{C}]\text{dpm})(\text{brain})/([\text{H}]\text{dpm}/[^{14}\text{C}]\text{dpm})(\text{injected solution}) * 100 \text{ with dpm where dpm} = \text{disintegrations per minute} \tag{8.7}$$

   BUI is a reliable method for labile and rapidly metabolizing compounds, but less exposure to the brain and cerebrovascular permeability cannot be correlated by BUI, as it depends on the brain region, blood flow, and time between injection and execution. So, compounds with high BBB permeability and radiolabel tagging are the most suitable to be measured by the BUI method. Examples of BUI values obtained in-vivo are 1.4% for sucrose and 90% for caffeine (van Rooy et al. 2011).

3. Permeability surface area product (PS) and permeability coefficient (PC) through in situ brain perfusion

   This is also called as an internal carotid artery perfusion technique, which is an advanced version of the BUI method to overcome the disadvantages of the latter. The in situ brain perfusion was first developed by Takasato et al. (Takasato, Rapoport, and Smith 1984), who derived an in situ brain perfusion method to circumvent the metabolism of the test compound, due to lack of systemic exposure of the test compound prior during entering the BBB. The compound is measured within the brain after 15–60 sec following perfusion of the carotid artery. The ligation is made into all the branches of the external and internal carotid artery (Figure 8.3), and the reference and test compounds are perfused retrograde via the external into the internal carotid artery. The animal is executed following infusion, the brain is collected, and reference and test compounds are measured to quantify the BBB-PS product (Mensch et al. 2009).

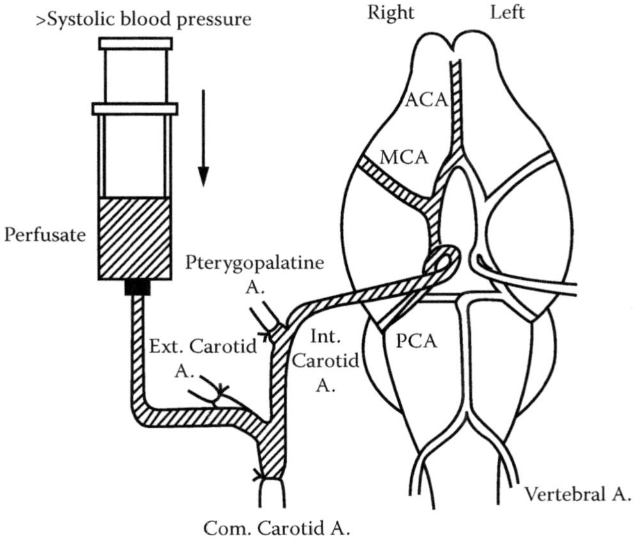

FIGURE 8.3
Schematic representation of in situ brain perfusion.

$$PS = (Vd - Vo)/t \tag{8.8}$$

where,

t = duration of the perfusion period (min),

Vd – Vo = the brain volume of distribution for the test and the reference compound, respectively, and calculated as the ratio of brain/perfusate concentration at time t.

The internal carotid artery perfusion at a rate of approximately 3.5–4.0 mL/min for 0.25–1.00 min is recommended by this technique. It is more sensitive compared to BUI method due to longer perfusion period (15–60 S). It allows analysis of saturable transport systems, hormones, neurotransmitter effects, and plasma protein binding due to control over the perfusate solute concentration. It allows the assessment of BBB entry of compounds in genetically modified organisms (GMOs) such as p-gp knockdown mice. The requirement of a high number of animals and additional labor are the only disadvantages of this method (Misra et al. 2003).

Both the PS product and PC are quantitative measures of the rate of transport obtained by in situ vascular perfusion technique. The three different groups as per following Table 8.5 can be classified by relating the octanol/water partition coefficient (PC) divided by

TABLE 8.5
PS and PC Product Correlation with Transport Mechanism

PS & PC Product Correlation level	Transport Mechanism
Highest correlation	Passive diffusion
Substrates exhibiting a significantly greater PS value than indicated by their lipophilicity	Facilitated transport
Substrates exhibiting a significantly smaller PS value than indicated by their lipophilicity	Active transport

the square root of the molecular weight (PC/Mw1/2) and the BBB permeability coefficient (PS).

4. Brain efflux index (BEX)

   The mechanism of the apparent restricted cerebral distribution of drugs after systemic administration is governed by the efflux transport process across the BBB. The intracerebral microinjection technique has been given which is coined as BEI. It is the relative percentage of drug effluxed from the ipsilateral (that is, they do not cross to the opposite hemisphere) cerebrum to the circulating blood across the BBB compared with the amount of drug injected into the cerebrum.

   $$\text{BEI\%} = \text{amount of drug effluxed at BBB/amount of drug injected into the brain} * 100$$

   It allows determination of the apparent in-vivo drug efflux rate constant across the BBB, monitoring the concentration dependency of the test drug, but only one data point can be obtained for a single intracerebral microinjection. Moreover, the drug concentration in the cerebrum cannot be accurately determined (Misra et al. 2003).

5. IV injection PS technique

   The IV injection PS technique (Figure 8.4A) is used to determine the BBB-PS product of a test compound. The cannulation of a femoral or tail-vein of animals for injection of the compound. Arterial blood is collected either by cannulation of a femoral or other vein or by executing the animals at single or multiple time points (0.25–60.0 min or longer) after injection. Animals must be executed in multiple time point experiments, whereas it is not necessary to execute in a single time point experiment. The aim of single time point study is to capture the unidirectional uptake phase. The single time point technique involves sampling the arterial blood at different intervals to derive a blood or plasma concentration-time curve ($AUC_{plasma(0-t)}$), which is done in this technique after executing the animals and measuring their brain concentrations to calculate the BBB-PS product. Brodie et al. (Brodie, Kurz, and Schanker 1960) derived multiple time

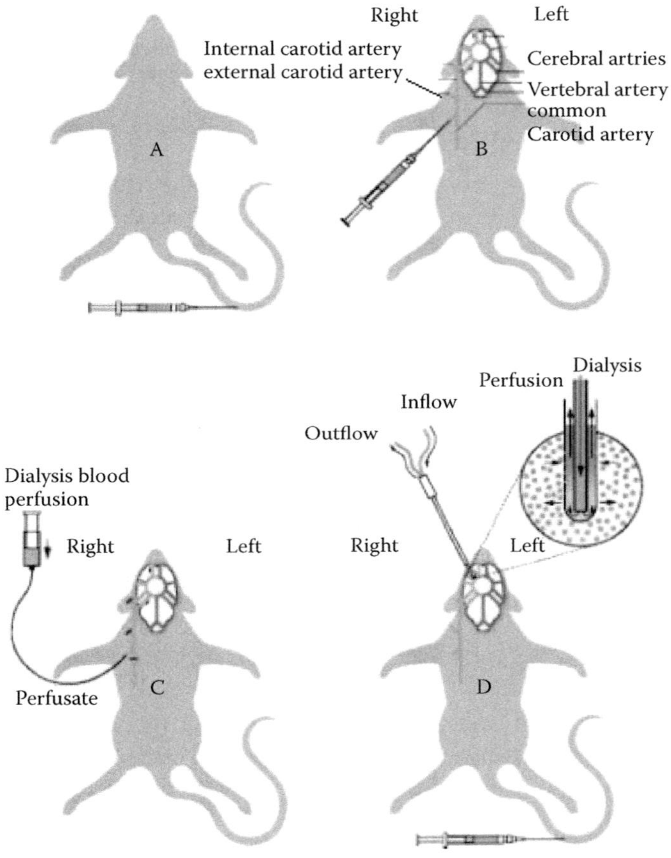

FIGURE 8.4
Schematic overview of the experimental setup of different invasive in-vivo brain uptake methods. (A) The intravenous injection PS technique as well as the logBB method with an intravenous injection of the compound into the tail vein; (B) the setup of the BUI method; (C) the experimental setup of the in situ brain perfusion technique characterized by the ligated branches of the internal and external carotid artery; (D) the intracerebral microdialysis technique with dialysis.

point uptake techniques that provided data about the time course of brain uptake. The BBB uptake can be calculated when the concentration difference between brain and plasma relative to the plasma concentration is plotted versus time and relating the Renkin–Crone equation to the BBB-PS product. There are several advantages with this technique that complete brain and plasma pharmacokinetic profiles, including high sensitivity due to longer exposure to cerebral micro-vessels and lesser complexity. The possibility of back diffusion from brain to plasma resulting in the faulty calculation of the BBB-PS product and the extensive metabolism of test compounds in the peripheral tissue are the few disadvantages associated with this method (Mensch et al. 2009; Misra et al. 2003).

6. Intracerebral microdialysis

   Intracerebral microdialysis is done by the implantation of a microdialysis probe in the brain. The SPM semipermeable membrane containing the probe is perfused constantly with a physiological solution. The oral, IV, or SC route may be selected to administer the drugs. Drugs can cross the semipermeable membrane by diffusion into the physiological buffer after crossing the BBB and entering brain interstitial fluid, and a concentration of the same is measured after buffer sampling from the probe that reflects the brain's free drug amount. The brain and blood level of the drugs can be determined at many time points in one animal, which is very useful to compute pharmacokinetic parameters. The main advantage of this method is the insertion of the probe in any region of the brain, which may be valuable when targeting a compound to a specific area of the brain. Like a brain tumor, however, it requires skilled person due to the possibility of damage to BBB (Mensch et al. 2009; van Rooy et al. 2011).

7. Quantitative autoradiography (QAR)

   The amount of radioactive test compound in different areas of brain like brain tumors and stroke-affected areas following oral, intravenous, or subcutaneous route can be traced by Quantitative autoradiography. The brain is sectioned into 20-μm thick sections after sampling the blood at different time points and uncovered to X-ray film along with radioactive standards. A BBB impermeable marker, such as radiolabeled sucrose, can be used to measure the intravascular volume. This technique gives very high spatial resolution which may be fruitful in the detection of tumor-affected area of the brain (Mensch et al. 2009).

8.4.1.2 Non-Invasive techniques

1. Positron Emission Tomography (PET)

   It is a non-invasive quantitative approach to measure a BBB-PS product in animals and humans (Mensch et al. 2009; Huang et al. 1998). A bolus injection (or infusion) into the body of a positron emitting radionuclide or a test compound labeled with an isotope which emits positrons. The following three main types of PET tracers are available (Table 8.6):

TABLE 8.6
PET Tracers and their use

PET Tracers	Use
Nutrient analogs like 18F-deoxy-glucose and several amino acid analogs (e.g., 11C-methionine, 18F-DOPA)	Metabolism measurement
Ligands	For neurotransmitter receptors or transporters measurement
68Ga-EDTA	BBB damage detection

The subject is placed in a counter that detects the emission of gamma photons emitted by the tracer in the brain as a result of devastation of the positrons subsequently periodic blood sampling allows to generate plasma tracer curve. The kinetic evaluation of brain uptake is carried out by the 2D images of the brain using computerized imaging techniques. The biological processes like glucose metabolism; P-gp transporter activity; blood flow; and distribution of receptors, enzymes, and neurotransmitters directly in-vivo can be easily analyzed, because it is noninvasive, rapid, and highly sensitive. The tracer transport can be visualized in the whole brain over time so compartment modeling in brain kinetic can be analyzed. However, PET is an expensive technique.

2. Magnetic Resonance Imaging (MRI)

   MRIs offer quantitative measurement of BBB permeability which is very useful in strokes, sclerosis, tumors, etc. Patlak et al. developed this technique to quantify the tracer, gadoliniumdiethylenetriamine-pentaacetic acid (Gd-DTPA) and its distribution in brain tissue after an IV bolus injection in a rat ischemic stroke brain model (Patlak, Blasberg, and Fenstermacher 1983).

3. Single-photon emission computed tomography (SPECT)

   SPECT is a nuclear medicine tomographic imaging method using gamma rays. A gamma-emitting radioisotope (such as an isotope of gallium[III]) is injected into the patient's bloodstream. Most of the time, though, a marker radioisotope is linked to a specific ligand to create a radioligand, whose properties bind it to certain types of tissues. This connection permits the combination of ligand and radiopharmaceutical to be carried and bound to a place of interest in the body, where the ligand concentration is seen by a gamma camera (Scuffham et al. 2012).

4. Visualization methods: Fluorescence microscopy and electron microscopy

   Microscopy is the most widely used qualitative method to study BBB uptake of drugs in-vivo. The nanoparticles are either loaded with fluorescent dyes (rhodamine-123, fluorescein, and 6-coumarin) or linked via covalent coupling, then IV injection is given followed by sectioning the brain. An endothelium staining marker, such as lectin, can be used to visualize the brain endothelium. A plasmid expressing a fluorescent protein can be combined into the particle, and visualization protein fluorescence in the brain sections is then utilized to evaluate gene expression in gene delivery experiment. Single particles can be visualized in targeted regions of the brain using electron microscopy. Zensi et al. (2009) (Mensch et al. 2009) showed that electron microscopy can detect human serum albumin nanoparticles in murine brain sections after iv administration.

8.4.2 In-Vivo Biodistribution by Radioisotopes

The compounds are mostly administered intravenously followed by executing the animals at a particular time point or at several time points. The brain, liver, spleen, etc. are taken out according to end accumulation of the particles. The radioactivity can be evaluated by liquid scintillation counting by dissolving the organs for a β-emitting label and organs can be directly calculated in a γ counter in case of a γ-emitting label. Tissue extraction can be performed in case of self-detection of the compound. The amount of drug in the brain and other organs can be analyzed by the HPLC determination of concentration. The drug targeting to the specific diseased-affected cells or area by a targeting ligand can be exactly judged by this technique by relating it with the % of the injected dose (% ID) which is the amount of drug taken up by the brain. Ke et al. (2009) carried out a dendrimer biodistribution study in mice in which they got the results as 0.03% ID in the brain for the unmodified dendrimers and 0.25% ID for the angiopep-2-modified dendrimers.

8.4.3 Pharmacodynamic Approach

Few drugs interfere with brain signaling and exert specific behavioral effects, so pharmacodynamic effects can be assessed utilizing this concept. The monitoring of the behavior of the animal with behavioral tests is carried out to assess the effect of the drug in the brain. There are mainly two type of behavioral tests: namely, the nociceptive test and motor function-learning-memory test (van Rooy et al. 2011).

The ability of an animal (usually a rodent) to detect a noxious stimulus, e.g., the feeling of pain due to stimulation of nociceptors, is analyzed in nociceptive test. It measures the existence of pain through behaviors such as withdrawal, vocalization, immobility, and licking. The antinociceptive effects by intracerebral administration of loperamide is used as an animal model mostly. The tail flick test and hot plate methods are used to measure nociception in rats and mice. The response to painful thermal stimulus is measured in both tests. A small area of the tail is heated by a light beam and latency (time) to flick the tail out of the path of the light beam is assessed in tail flick test. In the hot plate test, latency to lick the forepaws or to lift one of the hind paws is measured by putting animal onto the surface of the hot plate. The % maximum possible effect (% MPE) can be calculated as per following:

$$\begin{aligned}\% \text{ MPE} = &(\text{post drug latency-pre drug latency})/\\&(\text{cut off time-pre drug latency}) * 100\end{aligned} \tag{8.9}$$

The motor function-learning-memory test is carried out in patients with Alzheimer's disease and Parkinson's disease. Nerve growth factor (NGF) has been assessed for the treatment of Alzheimer's disease and Parkinson's disease which possess a low permeability through the BBB after IV administration. Kurakhmaeva et al. (Kurakhmaeva et al. 2009) assessed improvement in symptoms of parkinsonism in MPTP-induced Parkinsonism animals after IV administration of NGF nanoparticulate formulation. It showed a reduction in symptoms, including rigidity, tremor, locomotor activity, and orientation-research reaction. The behavioral skills can also be determined by transgenic mice like the APP/PS1 Alzheimer's disease (Scholtzova et al. 2008). The Y-maze test for improvement in cognitive behavioral skills may be also performed in the transgenic mice.

8.4.4 Capillary Depletion Studies

Capillary depletion studies are used to differentiate endothelial-associated particles and brain parenchyma–transcytosed particles (van Rooy et al. 2011). This happens due to the endothelium-bound nanoparticle present in endothelial cells or transcytosed cells. The dextran density centrifugation is used to deplete the homogenate of its vasculature after the homogenization of the brain. The compounds can be traced in both the parenchyma and the capillary fraction. The particles which are bound to the surface of the capillaries with low affinity may dissociate from the capillaries during centrifugation and reside into the parenchyma fraction. The flushing of brain vasculature after taking out the brain can prevent the above-mentioned action for low affinity-bound particles and particles bound with high affinity to the capillaries, should they remain associated. Robidas et al. (Paris-Robidas et al. 2016) performed a capillary depletion study and demonstrated endocytosis of Ri7-quantum dots by brain capillary endothelial cells and paved a way for brain endothelial cell drug delivery.

8.5 Anticipated Challenges and Future Perspectives

As largely apprehended from the diverse functions the brain carries out, it is of extreme importance that the functioning of brain be maintained, so that it does not get deregulated to critically hamper the homeostasis. This task is prominently carried out by BBB, which acts as a barrier regulating movement of range of molecules, ions, and therapeutics. Understanding the biology of a CNS disorder is quite complex and challenging. Further, the development of novel CNS therapeutics with a higher effectiveness, lower side effects, or that can modulate transport systems or the BBB can prove effective compared to the current treatment regime. A combination of strategies involving the destruction of the BBB or modification could be advantageous for targeting CNS tumors. The current in-vitro models used do not precisely mimic the in-vivo environment, and new precise models needs to be investigated. Endothelial cells cultured in-vitro lose some of the barrier properties during growth, and hence may provide a leaky platform for the therapeutics being tested. Several explorations are to be made in the following aspects of CNS: the initiation of pathogenic changes that occur in the human brain due to the breakdown of BBB lead to a disruption of brain function and structure in various CNS disorders is not clear. The use of advanced technology in determining complete symptoms prior to and during neuronal dysfunction and the breakdown of BBB are required. The molecular and imaging biomarkers employed for determination of CNS dysfunction in various neurodegenerative disorders need to be reliable tools for diagnostic and prognostic purposes.

With the advances made in the human genetics and the transgenic model for inherited disorders, it was investigated that each monogenic inherited disorder can be correlated with non-neuronal cell types of the neurovascular unit, and thus provide details of the link between BBB dysfunction and several neurological disorders. The use of advanced imaging techniques, particularly the brain-on-a-chip technique, has made it feasible to track the transport of biomolecules and gain deeper insight about the pathophysiological states in brain diseases such as Parkinson's disease and Alzheimer's. As far as the delivery aspects of therapeutics are concerned, strategies for their efficiency are being designed and investigated. One such widely investigated strategy is the use of nanocarriers. The cargo in nanocarriers may consist of drugs, genes, or biomolecules. Though nanocarriers have found their way for use in systemic delivery, their use for brain targeting have not seen the much clinical success, and many are at still being experimented at clinical stages. This is attributed to some limited details available for material transport mechanisms; the lack of models for prediction, quality checks, and industrial scalability for the manufacturing method; and storage stability and safety issues. As an alternative to the systemic or oral administration of drugs that increase the exposure to non-target organs, alternatives routes may be used for therapeutic benefits. Thus, in a nut shell, the increase in understanding and investigation of in-vitro and in-vivo approaches for precise prediction, and the correlation of performance of therapeutics, along with novel targeting strategies for accessing BBB, will enable to restore the normal functioning of the neurons and achieve remission of neurovascular diseases.

References

Abbott, N Joan. 2004. "Prediction of blood–brain barrier permeation in drug discovery from in vivo, in vitro and in silico models." *Drug Discovery Today: Technologies* no. 1 (4):407–416.

Abbott, N Joan. 2005. "Dynamics of CNS barriers: evolution, differentiation, and modulation." *Cellular and Molecular Neurobiology* no. 25 (1):5–23.

Abbott, N Joan, Diana EM Dolman, Svetlana Drndarski, and Sarah M Fredriksson. 2012. "An improved in vitro blood–brain barrier model: Rat brain endothelial cells co-cultured with astrocytes." *Astrocytes: Methods and Protocols* 415–430.

Alavijeh, Mohammad S, Mansoor Chishty, M Zeeshan Qaiser, and Alan M Palmer. 2005. "Drug metabolism and pharmacokinetics, the blood-brain barrier, and central nervous system drug discovery." *NeuroRx* no. 2 (4):554–571.

Aparicio-Blanco, Juan, Cristina Martín-Sabroso, and Ana-Isabel Torres-Suárez. 2016. "In vitro screening of nanomedicines through the blood brain barrier: A critical review." *Biomaterials* no. 103:229–255.

Aryal, Muna, Natalia Vykhodtseva, Yong-Zhi Zhang, Juyoung Park, and Nathan McDannold. 2013. "Multiple treatments with liposomal doxorubicin and ultrasound-induced disruption of blood–tumor and blood–brain barriers improve outcomes in a rat glioma model." *Journal of Controlled Release* no. 169 (1):103–111.

Atterwill, Christopher K, Arend Bruinink, Jorgen Drejer, Emilia Duarte, E McFarlane Abdulla, C Meridith, Pierluigi Nicotera, Ciaran Regan, and E Rodrigues-Farre. 1994. "In vitro neurotoxicity testing." *ATLA-NOTTINGHAM-* no. 22:350–350.

Ballabh, Praveen, Alex Braun, and Maiken Nedergaard. 2004. "The blood–brain barrier: An overview: Structure, regulation, and clinical implications." *Neurobiology of Disease* no. 16 (1):1–13.

Barbosa, Daniel José, João Paulo Capela, Maria de Lourdes Bastos, and Félix Carvalho. 2015. "In vitro models for neurotoxicology research." *Toxicology Research* no. 4 (4):801–842.

Becker, Stacey, and Xingrong Liu. 2006. "Evaluation of the utility of brain slice methods to study brain penetration." *Drug Metabolism and Disposition* no. 34 (5):855–861.

Bellavance, Marc-André, Marie Blanchette, and David Fortin. 2008. "Recent advances in blood–brain barrier disruption as a CNS delivery strategy." *AAPS Journal* no. 10 (1):166–177.

Berezowski, Vincent, Christophe Landry, Marie-Pierre Dehouck, Roméo Cecchelli, and Laurence Fenart. 2004. "Contribution of glial cells and pericytes to the mRNA profiles of P-glycoprotein and multidrug resistance-associated proteins in an in vitro model of the blood–brain barrier." *Brain Research* no. 1018 (1):1–9.

Bicker, Joana, Gilberto Alves, Ana Fortuna, and Amílcar Falcão. 2014. "Blood–brain barrier models and their relevance for a successful development of CNS drug delivery systems: A review." *European Journal of Pharmaceutics and Biopharmaceutics* no. 87 (3):409–432.

Bodor, Nicholas, and Peter Buchwald. 2003. "Brain-targeted drug delivery." *American Journal of Drug Delivery* no. 1 (1):13–26.

Borlongan, Cesario V, Dwaine F Emerich, Barry J Hoffer, and Raymond T Bartus. 2002. "Bradykinin receptor agonist facilitates low-dose cyclosporine-A protection against 6-hydroxydopamine neurotoxicity." *Brain Research* no. 956 (2):211–220.

Borlongan, CV, and DF Emerich. 2003. "Facilitation of drug entry into the CNS via transient permeation of blood brain barrier: Laboratory and preliminary clinical evidence from bradykinin receptor agonist, Cereport." *Brain Research Bulletin* no. 60 (3):297–306.

Boström, Emma, Ulrika SH Simonsson, and Margareta Hammarlund-Udenaes. 2006. "In vivo blood-brain barrier transport of oxycodone in the rat: Indications for active influx and implications for pharmacokinetics/pharmacodynamics." *Drug Metabolism and Disposition* no. 34 (9):1624–1631.

Brodie, Bernard B, Hehmann Kurz, and Lewis S Schanker. 1960. "The importance of dissociation constant and lipid-solubility in influencing the passage of drugs into the cerebrospinal fluid." *Journal of Pharmacology and Experimental Therapeutics* no. 130 (1):20–25.

Bussolari, Steven R, C Forbes Dewey Jr, and Michael A Gimbrone Jr. 1982. "Apparatus for subjecting living cells to fluid shear stress." *Review of Scientific Instruments* no. 53 (12):1851–1854.

Butt, Arthur M, and Hazel C Jones. 1992. "Effect of histamine and antagonists on electrical resistance across the blood-brain barrier in rat brain-surface microvessels." *Brain Research* no. 569 (1):100–105.

Calder, JA, and C Robin Ganellin. 1994. "Predicting the brain-penetrating capability of histaminergic compounds." *Drug Design and Discovery* no. 11 (4):259–268.

Cao, Yue, Christina I Tsien, Zhou Shen, Daniel S Tatro, Randall Ten Haken, Marc L Kessler, Thomas L Chenevert, and Theodore S Lawrence. 2005. "Use of magnetic resonance imaging to assess blood-brain/blood-glioma barrier opening during conformal radiotherapy." *Journal of Clinical Oncology* no. 23 (18):4127–4136.

Cecchelli, Romeo, Sezin Aday, Emmanuel Sevin, Catarina Almeida, Maxime Culot, Lucie Dehouck, Caroline Coisne, Britta Engelhardt, Marie-Pierre Dehouck, and Lino Ferreira. 2014. "A stable and reproducible human blood-brain barrier model derived from hematopoietic stem cells." *PloS one* no. 9 (6):e99733.

Cecchelli, Romeo, Vincent Berezowski, Stefan Lundquist, Maxime Culot, Mila Renftel, Marie-Pierre Dehouck, and Laurence Fenart. 2007. "Modelling of the blood–brain barrier in drug discovery and development." *Nature Reviews Drug Discovery* no. 6 (8):650–661.

Cecchelli, Roméo, Bénédicte Dehouck, Laurence Descamps, Laurence Fenart, Valérie Buée-Scherrer, C Duhem, Stefan Lundquist, M Rentfel, Gérard Torpier, and Marie-Pierre Dehouck. 1999. "In vitro model for evaluating drug transport across the blood–brain barrier." *Advanced Drug Delivery Reviews* no. 36 (2):165–178.

Cestelli, Alessandro, Caterina Catania, Stefania D'Agostino, Italia Di Liegro, Luana Licata, Gabriella Schiera, Giovanna Laura Pitarresi, Giovanni Savettieri, Viviana De Caro, and Giulia Giandalia. 2001. "Functional feature of a novel model of blood brain barrier: Studies on permeation of test compounds." *Journal of Controlled Release* no. 76 (1):139–147.

Chacko, Ann-Marie, Chunsheng Li, Daniel A Pryma, Steven Brem, George Coukos, and Vladimir Muzykantov. 2013. "Targeted delivery of antibody-based therapeutic and imaging agents to CNS tumors: Crossing the blood–brain barrier divide." *Expert Opinion on Drug Delivery* no. 10 (7):907–926.

Chekhonin, Vladimir P, Alexander V Kabanov, Yurii A Zhirkov, and Georgii V Morozov. 1991. "Fatty acid acylated Fab-fragments of antibodies to neurospecific proteins as carriers for neuroleptic targeted delivery in brain." *FEBS Letters* no. 287 (1–2):149–152.

Chen, Hongming, Susanne Winiwarter, Markus Fridén, Madeleine Antonsson, and Ola Engkvist. 2011. "In silico prediction of unbound brain-to-plasma concentration ratio using machine learning algorithms." *Journal of Molecular Graphics and Modelling* no. 29 (8):985–995.

Chen, Lei, Youyong Li, Huidong Yu, Liling Zhang, and Tingjun Hou. 2012. "Computational models for predicting substrates or inhibitors of P-glycoprotein." *Drug Discovery Today* no. 17 (7):343–351.

Chopineau, Joël, Stéphane Robert, Laurence Fenart, Roméo Cecchelli, Bernard Lagoutte, Stéphanie Paitier, Marie-Pierre Dehouck, and Dominique Domurado. 1998. "Monoacylation of ribonuclease A enables its transport across an in vitro model of the blood–brain barrier." *Journal of Controlled Release* no. 56 (1):231–237.

Clark, David E. 1999. "Rapid calculation of polar molecular surface area and its application to the prediction of transport phenomena. 1. Prediction of intestinal absorption." *Journal of Pharmaceutical Sciences* no. 88 (8):807–814.

Cory Kalvass, J, and Tristan S Maurer. 2002. "Influence of nonspecific brain and plasma binding on CNS exposure: implications for rational drug discovery." *Biopharmaceutics & Drug Disposition* no. 23 (8):327–338.

Culot, Maxime, Stefan Lundquist, Dorothée Vanuxeem, Stéphane Nion, Christophe Landry, Yannick Delplace, Marie-Pierre Dehouck, Vincent Berezowski, Laurence Fenart, and Roméo Cecchelli. 2008. "An in vitro blood-brain barrier model for high throughput (HTS) toxicological screening." *Toxicology In Vitro* no. 22 (3):799–811.

Daneman, Richard. 2012. "The blood–brain barrier in health and disease." *Annals of Neurology* no. 72 (5):648–672.

Del Pino, Manuel M Sánchez, Richard A Hawkins, and Darryl R Peterson. 1995. "Biochemical discrimination between luminal and abluminal enzyme and transport activities of the blood-brain barrier." *Journal of Biological Chemistry* no. 270 (25):14907–14912.

Deracinois, Barbara, Gwënaël Pottiez, Philippe Chafey, Tom Teerlink, Luc Camoin, Mariska Davids, Cedric Broussard, Pierre-Olivier Couraud, Marie-Pierre Dehouck, and Roméo Cecchelli. 2013. "Glial-cell-mediated re-induction of the blood-brain barrier phenotype in brain capillary endothelial cells: A differential gel electrophoresis study." *Proteomics* no. 13 (7):1185–1199.

Di, Li, Edward H Kerns, and Guy T Carter. 2008. "Strategies to assess blood–brain barrier penetration." *Expert Opinion on Drug Discovery* no. 3 (6):677–687.

Doran, Angela, R Scott Obach, Bill J Smith, Natilie A Hosea, Stacey Becker, Ernesto Callegari, Cuiping Chen, Xi Chen, Edna Choo, and Julie Cianfrogna. 2004. "The impact of P-glycoprotein on the disposition of drugs targeted for indications of the central nervous system: Evaluation using the MDR1A/1B knockout mouse model." *Drug Metabolism and Disposition.*

Eigenmann, Daniela E, Gongda Xue, Kwang S Kim, Ashlee V Moses, Matthias Hamburger, and Mouhssin Oufir. 2013. "Comparative study of four immortalized human brain capillary endothelial cell lines, hCMEC/D3, hBMEC, TY10, and BB19, and optimization of culture conditions, for an in vitro blood–brain barrier model for drug permeability studies." *Fluids and Barriers of the CNS* no. 10 (1):33.

Ekrami, Hossein M, Ann R Kennedy, and Wei-Chiang Shen. 1995. "Water-soluble fatty acid derivatives as acylating agents for reversible lipidization of polypeptides." *FEBS letters* no. 371 (3):283–286.

Elliott, Peter J, Neil J Hayward, Michael R Huff, Tricia L Nagle, Keith L Black, and Raymond T Bartus. 1996. "Unlocking the blood–brain barrier: A role for RMP-7 in brain tumor therapy." *Experimental Neurology* no. 141 (2):214–224.

Emerich, Dwaine F, Pamela Snodgrass, Melissa Pink, Floyd Bloom, and Raymond T Bartus. 1998. "Central analgesic actions of loperamide following transient permeation of the blood brain barrier with Cereport™(RMP-7)." *Brain Research* no. 801 (1):259–266.

Erdő, Franciska, László Denes, and Elizabeth de Lange. 2017. "Age-associated physiological and pathological changes at the blood–brain barrier: A review." *Journal of Cerebral Blood Flow & Metabolism* no. 37 (1):4–24.

Fan, Yi, Rayomand Unwalla, Rajiah A Denny, Li Di, Edward H Kerns, David J Diller, and Christine Humblet. 2010. "Insights for predicting blood-brain barrier penetration of CNS targeted molecules using QSPR approaches." *Journal of Chemical Information and Modeling* no. 50 (6):1123–1133.

Feher, Miklos, Elizabeth Sourial, and Jonathan M Schmidt. 2000. "A simple model for the prediction of blood–brain partitioning." *International Journal of Pharmaceutics* no. 201 (2):239–247.

Fike, John R, Glenn T Gobbel, Ali H Mesiwala, Hyung J Shin, Minoru Nakagawa, Kathleen R Lamborn, Theresa M Seilhan, and Peter J Elliott. 1998. "Cerebrovascular effects of the bradykinin analog RMP-7 in normal and irradiated dog brain." *Journal of Neuro-oncology* no. 37 (3):199–215.

Förster, Carola, Jens Waschke, Malgorzata Burek, Jörg Leers, and Detlev Drenckhahn. 2006. "Glucocorticoid effects on mouse microvascular endothelial barrier permeability are brain specific." *Journal of Physiology* no. 573 (2):413–425.

Franke, Helmut, Hans-Joachim Galla, and Carsten T Beuckmann. 2000. "Primary cultures of brain microvessel endothelial cells: A valid and flexible model to study drug transport through the blood–brain barrier in vitro." *Brain Research Protocols* no. 5 (3):248–256.

Fritsche, Ellen, Kathrin Gassmann, and Timm Schreiber. 2011. "Neurospheres as a model for developmental neurotoxicity testing." *In Vitro Neurotoxicology: Methods and Protocols* 99–114.

Fu, Xu-Chun, Chun-Xiao Chen, Wen-Quan Liang, and Qing-Sen Yu. 2001. "Predicting blood-brain barrier penetration of drugs by polar molecular surface area and molecular volume." *Acta Pharmacologica Sinica* no. 22 (7):663–668.

Fu, XC, GP Wang, WQ Liang, and QS Yu. 2004. "Predicting blood-brain barrier penetration of drugs using an artificial neural network." *Die Pharmazie-An International Journal of Pharmaceutical Sciences* no. 59 (2):126-130.

Gaillard, Pieter Jaap, Levina Helena Voorwinden, Jette Lyngholm Nielsen, Alexei Ivanov, Ryo Atsumi, Helena Engman, Carina Ringbom, Albertus Gerrit de Boer, and Douwe Durk Breimer. 2001. "Establishment and functional characterization of an in vitro model of the blood–brain barrier, comprising a co-culture of brain capillary endothelial cells and astrocytes." *European Journal of Pharmaceutical Sciences* no. 12 (3):215–222.

Goodwin, Jay T, and David E Clark. 2005. "In silico predictions of blood-brain barrier penetration: Considerations to "keep in mind"." *Journal of Pharmacology and Experimental Therapeutics* no. 315 (2):477–483.

Griep, LM, F Wolbers, B De Wagenaar, Paulus Martinus ter Braak, BB Weksler, Ignacio A Romero, PO Couraud, I Vermes, Andries Dirk van der Meer, and Albert van den Berg. 2013. "BBB on chip: Microfluidic platform to mechanically and biochemically modulate blood-brain barrier function." *Biomedical Microdevices* no. 15 (1):145–150.

Gumbleton, Mark, and Kenneth L Audus. 2001. "Progress and limitations in the use of in vitro cell cultures to serve as a permeability screen for the blood-brain barrier." *Journal of Pharmaceutical Sciences* no. 90 (11):1681–1698.

Harry, G Jean, Melvin Billingsley, Arend Bruinink, Iain L Campbell, Werner Classen, David C Dorman, Corrado Galli, David Ray, Robert A Smith, and Hugh A Tilson. 1998. "In vitro techniques for the assessment of neurotoxicity." *Environmental Health Perspectives* no. 106 (Suppl 1):131.

Hawkins, Brian T, and Thomas P Davis. 2005. "The blood-brain barrier/neurovascular unit in health and disease." *Pharmacological Reviews* no. 57 (2):173–185.

He, Yarong, Yao Yao, Stella E Tsirka, and Yu Cao. 2014. "Cell-culture models of the blood–brain barrier." *Stroke* no. 45 (8):2514–2526.

Hellinger, Éva, Szilvia Veszelka, Andrea E Tóth, Fruzsina Walter, Ágnes Kittel, Mónika Laura Bakk, Károly Tihanyi, Viktor Háda, Shinsuke Nakagawa, and Thuy Dinh Ha Duy. 2012. "Comparison of brain capillary endothelial cell-based and epithelial (MDCK-MDR1, Caco-2, and VB-Caco-2) cell-based surrogate blood–brain barrier penetration models." *European Journal of Pharmaceutics and Biopharmaceutics* no. 82 (2):340–351.

Helms, Hans C, and Birger Brodin. 2014. "Generation of primary cultures of bovine brain endothelial cells and setup of cocultures with rat astrocytes." *Cerebral Angiogenesis: Methods and Protocols* 365–382.

Helms, Hans C, N Joan Abbott, Malgorzata Burek, Romeo Cecchelli, Pierre-Olivier Couraud, Maria A Deli, Carola Förster, Hans J Galla, Ignacio A Romero, and Eric V Shusta. 2016. "In vitro models of the blood–brain barrier: An overview of commonly used brain endothelial cell culture models and guidelines for their use." *Journal of Cerebral Blood Flow & Metabolism* no. 36 (5):862–890.

Hoheisel, Dirk, Thorsten Nitz, Helmut Franke, Joachim Wegener, Ansgar Hakvoort, Thomas Tilling, and Hans-Joachim Galla. 1998. "Hydrocortisone reinforces the blood–brain barrier properties in a serum free cell culture system." *Biochemical and Biophysical Research Communications* no. 244 (1):312–316.

Hou, TJ, and XJ Xu. 2003. "ADME evaluation in drug discovery. 3. Modeling blood-brain barrier partitioning using simple molecular descriptors." *Journal of Chemical Information and Computer Sciences* no. 43 (6):2137–2152.

Hou, Zhonggang, Jue Zhang, Michael P Schwartz, Ron Stewart, C David Page, William L Murphy, and James A Thomson. 2013. "A human pluripotent stem cell platform for assessing developmental neural toxicity screening." *Stem Cell Research & Therapy* no. 4 (1):S12.

Huang, SC, C Hoh, JR Barrio, and ME Phelps. 1998. "Measurement of blood brain barrier permeability in humans with positron emission tomography." *Introduction to the Blood-Brain Barrier: Methodology, Biology and Pathology (Pardridge WM, ed)* 122–146.

Hutter, Michael C. 2003. "Prediction of blood–brain barrier permeation using quantum chemically derived information." *Journal of Computer-Aided Molecular Design* no. 17 (7):415–443.

Hynynen, K, N McDannold, N Vykhodtseva, and FA Jolesz. 2003. "Non-invasive opening of BBB by focused ultrasound." In *Brain Edema Xii*, 555–558. Springer.

Hynynen, Kullervo, Nathan McDannold, Nickolai A Sheikov, Ferenc A Jolesz, and Natalia Vykhodtseva. 2005. "Local and reversible blood–brain barrier disruption by noninvasive focused ultrasound at frequencies suitable for trans-skull sonications." *Neuroimage* no. 24 (1):12–20.

Jalali, Shahrzad, Yuexi Huang, Daniel J Dumont, and Kullervo Hynynen. 2010. "Focused ultrasound-mediated bbb disruption is associated with an increase in activation of AKT: Experimental study in rats." *BMC Neurology* no. 10 (1):114.

Jhala, DD, SS Chettiar, and JK Singh. 2012. "Optimization and validation of an in vitro blood brain barrier permeability assay using artificial lipid membrane." *J. Bioequiv. Availab.* no. 14:1–5.

Kabanov, Alexander V, Andrei V Levashov, and Karel Martinek. 1987. "Transformation of Water-Soluble Enzymes into Membrane Active Form by Chemical Modification." *Annals of the New York Academy of Sciences* no. 501 (1):63–66.

Kabanov, Alexander V, Andrey V Levashov, and Valery Yu Alakhov. 1989. "Lipid modification of proteins and their membrane transport." *Protein Engineering, Design and Selection* no. 3 (1):39–42.

Kaliszan, Roman, and Michał Markuszewski. 1996. "Brain/blood distribution described by a combination of partition coefficient and molecular mass." *International journal of pharmaceutics* no. 145 (1-2):9–16.

Ke, Weilun, Kun Shao, Rongqin Huang, Liang Han, Yang Liu, Jianfeng Li, Yuyang Kuang, Liya Ye, Jinning Lou, and Chen Jiang. 2009. "Gene delivery targeted to the brain using an Angiopep-conjugated polyethyleneglycol-modified polyamidoamine dendrimer." *Biomaterials* no. 30 (36):6976–6985.

Kelder, Jan, Peter DJ Grootenhuis, Denis M Bayada, Leon PC Delbressine, and Jan-Peter Ploemen. 1999. "Polar molecular surface as a dominating determinant for oral absorption and brain penetration of drugs." *Pharmaceutical Research* no. 16 (10):1514–1519.

Keserü, György M, and László Molnár. 2001. "High-throughput prediction of blood–brain partitioning: A thermodynamic approach." *Journal of Chemical Information and Computer Sciences* no. 41 (1):120–128.

King, Lester S. 1968. "The Human Brain and Spinal Cord: A Historical Study Illustrated by Writings From Antiquity to the 20th Century." *JAMA* no. 206 (6):1309–1309.

Klon, Anthony E. 2009. "Computational models for central nervous system penetration." *Current Computer-Aided Drug Design* no. 5 (2):71–89.

Kroll, Robert A, Edward A Neuwelt, and Edward A Neuwelt. 1998. "Outwitting the blood-brain barrier for therapeutic purposes: Osmotic opening and other means." *Neurosurgery* no. 42 (5):1083–1099.

Kurakhmaeva, Kamila B, Irma A Djindjikhashvili, Valery E Petrov, Vadim U Balabanyan, Tatiana A Voronina, Sergey S Trofimov, Jörg Kreuter, Svetlana Gelperina, David Begley, and Renad N Alyautdin. 2009. "Brain targeting of nerve growth factor using poly (butyl cyanoacrylate) nanoparticles." *Journal of Drug Targeting* no. 17 (8):564–574.

Lanevskij, Kiril, Justas Dapkunas, Liutauras Juska, Pranas Japertas, and Remigijus Didziapetris. 2011. "QSAR analysis of blood–brain distribution: The influence of plasma and brain tissue binding." *Journal of Pharmaceutical Sciences* no. 100 (6):2147–2160.

Larsen, Jessica M, Douglas R Martin, and Mark E Byrne. 2014. "Recent advances in delivery through the blood-brain barrier." *Current Topics in Medicinal Chemistry* no. 14 (9):1148–1160.

Lippmann, Ethan S, Samira M Azarin, Jennifer E Kay, Randy A Nessler, Hannah K Wilson, Abraham Al-Ahmad, Sean P Palecek, and Eric V Shusta. 2012. "Derivation of blood-brain barrier endothelial cells from human pluripotent stem cells." *Nature Biotechnology* no. 30 (8):783–791.

Liu, Hao-Li, Mu-Yi Hua, Pin-Yuan Chen, Po-Chun Chu, Chia-Hsin Pan, Hung-Wei Yang, Chiung-Yin Huang, Jiun-Jie Wang, Tzu-Chen Yen, and Kuo-Chen Wei. 2010. "Blood-brain barrier disruption with focused ultrasound enhances delivery of chemotherapeutic drugs for glioblastoma treatment." *Radiology* no. 255 (2):415–425.

Liu, Li-bo, Yi-xue Xue, and Yun-hui Liu. 2010. "Bradykinin increases the permeability of the blood-tumor barrier by the caveolae-mediated transcellular pathway." *Journal of Neuro-oncology* no. 99 (2):187–194.

Liu, Xingrong, Cuiping Chen, and Bill J Smith. 2008. "Progress in brain penetration evaluation in drug discovery and development." *Journal of Pharmacology and Experimental Therapeutics* no. 325 (2):349–356.

Lombardo, Franco, James F Blake, and William J Curatolo. 1996. "Computation of brain–blood partitioning of organic solutes via free energy calculations." *Journal of Medicinal Chemistry* no. 39 (24):4750–4755.

Mangas-Sanjuan, Victor, Isabel González-Álvarez, Marta González-Álvarez, Vicente G Casabó, and Marival Bermejo. 2013. "Innovative in vitro method to predict rate and extent of drug delivery to the brain across the blood–brain barrier." *Molecular Pharmaceutics* no. 10 (10):3822–3831.

Martin, Iain. 2004. "Prediction of blood–brain barrier penetration: Are we missing the point?" *Drug Discovery Today* no. 9 (4):161–162.

Mensch, Jurgen, Libuse Jaroskova, Wendy Sanderson, Anouche Melis, Claire Mackie, Geert Verreck, Marcus E Brewster, and Patrick Augustijns. 2010. "Application of PAMPA-models to predict BBB permeability including efflux ratio, plasma protein binding and physicochemical parameters." *International Journal of Pharmaceutics* no. 395 (1):182–197.

Mensch, Jurgen, Julen Oyarzabal, Claire Mackie, and Patrick Augustijns. 2009. "In vivo, in vitro and in silico methods for small molecule transfer across the BBB." *Journal of Pharmaceutical Sciences* no. 98 (12):4429–4468.

Mischeck, Uwe, Jörg Meyer, and Hans-Joachim Galla. 1989. "Characterization of γ-glutamyl transpeptidase activity of cultured endothelial cells from porcine brain capillaries." *Cell and Tissue Research* no. 256 (1):221–226.

Misra, Ambikanandan, S Ganesh, Aliasgar Shahiwala, and Shrenik P Shah. 2003. "Drug delivery to the central nervous system: A review." *J Pharm Pharm Sci* no. 6 (2):252–273.

Muehlbacher, Markus, Gudrun M Spitzer, Klaus R Liedl, and Johannes Kornhuber. 2011. "Qualitative prediction of blood–brain barrier permeability on a large and refined dataset." *Journal of Computer-Aided Molecular Design* no. 25 (12): 1095–1106.

Naik, Pooja, and Luca Cucullo. 2012. "In vitro blood–brain barrier models: Current and perspective technologies." *Journal of Pharmaceutical Sciences* no. 101 (4):1337–1354.

Nakagawa, Shinsuke, Mária A Deli, Shinobu Nakao, Masaru Honda, Kentaro Hayashi, Ryota Nakaoke, Yasufumi Kataoka, and Masami Niwa. 2007. "Pericytes from brain microvessels strengthen the barrier integrity in primary cultures of rat brain endothelial cells." *Cellular and Molecular Neurobiology* no. 27 (6):687–694.

Nakagawa, Shinsuke, Maria A Deli, Hiroko Kawaguchi, Takeshi Shimizudani, Takanori Shimono, Agnes Kittel, Kunihiko Tanaka, and Masami Niwa. 2009. "A new blood–brain barrier model using primary rat brain endothelial cells, pericytes and astrocytes." *Neurochemistry International* no. 54 (3):253–263.

Neuhaus, Winfried, Regina Lauer, Silvester Oelzant, Urs P Fringeli, Gerhard F Ecker, and Christian R Noe. 2006. "A novel flow based hollow-fiber blood–brain barrier in vitro model with immortalised cell line PBMEC/C1–2." *Journal of Biotechnology* no. 125 (1):127–141.

Neuwelt, Edward A, David L Goldman, Suellen A Dahlborg, John Crossen, Fred Ramsey, Simon Roman-Goldstein, Rita Braziel, and Bruce Dana. 1991. "Primary CNS lymphoma treated with osmotic blood-brain barrier disruption: Prolonged survival and preservation of cognitive function." *Journal of Clinical Oncology* no. 9 (9):1580–1590.

Nicolazzo, Joseph A, Susan A Charman, and William N Charman. 2006. "Methods to assess drug permeability across the blood-brain barrier." *Journal of Pharmacy and Pharmacology* no. 58 (3):281–293.

Nielsen, Peter Aadal, Olga Andersson, Steen Honoré Hansen, Klaus Bæk Simonsen, and Gunnar Andersson. 2011. "Models for predicting blood–brain barrier permeation." *Drug Discovery Today* no. 16 (11):472–475.

Norinder, Ulf, and Markus Haeberlein. 2002. "Computational approaches to the prediction of the blood–brain distribution." *Advanced Drug Delivery Reviews* no. 54 (3):291–313.

Norinder, Ulf, Per Sjöberg, and Thomas Österberg. 1998. "Theoretical calculation and prediction of brain–blood partitioning of organic solutes using MolSurf parametrization and PLS statistics." *Journal of Pharmaceutical Sciences* no. 87 (8):952–959.

Obermeier, Birgit, Richard Daneman, and Richard M Ransohoff. 2013. "Development, maintenance and disruption of the blood-brain barrier." *Nature Medicine* no. 19 (12):1584–1596.

Oldendorf, William H. 1970. "Measurement of brain uptake of radiolabeled substances using a tritiated water internal standard." *Brain Research* no. 24 (2):372–376.

Oldendorf, WILLIAM H. 1971. "Brain uptake of radiolabeled amino acids, amines, and hexoses after arterial injection." *American Journal of Physiology—Legacy Content* no. 221 (6):1629–1639.

Omidi, Yadollah, Lee Campbell, Jaleh Barar, David Connell, Saeed Akhtar, and Mark Gumbleton. 2003. "Evaluation of the immortalised mouse brain capillary endothelial cell line, b. End3, as an in vitro blood–brain barrier model for drug uptake and transport studies." *Brain Research* no. 990 (1):95–112.

Österberg, Thomas, and Ulf Norinder. 2000. "Prediction of polar surface area and drug transport processes using simple parameters and PLS statistics." *Journal of Chemical Information and Computer Sciences* no. 40 (6):1408–1411.

Paolinelli, Roberta, Monica Corada, Luca Ferrarini, Kavi Devraj, Cédric Artus, Cathrin J Czupalla, Noemi Rudini, Luigi Maddaluno, Eleanna Papa, and Britta Engelhardt. 2013. "Wnt activation of immortalized brain endothelial cells as a tool for generating a standardized model of the blood brain barrier in vitro." *PloS one* no. 8 (8):e70233.

Pardridge, William M. 1995. "Transport of small molecules through the blood-brain barrier: Biology and methodology." *Advanced Drug Delivery Reviews* no. 15 (1–3):5–36.

Pardridge, William M. 1998. "Isolated brain capillaries: An in vitro model of blood-brain barrier research." *Introduction to the Blood-Brain Barrier: Methodology, Biology and Pathology (Pardridge WM, ed)* 49–61.

Pardridge, William M. 2004. "Log (BB), PS products and in silico models of drug brain penetration." *Drug Discovery Today* no. 9 (9):392-393.

Pardridge, William M. 2007. "Drug targeting to the brain." *Pharmaceutical Research* no. 24 (9):1733–1744.

Paris-Robidas, Sarah, Danny Brouard, Vincent Emond, Martin Parent, and Frédéric Calon. 2016. "Internalization of targeted quantum dots by brain capillary endothelial cells in vivo." *Journal of Cerebral Blood Flow & Metabolism* no. 36 (4):731–742.

Patabendige, Adjanie, Robert A Skinner, and N Joan Abbott. 2013. "Establishment of a simplified in vitro porcine blood–brain barrier model with high transendothelial electrical resistance." *Brain Research* no. 1521:1–15.

Patabendige, Adjanie, Robert A Skinner, Louise Morgan, and N Joan Abbott. 2013. "A detailed method for preparation of a functional and flexible blood–brain barrier model using porcine brain endothelial cells." *Brain Research* no. 1521:16–30.

Patel, Rakesh R, and Minesh P Mehta. 2007. "Targeted therapy for brain metastases: Improving the therapeutic ratio." *Clinical Cancer Research* no. 13 (6):1675–1683.

Patlak, Clifford S, Ronald G Blasberg, and Joseph D Fenstermacher. 1983. "Graphical evaluation of blood-to-brain transfer constants from multiple-time uptake data." *Journal of Cerebral Blood Flow & Metabolism* no. 3 (1):1–7.

Pena, F. 2010. "Organotypic cultures as tool to test long-term effects of chemicals on the nervous system." *Current Medicinal Chemistry* no. 17 (10):987–1001.

Pereira, Cláudia V, Ana C Moreira, Susana P Pereira, Nuno G Machado, Filipa S Carvalho, Vilma A Sardão, and Paulo J Oliveira. 2009. "Investigating drug-induced mitochondrial toxicity: A biosensor to increase drug safety?" *Current Drug Safety* no. 4 (1):34–54.

Poller, Birk, Heike Gutmann, Stephan Krähenbühl, Babette Weksler, Ignacio Romero, Pierre-Olivier Couraud, Gerald Tuffin, Jürgen Drewe, and Jörg Huwyler. 2008. "The human brain endothelial cell line hCMEC/D3 as a human blood-brain barrier model for drug transport studies." *Journal of Neurochemistry* no. 107 (5):1358–1368.

Qin, De-Xing, Rong Zheng, Jin Tang, Jia-Xiu Li, and Yu-Hua Hu. 1990. "Influence of radiation on the blood-brain barrier and optimum time of chemotherapy." *International Journal of Radiation Oncology* Biology* Physics* no. 19 (6):1507–1510.

Radio, Nicholas M, and William R Mundy. 2008. "Developmental neurotoxicity testing in vitro: Models for assessing chemical effects on neurite outgrowth." *Neurotoxicology* no. 29 (3):361–376.

Ramsohoye, PV, and IB Fritz. 1998. "Preliminary characterization of glial-secreted factors responsible for the induction of high electrical resistances across endothelial monolayers in a blood-brain barrier model." *Neurochemical Research* no. 23 (12):1545–1551.

Rapoport, Stanley I. 2000. "Osmotic opening of the blood–brain barrier: Principles, mechanism, and therapeutic applications." *Cellular and Molecular Neurobiology* no. 20 (2):217–230.

Rascher, Gesa, and Hartwig Wolburg. 2002. "The blood-brain barrier in the aging brain." In *Neuroglia in the Aging Brain*, 305–320. Springer.

Rautio, Jarkko, Krista Laine, Mikko Gynther, and Jouko Savolainen. 2008. "Prodrug approaches for CNS delivery." *AAPS journal* no. 10 (1):92–102.

Rishton, Gilbert M, Kristen LaBonte, Anthony J Williams, Karim Kassam, and Eduard Kolovanov. 2006. "Computational approaches to the prediction of blood-brain barrier permeability: A comparative analysis of central nervous system drugs versus secretase inhibitors for Alzheimer's disease." *Current Opinion in Drug Discovery & Development* no. 9 (3):303–313.

Roux, F, O Durieu-Trautmann, N Chaverot, M Claire, P Mailly, J-M Bourre, AD Strosberg, and P-O Couraud. 1994. "Regulation of gamma-glutamyl transpeptidase and alkaline phosphatase activities in immortalized rat brain microvessel endothelial cells." *Journal of Cellular Physiology* no. 159 (1):101–113.

Sage, Michael R. 1982. "Blood-brain barrier: Phenomenon of increasing importance to the imaging clinician." *American Journal of Roentgenology* no. 138 (5):887–898.

Sanovich, Elena, Raymond T Bartus, Phillip M Friden, Reginald L Dean, Harrison Q Le, and Milton W Brightman. 1995. "Pathway across blood-brain barrier opened by the bradykinin agonist, RMP-7." *Brain Research* no. 705 (1):125–135.

Santaguida, Stefano, Damir Janigro, Mohammed Hossain, Emily Oby, Edward Rapp, and Luca Cucullo. 2006. "Side by side comparison between dynamic versus static models of blood–brain barrier in vitro: A permeability study." *Brain Research* no. 1109 (1):1–13.

Scholtzova, Henrieta, Youssef Z Wadghiri, Moustafa Douadi, Einar M Sigurdsson, Yong-Sheng Li, David Quartermain, Pradeep Banerjee, and Thomas Wisniewski. 2008. "Memantine leads to behavioral improvement and amyloid reduction in Alzheimer's-disease-model transgenic mice shown as by micromagnetic resonance imaging." *Journal of Neuroscience Research* no. 86 (12):2784–2791.

Scuffham, JW, MD Wilson, P Seller, MC Veale, PJ Sellin, SDM Jacques, and RJ Cernik. 2012. "A CdTe detector for hyperspectral SPECT imaging." *Journal of Instrumentation* no. 7 (08):P08027.

Shawahna, Ramzi, Xavier Declèves, and Jean-Michel Scherrmann. 2013. "Hurdles with using in vitro models to predict human blood-brain barrier drug permeability: A special focus on transporters and metabolizing enzymes." *Current Drug Metabolism* no. 14 (1):120–136.

Shityakov, S, E Salvador, and C Förster. 2013. "In silico, in vitro, and in vivo methods to analyse drug permeation across the blood–brain barrier: A critical review." *OA Anaesthetics* no. 1 (2):13.

Shityakov, Sergey, Norbert Roewer, Jens-Albert Broscheit, and Carola Förster. 2017. "In silico models for nanotoxicity evaluation and prediction at the blood-brain barrier level: A mini-review." *Computational Toxicology*.

Silva, RFM, AS Falcao, A Fernandes, AC Gordo, MA Brito, and D Brites. 2006. "Dissociated primary nerve cell cultures as models for assessment of neurotoxicity." *Toxicology Letters* no. 163 (1):1–9.

Slepnev, Vladimir I, Laurent Phalente, Hubert Labrousse, Nikolai S Melik-Nubarov, Veronique Mayau, Bruno Goud, Gerard Buttin, and Alexander V Kabanov. 1995. "Fatty acid acylated peroxidase as a model for the study of interactions of hydrophobically-modified proteins with mammalian cells." *Bioconjugate Chemistry* no. 6 (5):608–615.

Smith, Quentin R. 2003. "A review of blood-brain barrier transport techniques." *The Blood-Brain Barrier: Biology and Research Protocols* 193–208.

Stewart, Barbra H, and O Helen Chan. 1998. "Use of immobilized artificial membrane chromatography for drug transport applications." *Journal of Pharmaceutical Sciences* no. 87 (12):1471–1478.

Subramanian, Govindan, and Douglas B Kitchen. 2003. "Computational models to predict blood–brain barrier permeation and CNS activity." *Journal of Computer-Aided Molecular Design* no. 17 (10):643–664.

Summerfield, Scott G, and Kelly C Dong. 2013. "In vitro, in vivo and in silico models of drug distribution into the brain." *Journal of Pharmacokinetics and Pharmacodynamics* no. 40 (3):301–314.

Suzuki, Yasuhiro, Nobuo Nagai, and Kazuo Umemura. 2016. "A review of the mechanisms of blood-brain barrier permeability by tissue-type plasminogen activator treatment for cerebral ischemia." *Frontiers in Cellular Neuroscience* no. 10.

Takasato, Yoshio, Stanley I Rapoport, and Quentin R Smith. 1984. "An in situ brain perfusion technique to study cerebrovascular transport in the rat." *American Journal of Physiology-Heart and Circulatory Physiology* no. 247 (3):H484–H493.

Tolia, Evangelia, Ioannis P Fouyas, Paul AT Kelly, and Ian R Whittle. 2005. The blood–brain barrier in diabetes mellitus: A critical review of clinical and experimental findings. Paper read at International Congress Series.

Van Rooy, Inge, Serpil Cakir-Tascioglu, Wim E Hennink, Gert Storm, Raymond M Schiffelers, and Enrico Mastrobattista. 2011. "In vivo methods to study uptake of nanoparticles into the brain." *Pharmaceutical Research* no. 28 (3):456–471.

Van Wamel, Annemieke, Klazina Kooiman, Marcia Emmer, FJ Ten Cate, Michel Versluis, and Nico de Jong. 2006. "Ultrasound microbubble induced endothelial cell permeability." *Journal of Controlled Release* no. 116 (2):e100–e102.

Varsha, Awasarkar, Bagade Om, Ramteke Kuldeep, and Patel Bindiya Patel Riddhi. 2014. "Poles apart Inimitability of Brain Targeted Drug Delivery system in Middle of NDDS." *International Journal of Drug Development and Research.*

Wilhelm, Imola, Csilla Fazakas, and Istvan A Krizbai. 2011. "In vitro models of the blood-brain barrier." *Acta Neurobiol Exp (Wars)* no. 71 (1):113–128.

Wilhelm, Imola, and István A Krizbai. 2014. "In vitro models of the blood–brain barrier for the study of drug delivery to the brain." *Molecular Pharmaceutics* no. 11 (7):1949–1963.

Williams, Paul C, W David Henner, Roman-Goldstein Simon, Suellen A Dahlborg, Robert E Brummett, Mara Tableman, Bruce W Dana, and Edward A Neuwelt. 1995. "Toxicity and efficacy of carboplatin and etoposide in conjunction with disruption of the blood-brain tumor barrier in the treatment of intracranial neoplasms." *Neurosurgery* no. 37 (1):17–28.

Wilson, Melinda S, James R Graham, and Andrew J Ball. 2014. "Multiparametric High Content Analysis for assessment of neurotoxicity in differentiated neuronal cell lines and human embryonic stem cell-derived neurons." *Neurotoxicology* no. 42:33–48.

Wiranowska, Marzenna, Americo A Gonzalvo, Samuel Saporta, Orlando B Gonzalez, and Leon D Prockop. 1992. "Evaluation of blood-brain barrier permeability and the effect of interferon in mouse glioma model." *Journal of Neuro-oncology* no. 14 (3):225–236.

Yeon, Ju Hun, Dokyun Na, Kyungsun Choi, Seung-Wook Ryu, Chulhee Choi, and Je-Kyun Park. 2012. "Reliable permeability assay system in a microfluidic device mimicking cerebral vasculatures." *Biomedical Microdevices* no. 14 (6):1141–1148.

Young, Rodney C, Robert C Mitchell, Thomas H Brown, C Robin Ganellin, Robin Griffiths, Martin Jones, Kishore K Rana, David Saunders, and Ian R Smith. 1988. "Development of a new physicochemical model for brain penetration and its application to the design of centrally acting H2 receptor histamine antagonists." *Journal of Medicinal Chemistry* no. 31 (3):656–671.

Zensi, Anja, David Begley, Charles Pontikis, Celine Legros, Larisa Mihoreanu, Sylvia Wagner, Claudia Büchel, Hagen von Briesen, and Jörg Kreuter. 2009. "Albumin nanoparticles targeted with Apo E enter the CNS by transcytosis and are delivered to neurones." *Journal of Controlled Release* no. 137 (1):78–86.

Zhang, Yan, Cheryl SW Li, Yuyang Ye, Kjell Johnson, Julie Poe, Shannon Johnson, Walter Bobrowski, Rosario Garrido, and Cherukury Madhu. 2006. "Porcine brain microvessel endothelial cells as an in vitro model to predict in vivo blood-brain barrier permeability." *Drug Metabolism and Disposition* no. 34 (11):1935–1943.

Zlokovic, Berislav V. 2008. "The blood-brain barrier in health and chronic neurodegenerative disorders." *Neuron* no. 57 (2):178–201.

Zlokovic, Berislav V, and Michael LJ Apuzzo. 1997. "Cellular and molecular neurosurgery: Pathways from concept to reality-part I: Target disorders and concept approaches to gene therapy of the central nervous system." *Neurosurgery* no. 40 (4):789–804.

Zlokovic, Berislav V, Shigeyo Hyman, J Gordon McComb, Milo N Lipovac, Gordon Tang, and Hugh Davson. 1990. "Kinetics of arginine-vasopressin uptake at the blood-brain barrier." *Biochimica et Biophysica Acta (BBA)-Biomembranes* no. 1025 (2):191–198.

Zloković, Berislav V, Malcolm B Segal, David J Begley, Hugh Davson, and Lubšia Rakić. 1985. "Permeability of the blood-cerebrospinal fluid and blood-brain barriers to thyrotropin-releasing hormone." *Brain Research* no. 358 (1):191–199.

9

Parenteral Drug Delivery Systems

Aliasgar Shahiwala, Tejal A. Mehta, and Munira M. Momin

CONTENTS

9.1 Parenteral Drug Delivery Systems

Parenteral drug delivery is the most effective delivery for the drugs, as it has a narrow therapeutic index and is unstable in the gastrointestinal tract when administered orally. Also, there is a huge demand of protein- and biotechnology-based drugs in the treatment of certain diseases like cancer and viral diseases. These products are unstable when given orally and thus show poor bioavailability. However, conventional parenteral products have their own limitations in terms of specific targeting and delayed release. This leads to lots of research in novel parenteral drug delivery which is devoid of these side effects. Since last two decades, research in parenteral route of administration is increasingly been pursued. Even though extensive efforts were made for the systemic application of the drugs through alternative routes of administration, which includes mainly the nasal, buccal, pulmonary, and transdermal routes, poor and high variability in systemic absorption is still a major challenging issue. Major parenteral routes for drug administration are subcutaneous (s.c.), intramuscular (i.m.), and intravenous (i.v.). The intradermal and intra-arterial routes are used for specific purposes only.

Parenteral drug delivery systems can be classified broadly into two categories according to its purpose:

1. Controlled drug delivery systems
2. Targeted drug delivery systems

9.2 Controlled-Release Drug Delivery Systems

Conventional parenteral drug delivery show rapid action with rapid decline of systemic drugs. This requires frequent injections, which lead to discomfort, painful, and thereby poor patient compliance. Parenteral depot formulations, in contrast, are controlled-release drug delivery systems. Side effects due to fluctuations in the drug blood level caused by multiple injections are also avoided. Drug release from depot preparations can be modulated from days to years. The release from these systems can be either continuous through injections, or implanted into the muscle or subcutaneous tissue, thus providing a better option to i.v. infusions or chronic multiple injections in terms of better patient compliance, as frequency of administration is decreased. The release from these systems can be either continuous or pulsatile, depending on the type of the device and the polymer characteristics. Controlled-release parenteral drug products are broadly classified into injections and implants. Injections can be divided into polymer-based (polymer-based microspheres and polymer-based in situ gels) and lipid-based (oily solutions, oily suspensions, multivesicular liposomes, lipid microparticles, in situ solidifying organogels) categories. Other approaches used are low aqueous soluble salt (the largest injectable particle size with crystallinity, binding with adsorbents) and coadministration of vasoconstrictors. The characterizations of preformed depot formulations are based on type of dosage form, however, physicochemical characterizations such as size and shape, pKa, solubility at different pH, sterility, syringibility, visocosity, and rheology, Fourier transform infrared spectroscopy (FTIR), thermal analysis, drug content and drug release studies, histopathological studies, and stability studies are some of the most important parameters.

Compared to injections, in situ forming depots are easy and less costly to manufacture, since preformed depots, such as insoluble salts, larger drug crystals, and particulate carriers need to be isolated and washed after the preparation. With respect to parenterals, in situ gels are broadly classified into five categories: thermoplastic pastes, in situ crosslinked polymer systems, in situ polymer precipitation, thermally-induced gelling systems, and in situ solidifying organogels according to their mechanism of depot formation (Packhaeuser et al. 2004). The characterization of such systems are mainly based on characterizing the polymers for in situ gelling mechanisms. Common characterization tools for in situ gelling systems include sol to gel transition temperature and pH, gelling time, clarity, texture analysis (firmness, consistency, and cohesiveness), syringeability, gel strength, viscosity, and rheology, FTIR, thermalanalysis, drug content and drug release studies, histopathological studies, and stability studies (Packhaeuser et al. 2004).

It is not possible to discuss all the characterization tools (related to product quality) for controlled-release parenteral drug delivery systems; the most important characterization tool is the drug release study (related to product performance), which ultimately determines the clinical outcomes of such preparations.

9.2.1 In-Vitro Drug Release Studies for Controlled Drug Delivery Systems

It is crucial to ensure the quality and safety of controlled release parenteral drug delivery systems. In-vitro release testing is important for quality control purposes as well as to predict in-vivo performance, and is recommended as part of the demonstration of bioequivalence between test and reference products in the approval of most generic drugs. At present, no compendial guidelines are available for the dissolution testing of such products, which significantly delay the development and regulatory approval of these products (Schultz et al. 1997; Martinez et al. 2008; Burgess et al. 2004; Seidlitz and Weitschies 2012; Delplace et al. 2012; Pilaniya et al. 2011; Martinez et al. 2010). In recent years, considerable progress has been made towards the development of in-vitro release methods for parenteral drug delivery systems, which helps researchers to use the appropriate drug release method depending on the type of formulation.

9.2.1.1 Compendial Dissolution Methods

Out of United States Pharmacopeia (USP) apparatuses 1–7, only USP apparatuses 1, 2, 4, and 7 have been used for dissolution testing of controlled release formulations. Since compendial methods are mainly developed for oral (USP type 1 and 2 apparatus) and transdermal (USP 5, 6, and 7 apparatus) formulations, they pose different problems for their suitability as dissolution methods for parenteral-controlled release formulations such as:

1. For USP type 1 and 2 apparatus, large volumes of media are required which are not practical for estimation of low-dose parenteral products. Also, particulate systems need a dialysis bag/tube to place the sample. The evaporation of the media due to longer duration of testing of LAP may be another issue.
2. Flow-through cell apparatus is a widely accepted dissolution test by USP, European Pharmacopoeia as Apparatus 4 and Apparatus 3 by Japanese Pharmacopoeia for testing poorly soluble products or low-dose formulations with sustained release. Different types of cells are available for testing tablets, powders, suppositories, hard- and soft-gelatin capsules, implants, semisolids, suppositories, and drug-eluting stents (Fotaki 2011). Originally designed for poorly soluble compounds, different flow-through cell systems have been evolved for novel dosage forms such as drug-eluting stents and depot preparations. New cells have been developed and optimized in order to meet dosage form specific requirements, such as the introduction of injectables in the cell and the positioning of drug-eluting stents and implants. Cells with a volume of 1 mL are preferable for the dissolution testing of implants to provide the very slow flow rate required for such type of dissolution tests that can last for several weeks or even months (Looney 1996).

9.2.1.2 Custom-Design Methods

Custom-design methods are broadly classified into three categories:

1. Sample and separate method
2. Dialysis method
3. Continuous flow method

Comparisons of these methods are provided in Table 9.1 and the schematic diagram is provided in Figure 9.1.

1. Sample and separate (SS):

   This method is mainly used for microparticulate systems, gels, and in situ gels. In this method, particulate drug carriers are suspended in a vessel and the samples for the analysis are obtained by separating the particles using filtration or centrifugation. The volume of the release media may vary from less than 10 mL (Xu et al. 2009; Yang, Chia, and Chung 2000) or large volumes up to 400 ml (Yen et al.

TABLE 9.1

Comparison of Custom-Design Methods

	Sample and Separate Method	Dialysis Method	Continuous Flow Method
Sample container	Tube, vial, beaker, or flask	Dialysis bag	Flow-through cell
Sampling	Isolation of microparticles by filtration or centrifugation	Withdrawal of sample from the bulk media outside the membrane	Media is circulated through the cell containing the formulation and the media is sampled from the reservoir.
Advantages	Accurate measurement of the initial burst drug release and maintenance of sink conditions by replacement of the release media	Sampling and media replacement are convenient owing to a physical separation of the formulation from the outer media	Sampling can be continuously collected and analyzed via the automated process. Analysis of multiple time points can allow for complete characterization of the release profile.
Disadvantages	Cumbersome sampling process and undesirable withdrawal of microparticles from the media	Slow equilibration with the outer media limits an accurate measurement of initial drug levels	Variation in the flow rate due to clogging of the filter and difficulty in rapid replacement of the media.

Source: With kind permission from Springer Science+Business Media: Reprinted from Amatya, Sarmila, Eun Ji Park, Jong Hoon Park, Joon Sik Kim, Eunyoung Seol, Heeyong Lee, Hoil Choi, Young-Hee Shin, and Dong Hee Na. 2013. Drug Release Testing Methods of Polymeric Particulate Drug Formulations. *Journal of Pharmaceutical Investigation* 43 (4):259–66. https://doi.org/10.1007/s40005-013-0072-5.

2001; Jeong et al. 2003) that are necessary to maintain sink conditions and according to the sensitivity of the drug assay. Depending on the volume, either tubes or vials for small volumes or bottle or flasks for large volumes can be selected. The movement of release media is provided either by intermittent or continuous agitation. Different types of shakers have been used in different studies, such as, paddle (Bain, Munday, and Smith 1999), magnetic stirrer (Negrin et al. 2001), wrist shaker (Murty et al. 2003), incubator shaker (Latha et al. 2000), shaking water bath (Mi et al. 2003; Yen et al. 2001; Kim and Burgess 2002), tumbling end-over-end (Liggins and Burt 2001), or high-speed stirring/revolution of bottles (Latha et al. 2000). At different time points, the media is separated from the particulate carriers either by filtration using membrane filters with the appropriate pore size

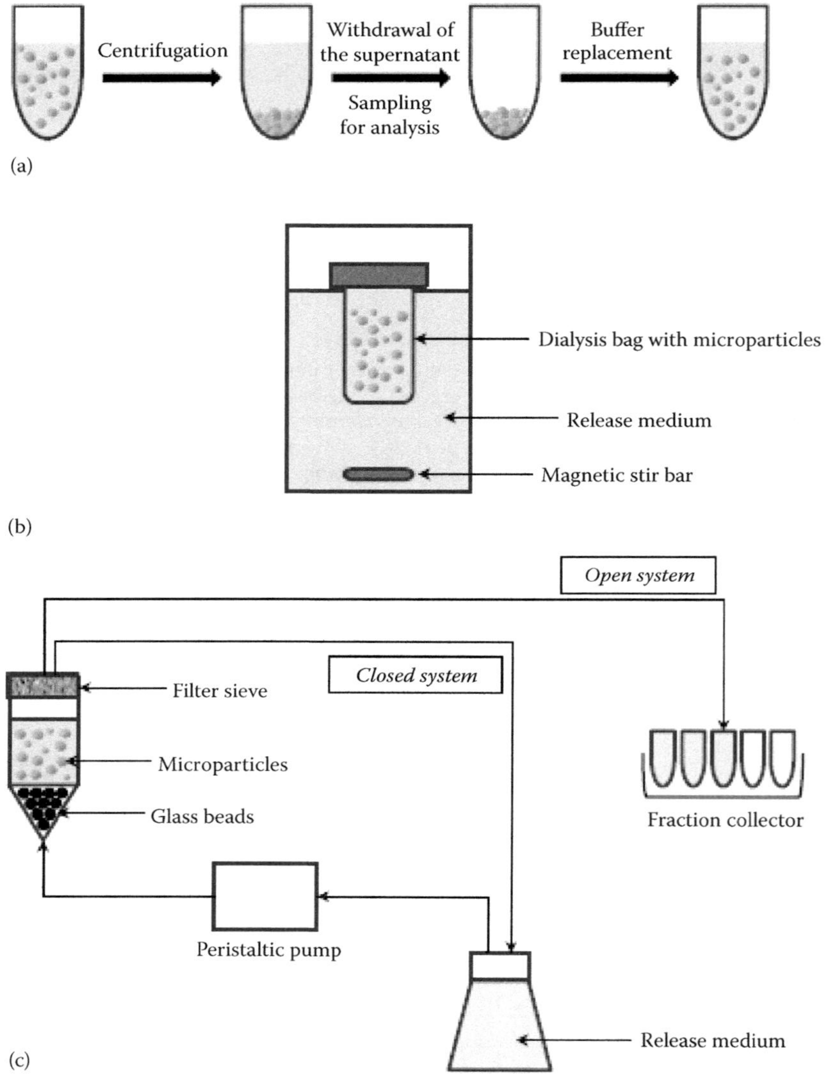

FIGURE 9.1
Basic principles of in-vitro drug release testing methods, (a) sample and separate method, (b) dialysis method, and (c) continuous flow method with open or closed system. (Modified from Larsen, Claus, Susan Weng Larsen, Henrik Jensen, Anan Yaghmur, and Jesper Ostergaard. 2009a. Role of in Vitro Release Models in Formulation Development and Quality Control of Parenteral Depots. Expert Opinion on Drug Delivery 6 (12):1283–95. https://doi.org/10.1517/17425240903307431; With kind permission from Springer Science+Business Media: Reprinted from Amatya, Sarmila, Eun Ji Park, Jong Hoon Park, Joon Sik Kim, Eunyoung Seol, Heeyong Lee, Hoil Choi, Young-Hee Shin, and Dong Hee Na. 2013. Drug Release Testing Methods of Polymeric Particulate Drug Formulations. *Journal of Pharmaceutical Investigation* 43 (4):259–66. https://doi.org/10.1007/s40005-013-0072-5.)

(Liu et al. 2003; Yen et al. 2001) or using centrifugation at sufficient speed (Schaefer and Singh 2002; Park, Yong, and Sung 1998; Jiang et al. 2002; Lacasse et al. 1997), to assess the drug release. If the filtration is used to separate the released drug from the particulate carriers, the uniform distribution of the particulate carriers is required, as the collected sample should contain same concentration of the carriers as the remaining media. Otherwise fresh dispersion should be used in the study for each sampling point. Centrifugation is more common since it is a more user-friendly since supernatant that can be easily separated by centrifugal force and the withdrawn sample can be easily replaced with a fresh buffer to maintain sink conditions (Wei et al. 2004). The particles need to be resuspended again after collecting the samples. In some cases, instead of collecting the supernatant, the remaining drug in the dosage form has been determined (Park, Na, and Lee 2007). However, this requires fresh formulation each time. Care must be taken since the particulate systems start degrading and changing its dimensions during the release study, and the time for sedimentation in centrifugation and pore size in filtration should be carefully selected.

In order to study drug release from oily formulations, different variations in the sample and separate method were applied. These include formulation floating on the top of the release media with stirring of both the phases in a jacketed vessel (Kakemi et al. 1972), formulation placed in the open tube on the top of the release medium with constant stirring of the media (Crommelin and de Blaey 1980), formulation placed in the inverted cup which is placed in the release medium (Söderberg et al. 2006a), and formulation held as a single drop in a continuously rotating downward flow of release media inside the tube (Söderberg et al. 2002, 2006b).

2. Continuous flow (CF) method

As the name suggests, the release media is circulated through a column containing the formulation and drug release is assessed over a period of time. While selecting the apparatus in CF method, consideration should be given to its agitation characteristics, flow rate and choice of medium. Mainly USP Apparatus IV with different setups, pumps, and flow rates are mainly used in this method. The release media is circulated through the column containing formulation either in closed or open fashion. Oily formulations usually floats on the release media inside the column (Janicki et al. 2001; Lootvoet et al. 1992). In closed system, the same media is circulated through the formulation (Wagenaar and Müller 1994; Wang, Wang, and Schwendeman 2002), while in open system, fresh media is circulated through the formulation each time (Rawat, Bhardwaj, and Burgess 2012). Several pumps have been utilized based on the desired flow rate to circulate the release media through the column containing microparticles, such as a peristaltic pump (3–30 ml/min) (Wagenaar and Müller 1994; Vandelli et al. 2001), high performance liquid chromatography (HPLC) pump (0.4 ml/min) (Wang, Wang, and Schwendeman 2002) and syringe pump (5 µl/min) (Aubert-Pouëssel et al. 2004). The choice flow rate is an important parameter in this method, since the low flow rates can result in slow and incomplete drug release due to the insufficient transfer of the release media. D'Souza and DeLuca (2006) recommended HPLC pumps since they provide necessary accuracy and precision at very low flow rates.

Compare to the SS method, the CF method offers several advantages such as standardized setup, automation, ease of operation, separation of microspheres from the release medium, minimum evaporation of the media, and flexibility of monitoring release via inline drug analysis, and automation (Voisine, Zolnik, and Burgess 2008; Rawat et al. 2011; Rawat and Burgess 2011; Shen and Burgess 2012a).

3. Dialysis Method

Dialysis methods are widely used for in-vitro release testing of particulate carriers specially for oily suspensions/solutions (Schultz et al. 1997; Laresen et al. 2002), nanoparticles (Asadishad, Vossoughi, and Alamzadeh 2010) and lipid carriers (Xu, Khan, and Burgess 2012b). In this method, the dialysis bag with the suitable molecular weight cutoff (MWCO) is utilized to separate the formulation from the release media. The formulation is usually filled inside the dialysis sac. The schematic diagram is provided in Figure 9.1b. D'Souza and DeLuca (2006) suggested that the dialysis membrane should have MWCO of 100 times of the size of drug molecules in order not to be a limiting factor for drug diffusion, and the volume of the release media should be 5–10 times more than the formulation volume in order to have a sufficient driving force for drug to diffuse outside and to maintain sink conditions.

An alternative to the dialysis method is reverse dialysis, where the drug is placed outside the dialysis sac and release media is kept inside the dialysis bag. The sampling is done either by opening the dialysis sac or by replacing the dialysis sac with the new one (Maestrelli, Mura, and Alonso 2004; Xu, Khan, and Burgess 2012b). Since the formulation is agitated, the aggregation of the particles can be avoided. However, loss of solvent can be a major concern.

Different variations to this method are reported in the literature, which include rotating dialysis cell (Fredholt, Larsen, and Larsen 2000; Larsen, Fredholt, and Larsen 2001; Larsen et al. 2002; Larsen, Fredholt, and Larsen 2000; Schultz et al. 1997; Larsen et al. 2006,

2008; Pedersen et al. 2008; Dibbern and Wirbitzki 1983), formulation inside the dialysis bag (Float A Lyzer®) (Larsen et al. 2008; D'Souza and DeLuca 2005), and reverse dialysis method (Chidambaram and Burgess 1999; Levy and Benita 1990). Some of the studies have also performed drug release studies using dialysis bags in either a USP I (basket) apparatus (Abdel-Mottaleb and Lamprecht 2011; Gao et al. 2013) or a USP II apparatus (Cao et al. 2012, 2013). Results suggested that a modified USP I setup is more discriminatory in nature due to reduced surface area of the dialysis membrane as compared to common dialysis techniques (Abdel-Mottaleb and Lamprecht 2011).

9.2.2 *In Situ* Methods

In the sample-and-separate method, it is difficult to preserve the physical integrity of emulsion since filtration and centrifugation is used to separate sample from the formulation (Ammoury et al. 1990). A dialysis technique due to the limited membrane surface area may lead to a violation of sink conditions. An additional constraint of the barrier methods is the limited volume of the continuous phase available to solubilize the released drug in the donor chamber (Armoury et al. 1989; Gupta, Hung, and Perrier 1987; Sasaki et al. 1984). In order to avoid these difficulties, novel in situ methods were developed, which involve the analysis of the released drug under real-time conditions without the need of separating them from the dosage form. These methods mainly utilized electrochemical analysis for electroactive drugs. For example, Tan et al. used a potentiometric drug selective electrode to monitor the real-time release of procaine hydrochloride from pH-responsive nanogels (Tan, Goh, and Tam 2007) and Mora et al. used a square-wave voltammetric technique for the real-time measurement of doxorubicin hydrochloride (Mora et al. 2009). Similarly, differential pulse polarography was used to study the real-time release of piroxicam, chloramphenicol, and diazepam (Rosenblatt, Douroumis, and Bunjes 2007; Charalampopoulos, Avgoustakis, and Kontoyannis 2003). Since the formulation is in direct contact with the release media, the sink conditions can be easily maintained. However, only electroactive drugs are suitable candidates for in situ methods, and also the sensitivity of this method depends on the sensors used.

9.2.2.1 Accelerated In-Vitro Release Testing

Depot preparations or controlled release formulations are typically designed to release the drug over period of days to months and sometimes up to years; therefore, real time drug release testing is not practical. Accelerated in-vitro release testing conditions such as temperature and use of solvents have been used to mitigate this issue. For example, generally, in-vitro release studies are performed at 37°C (physiological temperature), testing at elevated temperatures has been performed to characterize drug release from variety of dosage forms (Zackrisson et al. 1995; D'Souza, Faraj, and DeLuca 2005). However, the accelerated drug release studies only serve as quality control purposes and batch release testing, the real-time testing method to predict in-vivo product performance is still necessary. Moreover, accelerated conditions can affect both the rate and mechanism of drug release, and therefore, careful selection of accelerated conditions is required. As an example, elevated temperature and pH-affect drug diffusion due to increased polymer mobility (Zolnik, Leary, and Burgess 2006) and polymer erosion due to accelerated hydrolysis (Zolnik and Burgess 2007) in case of biodegradable polymer-based dosage forms. A detailed review on "accelerated in-vitro release testing methods for extended release parenteral dosage forms" is available, and can be used to further develop understanding on factors to be considered while designing the accelerated in-vitro release testing methods of such dosage forms (Shen and Burgess 2012b).

9.2.3 Implications of In-Vitro Release Testing

The development of a suitable in-vitro release model is a key activity in the development of new formulations, since it serves as a valuable tool for quality control and batch release, for further formulation developments and product regulatory approvals. Much attention has been provided to the development of suitable methods to assess in-vitro release from parenteral drug delivery systems. Success has been reported with the use of a modified rotating paddle for suspensions, a Franz cell diffusion system for gels, a flow-through cell for implants, and a floatable dialysis bag for microspheres or nanoparticles.

It is also important to understand the in-vitro drug release mechanism for parenteral depot formulation in developing a suitable in-vitro release testing method. For example, drug release from biodegradable polymer-based dosage forms (microparticles, in situ forming systems, and implants) usually involve two phases; an initial burst release phase involves drug diffusion from a polymer matrix, followed by prolonged drug release phase which is primarily govern by polymer erosion (Wang, Venkatraman, and Kleiner 2004; Bhardwaj et al. 2010; Alexis 2005). Polymer erosion occurs through the hydrolysis of polymer chains, and can be surface (heterogeneous) erosion at the matrix boundary (in case of polyortho esters or polyanhydrides) or bulk (homogeneous) erosion throughout the entire microsphere structure (in case of poly lactide/glycolide polymers) (Burkersroda, Schedl, and Göpferich 2002).

For oily formulations, the major rate-limiting in-vivo release mechanism is the drug partitioning between the oil vehicle and the tissue fluid (Chien 1981; Zuidema et al. 1994; Armstrong and James 1980; Luo, Hubbard, and Midha 1998; Minto et al. 1997). Therefore, linear relationships between the log absorption rate and the log oil–water distribution coefficient was observed (Al-Hindawi, James, and Nicholls 1987; Hirano, Ichihashi, and Yamada 1982; James, Nicholls, and Roberts 1969; Hirano, Ichihashi, and Yamada 1981). Therefore, it is possible to obtain desired rate of drug release by proper selection of transport group for prodrug design and/or vehicle composition (Weng Larsen and Larsen 2009). However, the in-vivo release process may become more complex with more lipophilic drugs and a longer duration of action (Weng Larsen and Larsen 2009).

9.2.4 In-Vivo Studies for Controlled Drug Delivery Systems

Preclinical in-vivo animal studies are required to predict the drug safety, efficacy, and initial dose range in humans. However, the selection of an animal species can have a pronounced effect on the outcome and potential fate of the long acting parenterals. Martinez has reviewed the factors influencing the use and interpretation of animals models in the development of parenteral drug delivery systems (Martinez 2011) and some of the differences are provided in Table 9.2.

TABLE 9.2

Important Interspecies Differences and Its Implications in Clinical Trials

Parameter	Implication	Reference
Tissue fluid volume	Difference in fluid volume at the injection site may influence in-vivo drug dissolution and absorption. For example, in rats, fluid volume ranges from about 0.05 mL/g wet weight for muscles to over 0.4 mL/g wet weight of skin.	(Hirano, Ichihashi, and Yamada 1982; Larsen et al. 2009b)
The interspecies difference in HA composition in interstitial matrix may delay the drug absorption	Miligram amount of HA in pigs, dogs, and ferrets in lamina propria of the vocal folds was approximately 2, 3, and 4-fold greater than that observed in humans. The corresponding distribution of HA across different layers of the lamina propria varies in dogs and ferrets, while HA content is relatively evenly distributed across all three layers in people.	(Hahn et al. 2006)
Movement of drug from the interstitium to blood	Diffusion of drug from interstitial fluid and blood was approximately 2.6 times greater in the rat as compared to cats. Less dispersion in interstitial tissue in rats, compared to higher animals which build pressure due to volume of the injection, forces the drug to move into the blood.	(Watson 1998)
Lymphatic drug absorption	Macromolecules absorb primarily through capillaries, while higher animal species, such as sheep, dogs, and pigs, absorb mainly through lymphatic transport.	(Porter, Edwards, and Charman 2001)
Subcutaneous fat	Similarity between lipid structure between pig and humans; lower amount of SC fat in dogs and rats allowed faster absorption of drugs.	(Plum, Agerso, and Andersen 2000)

Abbreviation: HA: Hyaluronic acid.

9.2.5 In-Vitro–In-Vivo Correlations (IVIVC)

In-vitro release studies are commonly used as predictors of in-vivo behavior, historically with traditional dosage forms like capsules and tablets (i.e., dissolution), and more recently with parenteral-controlled drug delivery systems (Amann et al. 2010; D'Souza et al. 2014; Buch et al. 2010). Ideally, in-vitro drug release model development should be based on the knowledge of the in-vivo drug release mechanisms in order to develop better IVIVC (Burgess et al. 2004, 2002; Martinez et al. 2008). IVIVC mainly describes a relationship between the in-vitro drug release rates to the in-vivo absorption rates. Thus, for parenteral delivery systems, IVIVC can be applied to when drug need to partition in the blood such as s.c. or i.m. route (though not for i.v. or i.a. routes). Developing a predictable IVIVC for s.c. or i.m. routes is really challenging due to multiple factors governing the drug diffusion in the surrounding tissues, the change in the dimension of drug delivery systems, lymphatic drug uptake, interstitial tissue composition, the partition of drug between interstitium to the blood. Many authors have tried to establish IVIVC for controlled release parenteral formulations, as listed in Table 9.3.

9.3 Targeted Drug Delivery Systems

Drug targeting to the specific part of the body can be achieved by either passive, physical, and active targeting approaches. Passive targeting, or enhanced permeability and retention (EPR) is resulted mainly due to the leaky vasculature of a diseased area such as tumors, infarcts, and inflammation. In physical targeting, the carrier itself does not have targeting affinity, it only degrades at the target sited based on pH, temperature, redox potential, ultrasound, or magnetic properties and, as a result, the drug will release and accumulate at the target site. Active targeting uses the pairing of drug carriers with site-specific ligands. Different nanocarriers have been developed to prolong the half-life of drugs in systemic circulation, deliver drugs in a targeted manner to minimize systemic side effects, as well as release drugs at favorable rates (fast or sustained) or in an environmentally responsive manner.

The characterization of targeted drug delivery systems depends on type of nanocarriers, and are presented in Table 9.4. The most common tools are described in sections from 9.3.1 to 9.3.6.

Although, some of the characterization parameters are applied to all nanocarriers (for both controlled and targeted drug delivery), the emphasis is given to how these parameters will affect the targeting of these nanocarriers.

TABLE 9.3
Example Studies Involving IVIVC for Different Formulation Types

Formulation	Drug	In-Vitro Drug Release Method	In-Vitro Drug Release Period	In-Vivo Study Period	Drug Release Media	References
Oily solution	Buprivacaine	Rotating dialysis cell method	2 days	2 days in rats	pH 6.0, 0.05 M PB	(Dorrit Bjerg Larsen, Joergensen, et al. 2002)
In situ gel	Insulin	Shake flask method	55 days	30 days in rats	pH 7.4, PBS	(Al-Tahami, Oak, and Singh 2011)
	Doxorubicin hydrochloride	Incubation method	14 days	4 days in mice	pH 7.4, 0.1 M PBS	(Kakinoki and Taguchi 2007)
Microspheres	Ornitide acetate	Dialysis bag (Tube-o-dialyzer)	35 days	38 days in rats	pH 7.4, 0.1 M acetate buffer	(Kostanski and DeLuca 2000)
	Tolterodine	Shake flask method	28 days	18 days in dogs	pH 7.4, PBS, 0.02% sodium azide, 0.01% Tween 80	(Sun et al. 2010)
	Triptorelin	Shake flask method	180 days	168 days in rats	pH 7.4, tris buffer	(Asmus et al. 2013)
	Clozapine	Dialysis bag	24 days	21 days in rats	pH 7.4, PB, 25% (v/v) methanol	(Ishak et al. 2014)
	Leuprolide	Dialysis bag (Tube-o-dialyzer)	150 days	120 days in rats	pH 7.4, 0.1 M PBS, 0.02% sodium azide	(Woo et al. 2001)
	Exenatide	Shake flask method	30 days	30 days in rats	pH 7.4, PBS	(Qi et al. 2014)
Oily Suspension	Drospirenone	Shake flask method	9 days	28 days in rats and monkeys	pH 7.4, PB, 0.05% (w/v) sodium azide, 8% (w/w) HP-β-CD	(Plourde et al. 2005)
	Doxepin (Doxipine pamoate	Continuous -flow method	5 days	5 days in dogs	Human plasma, pH 7.4, 10 mM PB	(Gido, Langguth, and Mutschler 1994)
Implant	Risperidone	Shake flask method	112 days	112 days in rats	pH 7.4, PBS	(Amann et al. 2010)
	S-nitrosothiols	Shake flask method	37 days	28 days in rats	PBS	(Parent et al. 2013)
	Buserelin	Incubation method	70 days	35 days in dogs	pH 7.4, 0.05M PB, 0.05% benzalkonium chloride, 0.1% sodium azide	(Schliecker et al. 2004)

TABLE 9.4

Characterization of Different Targeted Drug Delivery Systems

Name of the Parenteral Drug Delivery System	Characterization Parameter	Equipment Used	Ref.
PEGylated Liposomes Radiofrequency Ablasion Therapy (RFA)	Tumor size	Direct caliper	(Andriyanov et al. 2017)
	Pharmacokinetic analysis	Phoenix WinNonlin	
	Imaging	IVIS kinetic in-vivo imaging system	
Liposomes	Solubility	Atomic absorption spectroscopy	(Wehbe et al. 2017)
	Imaging	INCELL Analyzer 2200	
	Tumor size	Digital callipers	
	Lipid concentration	Liquid scintillation counting	
Secretory phospholipase A2 ($sPLA_2$ sensitive liposomes)	Concentration of platinum in Oxaliplatin	Inductively coupled plasma mass spectroscopy (ICP-MS)	(Pourhassan et al. 2017)
	Concentration of sPLA2	Microplate reader	
Liposomes	Tumor implantation	Stereotaxic apparatus	(Nordling-David et al. 2017)
	Brain biodistribution and therapeutic effect	Clinical Optima MR450w 1.5 T MRI system	
Cationic liposomes	Cellular uptake	Flow cytometry	(Chi et al. 2017)
	Pharmacokinetic study	High performance HPLC analysis	
	In-vivo imaging and antitumor study	Carestream Molecular Imaging FX PRO	
Liposomes Lipid nanocapsule Polymeric nanocapsule	Zeta potential, particle size and polydispersity index	Zetasizer Nano ZS	(Karim et al. 2017)
	In-vitro drug release profile	HPLC analysis	
Nanoemulsion	Biodistribution study	Small animal IVIS®, spectrum In-Vivo Imaging System	(Afzal, Shareef, and Kishan 2016)
	In-vivo measurement of tumor volume	Digital screw gauge	
Lipoplexes	Confirmation of conjugation	^{1}H-NMR	(Martens et al. 2017)
	Uptake and transfection efficiency	Flow cytometer	
	Analysis of cell associated fluorescence	FACSCalibur	

(Continued)

TABLE 9.4 (CONTINUED)

Characterization of Different Targeted Drug Delivery Systems

Name of the Parenteral Drug Delivery System	Characterization Parameter	Equipment Used	Ref.
Electrostatic polymer–DNA complex (polyplexes)	Dynamics of polyplexes	Fluorolog Yobin Yvon-SPEX Fluorometer	(Vuorimaa-Laukkanen et al. 2017)
Eudragit (50:50) RL and RS mixture cationic polymeric NPs	Drug release profile	Logan Franz cell	(Duxfield et al. 2016)
	Cytotoxicity testing	Beckmann DU-600 Spectrophotometer	
Hydrophilic Polymeric NPs	Osmolarity	Osmometer	(Ogunjimi et al. 2017)
	PDI, zeta potential, particle size	Nano ZS zetasizer	
	In-vitro release study	Franz diffusion cell	
	In-vitro cell cytotoxicity study	Microplate reader	
	In-vitro cellular uptake study	Flow cytometry and confocal microscopy	
Solid lipid NPs (SLN), nanostructured lipid NPs (NLC)	Entrapment efficiency and in-vitro release of drug	HITACHI 4010 inductively coupled plasma mass spectrometry (ICP-MS)	(Wu et al. 2016)
Nanoemulsion	Rheological behavior	Visco Elite-R Rotational viscometer	(Đorđević et al. 2017)
	Electrical conductivity measurement	sensION™ + EC71 conductivity meter	
	pH value measurement	HI9321 pH meter	
	In-vivo pharmacodynamic study (Basal Locomotor activity and amphetamine-induced locomotor activity)	Ceiling-mounted camera connected to ANY-maze video tracking system software	
	In-vivo pharmacokinetic and biodistribution study	Ultra high-performance liquid chromatography-tandem mass spectrometry (UHPLC-MS/MS)	
Lipid nanoemulsion	Quantification of paclitaxel and in-vitro release of drug	HPLC analysis	(Chen et al. 2017)
	In-vitro cytotoxicity assay	Microplate reader	

(Continued)

TABLE 9.4 (CONTINUED)

Characterization of Different Targeted Drug Delivery Systems

Name of the Parenteral Drug Delivery System	Characterization Parameter	Equipment Used	Ref.
O/W nanoemulsion	Isothermal titration calorimetry analysis	Nano ITC Low Volume Isothermal Titration Calorimetry	(Fotticchia et al. 2017)
	Cell uptake study (fluorescence analysis)	Confocal and multiphoton microscope system	
Nanoemulsion	Turbidity measurement	Turbiscan LAB Expert device	(de Oliveira et al. 2017)
Quantum dots and paclitaxel co-loaded NLC	Morphology analysis	Transmission electron microscopy	(Olerile et al. 2017)
	Drug loading and in-vitro drug release	HPLC	
	In-vivo biodistribution and ex vivo study	Near infrared fluorescence imaging	
Quantum dot–engineered polymeric nanoparticles	Morphology and surrounding properties of the NPs	Atomic force microscope	(Belletti et al. 2017)
	Quantum dot binding efficiency	Spectrofluorimeter	
	Excitation–emission profile of sample	Confocal DM-IRE2 microscope	
Hollow Mesoporous silica NPs	Morphology and composition of NPs	TEM on JEM-100CX II electron microscope	(Li et al. 2017)
	Quantification of drug and drug release	FLS-920Edinburg Fluorescence Spectrometer	
	Cytotoxicity of NPs	TCS SP5 confocal laser scanning microscope	
Gelatin-functionalized mesoporous silica NPs	Structural and porous properties	TEM	(Liao et al. 2017)
	Determination of functional groups on the sample	FTIR spectrometer	
Penetratin-modified poly(amidoamine/Hyaluronic acid) complex	Authenticate proportion of HA and penetratin complexes	Fluorescence microscope	(Tai et al. 2017)
	Particle size	Zeta sizer Nano-ZS	
	Cellular uptake	Flow cytometer	
HA-coated albumin NPs	Illuminate the fluorescence NPs	Burton lamp	(Huang et al. 2017)
	Cross-sectioning of tissue	Cryostat microtome	

(Continued)

TABLE 9.4 (CONTINUED)

Characterization of Different Targeted Drug Delivery Systems

Name of the Parenteral Drug Delivery System	Characterization Parameter	Equipment Used	Ref.
Novel dual DNA nanocomplex in a NPs system	Analysis of GFP (green fluorescence protein) expression	Fluorescence microscope and ARIA-III flow cytometer	(Shukla et al. 2017)
	Analysis of apoptosis	ARIA-III flow cytometer	
	Immunocytochemistry	Confocal laser scanning microscope	

Abbreviation: TEM: Transmission electron microscopy.

9.3.1 Surface Morphology

Surface morphology and shape of nanoparticles play important role in the interaction with the cells and resulting uptake. Owing to the nanosize of the particulate system, they are usually not detected by the optical microscopy; hence, higher resolution techniques, such as electron microscopy, are required for the determination of surface morphology of nanoparticles. The TEM employs a beam of electron which is transmitted through an ultra-thin specimen, interacting with the specimen as it passes through. This interaction of electrons forms an image which is further magnified and can be focused onto an imaging device, such as fluorescent screen or charged coupled device (CCD) camera. The TEM provides information not only about the size of nanoparticles but also about the surface morphology (Nimesh 2013).

Atomic force microscopy (AFM) or scanning force microscopy (SFM) is a high-resolution scanning probe microscope with the resolution in the range of nanometers. The instrumental setup of AFM consists of cantilever with very fine tip (probe) at its end, which is typically silicon with a radius of few nanometers, and is employed to scan the sample surface. At the time of scanning, the forces between the cantilever tip and the sample lead to deflection of the cantilever, which can be detected by using a laser spot reflected from the top surface of the cantilever. AFM is preferred for the characterization of nanoparticles, as it possesses three-dimensional (3D) visualization capability and provides both qualitative and quantitative information about the sample topology, including morphology, surface texture and roughness, and more importantly in this case, the size of the particles (Montasser, Fessi, and Coleman 2002; Belletti et al. 2017).

Scanning electron microscopy (SEM) is another high-resolution scanning microscope, which employs a high energy beam of electrons to scan the sample surfaces. On this instrument, samples are analyzed after drying and coating with a thin layer of gold or platinum. The SEM provides direct picture of the surface of nanoparticles (Nimesh 2013; Martins et al. 2010).

9.3.2 Size and Size Distribution

The transfection and targeting efficiency of the nanoparticles are significantly influenced by the particle size. Hence, most of the formulations, such as polymeric nanoparticles, liposomes, solid lipid nanoparticles, and nanostructured lipid carriers are prepared with special attention to particle size (Dauty et al. 2001; Lee et al. 2001; Sakurai et al. 2000). Various publications have revealed that particle size of nanoparticles significantly dictates their cellular and tissue uptake; in many cell lines, only the nanosized particles are taken up efficiently, but not the larger sized microparticles (Desai et al. 1997; Zauner, Farrow, and Haines 2001). The characterization of nanoparticles with multiple techniques, such as TEM and dynamic light scattering (DLS), provides information about size of nanoparticles which may be relevant to their therapeutic potential. DLS measures the temporal fluctuations of the light scattered due to the Brownian motion of the particles, when a solution containing the particles is placed in the path of a monochromatic beam of light. It is also known as photon correlation spectroscopy or quasi-elastic light scattering. This technique provides particle size information in terms of hydrodynamic diameter along with the polydispersity index of the sample. DLS is a sensitive, nonintrusive, and powerful analytical tool, routinely employed for characterization of macromolecules and colloids in solution. To perform DLS, nanoparticles are suspended in water or a buffer, making them completely hydrated, whereas AFM or TEM samples are dried on a glass slide or copper grid surface. AFM and TEM have the limitation of visualizing a small number of nanoparticles, while the DLS technique provides an average picture of the sample by determining the size of millions of particles. The properties of the nanoparticles are highly influenced by the surrounding environment; for instance, the size distribution at physiological conditions may differ from that in water or in the dry state. In this regard, DLS seems to be the more suitable method, as it provides measurements in physiological buffers or biological fluids, such as blood plasma (Nimesh 2013).

9.3.3 Surface Charge

Surface charge plays a significant role in determining the interaction of nanoparticles with the cell surface. Zeta potential is used to describe the surface charge of nanoparticles, which can be defined as a measure of the magnitude of repulsion or attraction between particles. A particle suspended in a solution containing ions is surrounded by an electric double layer of ions and counterions. The potential that exists at the hydrodynamic boundary of the particle is known as the zeta potential. The stability of the colloidal system can be predicted by the magnitude of the zeta potential values. Particles in a suspension with a high negative or positive (i.e., above or ~±30 mV) zeta potential will tend to repel each other, thereby resulting in a stable colloid. However, in the contrary, particles with low zeta potential values lack sufficient repulsive force, which results in particle aggregation (Ogunjimi et al. 2017; Karim et al. 2017).

9.3.4 Entrapment/Drug Loading Efficiency

Entrapment efficiency (EE) is the percent drug that is successfully entrapped/adsorbed into the nanoparticles. The EE of nanoparticles can be influenced by the following factors: 1) type of polymer used; 2) concentration of polymer; 3) drug-to-polymer ratio; 4) solubility of the drug; 5) dispersed phase to continuous phase ratio, etc. (Dhakar 2012; Afzal, Shareef, and Kishan 2016; Karim et al. 2017). To determine the EE, the separation of free and encapsulated drugs is required. Different methods are used to separate the free drugs from the encapsulated drugs, which are dialysis-based methods, ultracentrifugation (Ricci et al. 2006; Zheng et al. 2006), centrifugal ultrafiltration (Ozer and Talsma H. 1989; Wang et al. 2009; Magalhaes et al. 1995), and pressure ultrafiltration (Boyd 2003; Cui et al. 2006). The EE is calculated as % EE = [(drug added- free drug)/drug added]* 100. The drug loading efficiency is calculated as ratio of drug to weight of the nanoparticles.

9.3.5 Drug Release Studies

A rapid and efficient physical separation of free drugs from encapsulated drugs is the main requirement for determination of both EE and in-vitro drug release from nanocarriers. Therefore, in-vitro performance testing of nanoparticulate systems is more complex and challenging than other modified release dosage forms (e.g., microspheres). Therefore, the dialysis method is the most versatile and popular method used to assess drug release from nanosized dosage forms, and physical separation of the dosage forms is achieved by the use of a dialysis membrane which allows for the ease of sampling at periodic intervals. In some instances, the sample and separate method (Wallace et al. 2012; Yue et al. 2009; Juenemann et al. 2011) and continuous flow method (USP apparatus 4) (Bhardwaj and Burgess 2010; Heng et al. 2008) are used. All these methods are discussed in earlier sections and detailed reviews are available for in-vitro drug release testing with respect to nanoparticulate drug delivery systems (Shen and Burgess 2013; S. D'Souza 2014).

In dialysis techniques, nanoparticle suspension (inner compartment) is placed in the dialysis bag that is subsequently sealed and placed in a larger vessel containing release media (outer compartment) and agitated to minimize the unstirred water layer effect. In general, the volume enclosed in the dialysis bag is much smaller than the outer media. For instance, the inner media volume reported in the literature ranges from 1 to 10 ml, whereas the outer media volume is much greater, typically around 40 to 90 ml, thus the container size will depend on the total volume of release media required for the in-vitro release study. Furthermore, the drug release from the nanoparticles diffuses through the dialysis membrane to the outer compartment from where it is sampled for analysis (Afzal, Shareef, and Kishan 2016; Karim et al. 2017; Chi et al. 2017). However, in the dialysis method, time is required for the free drug to equilibrate across the membrane. Therefore, the concentration of the drug in the release media may not necessarily reflect the actual free/release drug concentration in the donor compartment (Wallace et al. 2012). In such cases, the true release rate must be deconvoluted from a priori knowledge of kinetics of free drug diffusion across the membrane using a control experiment (Washington 1990). As an alternative to the dialysis method, a column-switching high-performance liquid chromatography (HPLC) system comprising of an online solid-phase extraction (SPE) system comprising a Diol SPE column and an octadecylsilane bonded silica (ODS) SPE column connected in series that can evaluate free (released) and encapsulated drug was developed (Ohnishi et al. 2013). In comparison to the dialysis method, this method is fast, requires less volume and does not include a rate-limiting barrier, which is the dialysis membrane.

Developing biorelevant in-vitro drug release methods for targeted nanocarriers is more challenging, as it requires an estimate of the absence of drug release prior to reaching the target and triggered drug release at the target site. Xu et al. have developed a two-stage reverse dialysis in-vitro release method to demonstrate the two-stage drug release characteristics of passively targeted liposomes using the reverse dialysis method (Xu, Khan, and Burgess 2012a). Liposomal dispersion was added to the external release medium and release samples were collected from the interior of the dialysis tubes. Buffer solution (pH 7.4 HEPES buffer) and a buffered surfactant solution (1% (v/v) Triton X-100 in HEPES buffer) were used as external release mediums during the first and second stages to mimic the two-phase drug release behavior of targeted liposome formulations.

9.3.6 Biodistribution Studies

The pharmacokinetic behavior of drug delivery systems with various targeting functions and controlled drug release capabilities is crucial for the successful transition of this research into clinical practice.

Biodistribution studies of developed targeted drug delivery systems should be performed to determine the targeting efficiency of the carrier in experimental animals. These studies should be performed in accordance with the general guidelines for animal experiments and consideration should be given to the appropriateness of experimental procedures, species, and number of animals. Generally, rats or mice are more common than other species (such as rabbits, which are also used). Formulation is usually injected intravenously into a group of 16–20 rodents (usually mice or rats) and at different time intervals, and smaller groups (4–5) of the animals are euthanized, then dissected. The organs of interest (usually: blood, liver, spleen, kidney, muscle, fat, adrenals, pancreas, brain, bone, stomach, small intestine, upper and lower large intestine, etc.) are collected, weighed, and analyzed for drug content by a suitable analytical technique. The drug content can be analyzed by two means: In the first method, the organs are weighed, homogenized, and then extracted with suitable solvent followed by estimation of drug content by suitable analytical technique like HPLC or LC-MS (liquid chromatography–mass spectrometry). In second approach, a targeted drug delivery system with either a radiolabeled drug or carrier (with 14C, 99mTc, etc.) is injected in the animal being tested. The amount of radioactive drug in each organ is then analyzed for radioactivity with a scintillation counter/detector at specific time points after administration. The advantage of second approach is the non-invasive imaging of the animal possible with the availability of the suitable scanners, and same study can be performed with fewer animals. The most common non-invasive imaging modalities routinely used are positron emission tomography (PET), single photon emission computed tomography (SPECT), magnetic resonance imaging (MRI), and optical imaging (OI) (Lammers et al. 2011; Kunjachan et al. 2012) Among these, OI has become extremely popular recently in the drug delivery field because of its time- and cost-effectiveness, its user-friendliness and its ability to be used for high-throughput analyses (Kunjachan et al. 2013). Detailed information on OI procedures, their applications, and limitations is beyond the scope of this book, and is readily available in the published literature (Appel et al. 2013; Rana et al. 2015; Etrych et al. 2016; Kunjachan et al. 2013).

References

Abdel-Mottaleb, Mona M A, and Alf Lamprecht. 2011. "Standardized in Vitro Drug Release Test for Colloidal Drug Carriers Using Modified USP Dissolution Apparatus I." *Drug Development and Industrial Pharmacy* 37 (2):178–84. https://doi.org/10.3109/03639045.2010.502534.

Afzal, Syed Muzammil, Mohammad Zubair Shareef, and Veerabrahma Kishan. 2016. "Transferrin Tagged Lipid Nanoemulsion of Docetaxel for Enhanced Tumor Targeting." *Journal of Drug Delivery Science and Technology* 36 (Supplement C):175–82. https://doi.org/10.1016/j.jddst.2016.10.008.

Al-Hindawi, M K, K C James, and P J Nicholls. 1987. "Influence of Solvent on the Availability of Testosterone Propionate from Oily, Intramuscular Injections in the Rat." *The Journal of Pharmacy and Pharmacology* 39 (2):90–95. http://www.ncbi.nlm.nih.gov/pubmed/2882010.

Al-Tahami, Khaled, Mayura Oak, and Jagdish Singh. 2011. "Controlled Delivery of Basal Insulin from Phase-Sensitive Polymeric Systems after Subcutaneous Administration: In Vitro Release, Stability, Biocompatibility, in Vivo Absorption, and Bioactivity of Insulin." *Journal of Pharmaceutical Sciences* 100 (6):2161–71. https://doi.org/10.1002/jps.22433.

Alexis, Frank. 2005. "Factors Affecting the Degradation and Drug-Release Mechanism of Poly(lactic Acid) and Poly[(lactic Acid)-Co-(Glycolic Acid)]." *Polymer International* 54 (1). John Wiley & Sons, Ltd.:36–46. https://doi.org/10.1002/pi.1697.

Amann, Laura C, Michael J Gandal, Robert Lin, Yuling Liang, and Steven J Siegel. 2010. "In Vitro-in Vivo Correlations of Scalable PLGA-Risperidone Implants for the Treatment of Schizophrenia." *Pharmaceutical Research* 27 (8):1730–37. https://doi.org/10.1007/s11095-010-0152-4.

Amatya, Sarmila, Eun Ji Park, Jong Hoon Park, Joon Sik Kim, Eunyoung Seol, Heeyong Lee, Hoil Choi, Young-Hee Shin, and Dong Hee Na. 2013. "Drug Release Testing Methods of Polymeric Particulate Drug Formulations." *Journal of Pharmaceutical Investigation* 43 (4):259–66. https://doi.org/10.1007/s40005-013-0072-5.

Ammoury, N, H Fessi, J P Devissaguet, F Puisieux, and S Benita. 1990. "In Vitro Release Kinetic Pattern of Indomethacin from poly(D,L-Lactide) Nanocapsules." Journal of Pharmaceutical Sciences 79 (9):763–67. http://www.ncbi.nlm.nih.gov/pubmed/2273454.

Andriyanov, Alexander V, Emma Portnoy, Erez Koren, Semenenko Inesa, Sara Eyal, S Nahum Goldberg, and Yechezkel Barenholz. 2017. "Therapeutic Efficacy of Combined PEGylated Liposomal Doxorubicin and Radiofrequency Ablation: Comparing Single and Combined Therapy in Young and Old Mice." *Journal of Controlled Release* 257 (November):2–9. https://doi.org/10.1016/j.jconrel.2017.02.018.

Appel, Alyssa A, Mark A Anastasio, Jeffery C Larson, and Eric M Brey. 2013. "Imaging Challenges in Biomaterials and Tissue Engineering." *Biomaterials* 34 (28). NIH Public Access:6615–30. https://doi.org/10.1016/j.biomaterials.2013.05.033.

Armoury, N, H Fessi, JP Devissauget, F Puisieux, and S Benita. 1989. "Physicochemical Characterization of Polymeric Nanocapsules and in Vitro Release Evaluation of Indomethacin as a Drug Model." Sci Technol Pract Pharm 5:647–51.

Armstrong, N A, and K C James. 1980. "Drug Release from Lipid-Based Dosage Forms. I." *International Journal of Pharmaceutics* 6 (3):185–93. https://doi.org/10.1016/0378-5173(80)90103-9.

Asadishad, B, M Vossoughi, and I Alamzadeh. 2010. "In Vitro Release Behavior and Cytotoxicity of Doxorubicin-Loaded Gold Nanoparticles in Cancerous Cells." *Biotechnology Letters* 32 (5):649–54. https://doi.org/10.1007/s10529-010-0208-x.

Asmus, Lutz R., Jean-Christophe Tille, Béatrice Kaufmann, Louise Melander, Torsten Weiss, Kerstin Vessman, Wolfgang Koechling, Grégoire Schwach, Robert Gurny, and Michael Möller. 2013. "In Vivo Biocompatibility, Sustained-Release and Stability of Triptorelin Formulations Based on a Liquid, Degradable Polymer." *Journal of Controlled Release* 165 (3):199–206. https://doi.org/10.1016/j.jconrel.2012.11.014.

Aubert-Pouëssel, Anne, Marie-Claire Venier-Julienne, Anne Clavreul, Michelle Sergent, Christophe Jollivet, Claudia N Montero-Menei, Emmanuel Garcion, David C Bibby, Philippe Menei, and Jean-Pierre Benoit. 2004. "In Vitro Study of GDNF Release from Biodegradable PLGA Microspheres." *Journal of Controlled Release: Official Journal of the Controlled Release Society* 95 (3):463–75. https://doi.org/10.1016/j.jconrel.2003.12.012.

Bain, D F, D L Munday, and A Smith. 1999. "Modulation of Rifampicin Release from Spray-Dried Microspheres Using Combinations of Poly-(DL-Lactide)." *Journal of Microencapsulation* 16 (3):369–85. https://doi.org/10.1080/026520499289086.

Belletti, D, G Riva, M Luppi, G Tosi, F Forni, M A Vandelli, B Ruozi, and F Pederzoli. 2017. "Anticancer Drug-Loaded Quantum Dots Engineered Polymeric Nanoparticles: Diagnosis/therapy Combined Approach." *European Journal of Pharmaceutical Sciences: Official Journal of the European Federation for Pharmaceutical Sciences* 107:230–39. https://doi.org/10.1016/j.ejps.2017.07.020.

Bhardwaj, Upkar, and Diane J Burgess. 2010. "A Novel USP Apparatus 4 Based Release Testing Method for Dispersed Systems." *International Journal of Pharmaceutics* 388 (1–2):287–94. https://doi.org/10.1016/j.ijpharm.2010.01.009.

Bhardwaj, Upkar, Radhakrishana Sura, Fotios Papadimitrakopoulos, and Diane J. Burgess. 2010. "PLGA/PVA Hydrogel Composites for Long-Term Inflammation Control Following S.c. Implantation." *International Journal of Pharmaceutics* 384 (1–2):78–86. https://doi.org/10.1016/j.ijpharm.2009.09.046.

Boyd, Ben J. 2003. "Characterisation of Drug Release from Cubosomes Using the Pressure Ultrafiltration Method." *International Journal of Pharmaceutics* 260 (2):239–47. http://www.ncbi.nlm.nih.gov/pubmed/12842343.

Buch, Philipp, Per Holm, Jesper Qvist Thomassen, Dieter Scherer, Robert Branscheid, Ute Kolb, and Peter Langguth. 2010. "IVIVC for Fenofibrate Immediate Release Tablets Using Solubility and Permeability as in Vitro Predictors for Pharmacokinetics." *Journal of Pharmaceutical Sciences* 99 (10):4427–36. doi:10.1002/jps.22148.

Burgess, Diane J, Daan J A Crommelin, Ajaz S Hussain, and Mei-Ling Chen. 2004. "Assuring Quality and Performance of Sustained and Controlled Released Parenterals." *European Journal of Pharmaceutical Sciences: Official Journal of the European Federation for Pharmaceutical Sciences* 21 (5):679–90. http://www.ncbi.nlm.nih.gov/pubmed/15188769.

Burgess, Diane J, Ajaz S Hussain, Thomas S Ingallinera, and Mei-Ling Chen. 2002. "Assuring Quality and Performance of Sustained and Controlled Release Parenterals: Workshop Report." *AAPS pharmSci* 4 (2). Springer:E7. https://doi.org/10.1208/PS040205.

Burkersroda, Friederike von, Luise Schedl, and Achim Göpferich. 2002. "Why Degradable Polymers Undergo Surface Erosion or Bulk Erosion." *Biomaterials* 23 (21). Elsevier:4221–31. https://doi.org/10.1016/S0142-9612(02)00170-9.

Cao, Xia, Wen-Wen Deng, Min Fu, Liang Wang, Shan-Shan Tong, Ya-Wei Wei, Ying Xu, Wei-Yan Su, Xi-Ming Xu, and Jiang-Nan Yu. 2012. "In Vitro Release and in Vitro-in Vivo Correlation for Silybin Meglumine Incorporated into Hollow-Type Mesoporous Silica Nanoparticles." *International Journal of Nanomedicine* 7:753–62. https://doi.org/10.2147/IJN.S28348.

Cao, Xia, Wenwen Deng, Min Fu, Yuan Zhu, Hongfei Liu, Li Wang, Jin Zeng, Yawei Wei, Ximing Xu, and Jiangnan Yu. 2013. "Seventy-Two-Hour Release Formulation of the Poorly Soluble Drug Silybin Based on Porous Silica Nanoparticles: In Vitro Release Kinetics and in Vitro/in Vivo Correlations in Beagle Dogs." *European Journal of Pharmaceutical Sciences* 48 (1):64–71. https://doi.org/10.1016/j.ejps.2012.10.012.

Charalampopoulos, Nikolaos, Konstantinos Avgoustakis, and Christos G Kontoyannis. 2003. "Differential Pulse Polarography: A Suitable Technique for Monitoring Drug Release from Polymeric Nanoparticle Dispersions." Analytica Chimica Acta 491 (1):57–62. doi:10.1016/S0003-2670(03)00788-8.

Chen, Lina, Bingchen Chen, Li Deng, Baoan Gao, Yuansheng Zhang, Chan Wu, Nong Yu, Qinqin Zhou, Jianzhong Yao, and Jianming Chen. 2017. "An Optimized Two-Vial Formulation Lipid Nanoemulsion of Paclitaxel for Targeted Delivery to Tumor." *International Journal of Pharmaceutics* 534 (1–2):308–15. https://doi.org/10.1016/j.ijpharm.2017.10.005.

Chi, Yingying, Xuelei Yin, Kaoxiang Sun, Shuaishuai Feng, Jinhu Liu, Daquan Chen, Chuanyou Guo, and Zimei Wu. 2017. "Redox-Sensitive and Hyaluronic Acid Functionalized Liposomes for Cytoplasmic Drug Delivery to Osteosarcoma in Animal Models." *Journal of Controlled Release: Official Journal of the Controlled Release Society* 261:113–25. https://doi.org/10.1016/j.jconrel.2017.06.027.

Chidambaram, N, and D J Burgess. 1999. "A Novel in Vitro Release Method for Submicron-Sized Dispersed Systems." *AAPS PharmSci* 1 (3):32–40. https://doi.org/10.1208/ps010311.

Chien, Y W. 1981. "Long-Acting Parenteral Drug Formulations." *Journal of Parenteral Science and Technology: A Publication of the Parenteral Drug Association* 35 (3):106–39. http://www.ncbi.nlm.nih.gov/pubmed/6113276.

Crommelin, D J A , and C J de Blaey. 1980. "In Vitro Release Studies on Drugs Suspended in Non-Polar Media I. Release of Sodium Chloride from Suspensions in Liquid Paraffin." *International Journal of Pharmaceutics* 5 (4):305–16. https://doi.org/10.1016/0378-5173(80)90038-1.

Crommelin, D J A, and C J de Blaey. 1980. "In Vitro Release Studies on Drugs Suspended in Non-Polar Media II. The Release of Paracetamol and Chloramphenicol from Suspensions in Liquid Paraffin." *International Journal of Pharmaceutics* 6 (1):29–42. https://doi.org/10.1016/0378-5173(80)90027-7.

Cui, JingXia, ChunLei Li, YingJie Deng, YongLi Wang, and Wei Wang. 2006. "Freeze-Drying of Liposomes Using Tertiary Butyl Alcohol/water Cosolvent Systems." *International Journal of Pharmaceutics* 312 (1–2):131–36. https://doi.org/10.1016/j.ijpharm.2006.01.004.

D'Souza, Susan. 2014. "A Review of In Vitro Drug Release Test Methods for Nano-Sized Dosage Forms." *Advances in Pharmaceutics* 2014 (October):1–12. https://doi.org/10.1155/2014/304757.

D'Souza, Susan S, Jabar A Faraj, and Patrick P DeLuca. 2005. "A Model-Dependent Approach to Correlate Accelerated with Real-Time Release from Biodegradable Microspheres." *AAPS PharmSciTech* 6 (4):E553–564. doi:10.1208/pt060470.

D'Souza, Susan S, and Patrick P DeLuca. 2005. "Development of a Dialysis in Vitro Release Method for Biodegradable Microspheres." *AAPS PharmSciTech* 6 (2):E323–28. https://doi.org/10.1208/pt060242.

D'Souza, Susan S, and Patrick P DeLuca. 2006. "Methods to Assess in Vitro Drug Release from Injectable Polymeric Particulate Systems." *Pharmaceutical Research* 23 (3):460–74. https://doi.org/10.1007/s11095-005-9397-8.

Dauty, E, J S Remy, T Blessing, and J P Behr. 2001. "Dimerizable Cationic Detergents with a Low Cmc Condense Plasmid DNA into Nanometric Particles and Transfect Cells in Culture." *Journal of the American Chemical Society* 123 (38):9227–34. http://www.ncbi.nlm.nih.gov/pubmed/11562201.

Davies, B, and T Morris. 1993. "Physiological Parameters in Laboratory Animals and Humans." *Pharmaceutical Research* 10 (7):1093–95. http://www.ncbi.nlm.nih.gov/pubmed/8378254.

Delplace, C, F Kreye, D Klose, F Danède, M Descamps, J Siepmann, and F Siepmann. 2012. "Impact of the Experimental Conditions on Drug Release from Parenteral Depot Systems: From Negligible to Significant." International Journal of Pharmaceutics, October. http://agris.fao.org/agris-search/search.do?recordID=US201500125013.

Desai, M P, V Labhasetwar, E Walter, R J Levy, and G L Amidon. 1997. "The Mechanism of Uptake of Biodegradable Microparticles in Caco-2 Cells Is Size Dependent." *Pharmaceutical Research* 14 (11):1568–73. http://www.ncbi.nlm.nih.gov/pubmed/9434276.

Dhakar, Ram Chand. 2012. "From Formulation Variables to Drug Entrapment Efficiency of Microspheres: A Technical Review." *Journal of Drug Delivery and Therapeutics* 2 (6). doi:10.22270/jddt.v2i6.160.

Dibbern, H W, and E Wirbitzki. 1983. "Möglikeiten Zur Bestimmung Der Wirkstoffreigabe Aus Hydrophoben Trägern. Insbesondere Aus Suppositorien." *Pharmazeutische Industrie* 45 (10). Edito Cantor fuṙ Medizin und Naturwissenschaften KG:985–90. http://www.refdoc.fr/Detailnotice?idarticle.

Đorđević, Sanela M, Anja Santrač, Nebojša D Cekić, Bojan D Marković, Branka Divović, Tanja M Ilić, Miroslav M Savić, and Snežana D Savić. 2017. "Parenteral Nanoemulsions of Risperidone for Enhanced Brain Delivery in Acute Psychosis: Physicochemical and in Vivo Performances." *International Journal of Pharmaceutics* 533 (2):421–30. https://doi.org/10.1016/j.ijpharm.2017.05.051.

Duxfield, Linda, Rubab Sultana, Ruokai Wang, Vanessa Englebretsen, Samantha Deo, Simon Swift, Ilva Rupenthal, and Raida Al-Kassas. 2016. "Development of Gatifloxacin-Loaded Cationic Polymeric Nanoparticles for Ocular Drug Delivery." *Pharmaceutical Development and Technology* 21 (2):172–79. https://doi.org/10.3109/10837450.2015.1091839.

Etrych, Tomáš, Henrike Lucas, Olga Janoušková, Petr Chytil, Thomas Mueller, and Karsten Mäder. 2016. "Fluorescence Optical Imaging in Anticancer Drug Delivery." *Journal of Controlled Release* 226 (March):168–81. https://doi.org/10.1016/j.jconrel.2016.02.022.

Fotaki, Nikoletta. 2011. "Flow-Through Cell Apparatus (USP Apparatus 4): Operation and Features." *Dissolution Technologies*, 46–49. https://doi.org/10.14227/DT180411P46.

Fotticchia, Teresa, Raffaele Vecchione, Pasqualina Liana Scognamiglio, Daniela Guarnieri, Vincenzo Calcagno, Concetta Di Natale, Chiara Attanasio et al. 2017. "Enhanced Drug Delivery into Cell Cytosol *via* Glycoprotein H-Derived Peptide Conjugated Nanoemulsions." *ACS Nano* 11 (10):9802–13. https://doi.org/10.1021/acsnano.7b03058.

Fredholt, K, D H Larsen, and C Larsen. 2000. "Modification of in Vitro Drug Release Rate from Oily Parenteral Depots Using a Formulation Approach." *European Journal of Pharmaceutical Sciences: Official Journal of the European Federation for Pharmaceutical Sciences* 11 (3):231–37. http://www.ncbi.nlm.nih.gov/pubmed/11042229.

Gao, Yuan, Jieyu Zuo, Nadia Bou-Chacra, Terezinha de Jesus Andreoli Pinto, Sophie-Dorothee Clas, Roderick B Walker, L&#xf6, and Raimar Benberg. 2013. "In Vitro Release Kinetics of Antituberculosis Drugs from Nanoparticles Assessed Using a Modified Dissolution Apparatus." *BioMed Research International*. https://www.hindawi.com/journals/bmri/2013/136590/.

Gido, C, P Langguth, and E Mutschler. 1994. "Predictions of in Vivo Plasma Concentrations from in Vitro Release Kinetics: Application to Doxepin Parenteral (I.m.) Suspensions in Lipophilic Vehicles in Dogs." *Pharmaceutical Research* 11 (6):800–808. http://www.ncbi.nlm.nih.gov/pubmed/7937517.

Goldbach, Pierre, Hervé Brochart, Pascal Wehrlé, and André Stamm. 1995. "Sterile Filtration of Liposomes: Retention of Encapsulated Carboxyfluorescein." *International Journal of Pharmaceutics* 117:225–30. https://doi.org/10.1016/0378-5173(94)00346-7.

Green, R H. 1974. "The Association of Viral Activation with Penicillin Toxicity in Guinea Pigs and Hamsters." *The Yale Journal of Biology and Medicine* 47 (3):166–81. http://www.ncbi.nlm.nih.gov/pubmed/4446629.

Gupta, P K, C T Hung, and D G Perrier. 1987. "Quantitation of the Release of Doxorubicin from Colloidal Dosage Forms Using Dynamic Dialysis." *Journal of Pharmaceutical Sciences* 76 (2):141–45. http://www.ncbi.nlm.nih.gov/pubmed/3572752.

Hahn, Mariah S, James B Kobler, Barry C Starcher, Steven M Zeitels, and Robert Langer. 2006. "Quantitative and Comparative Studies of the Vocal Fold Extracellular Matrix. I: Elastic Fibers and Hyaluronic Acid." *The Annals of Otology, Rhinology, and Laryngology* 115 (2):156–64. https://doi.org/10.1177/000348940611500213.

Heng, Desmond, David J Cutler, Hak-Kim Chan, Jimmy Yun, and Judy A. Raper. 2008. "What Is a Suitable Dissolution Method for Drug Nanoparticles?" *Pharmaceutical Research* 25 (7):1696–1701. https://doi.org/10.1007/s11095-008-9560-0.

Hirano, K, T Ichihashi, and H Yamada. 1982. "Studies on the Absorption of Practically Water-Insoluble Drugs Following Injection V: Subcutaneous Absorption in Rats from Solutions in Water Immiscible Oils." *Journal of Pharmaceutical Sciences* 71 (5):495–500. http://www.ncbi.nlm.nih.gov/pubmed/7097491.

Hirano, Koichiro, Teruhisa Ichihashi, and Hideo Yamada. 1981. "Studies on the Absorption of Practically Water-Insoluble Drugs Following Injection. I. Intramuscular Absorption from Water-Immiscible Oil Solutions in Rats." *Chemical & Pharmaceutical Bulletin*. Vol. 29. https://doi.org/10.1248/cpb.29.519.

Huang, Di, Ying-Shan Chen, Sachin S Thakur, and Ilva D Rupenthal. 2017. "Ultrasound-Mediated Nanoparticle Delivery across Ex Vivo Bovine Retina after Intravitreal Injection." *European Journal of Pharmaceutics and Biopharmaceutics* 119 (November):125–36. https://doi.org/10.1016/j.ejpb.2017.06.009.

Ishak, Rania A H, Nahed D Mortada, Noha M Zaki, Abd El-Hamid A El-Shamy, and Gehanne A S Awad. 2014. "Impact of Microparticle Formulation Approaches on Drug Burst Release: A Level A IVIVC." *Journal of Microencapsulation* 31 (7):674–84. https://doi.org/10.3109/02652048.2014.913724.

James, K C, P J Nicholls, and M Roberts. 1969. "Biological Half-Lives of [4-14C] testosterone and Some of Its Esters after Injection into the Rat." *The Journal of Pharmacy and Pharmacology* 21 (1):24–27. http://www.ncbi.nlm.nih.gov/pubmed/4388185.

Janicki, S, M Sznitowska, W Zebrowska, H Gabiga, and M Kupiec. 2001. "Evaluation of Paracetamol Suppositories by a Pharmacopoeial Dissolution Test—Comments on Methodology." *European Journal of Pharmaceutics and Biopharmaceutics: Official Journal of Arbeitsgemeinschaft Fur Pharmazeutische Verfahrenstechnik e.V* 52 (2):249–54. http://www.ncbi.nlm.nih.gov/pubmed/11522493.

Jeong, Young-Il, Jin-Gyu Song, Sam-Suk Kang, Hyang-Hwa Ryu, Young-Hwa Lee, Chan Choi, Boo-Ahn Shin, Kyung-Keun Kim, Kyu-Youn Ahn, and Shin Jung. 2003. "Preparation of poly(DL-Lactide-Co-Glycolide) Microspheres Encapsulating All-Trans Retinoic Acid." *International Journal of Pharmaceutics* 259 (1–2):79–91. http://www.ncbi.nlm.nih.gov/pubmed/12787638.

Jiang, Ge, Byung H Woo, Feirong Kang, Jagdish Singh, and Patrick P DeLuca. 2002. "Assessment of Protein Release Kinetics, Stability and Protein Polymer Interaction of Lysozyme Encapsulated poly(D,L-Lactide-Co-Glycolide) Microspheres." *Journal of Controlled Release: Official Journal of the Controlled Release Society* 79 (1–3):137–45. http://www.ncbi.nlm.nih.gov/pubmed/11853925.

Juenemann, Daniel, Ekarat Jantratid, Christian Wagner, Christos Reppas, Maria Vertzoni, and Jennifer B. Dressman. 2011. "Biorelevant in Vitro Dissolution Testing of Products Containing Micronized or Nanosized Fenofibrate with a View to Predicting Plasma Profiles." *European Journal of Pharmaceutics and Biopharmaceutics* 77 (2):257–64. https://doi.org/10.1016/j.ejpb.2010.10.012.

Kakemi, K, H Sezaki, S Muranishi, H Ogata, and K Giga. 1972. "Mechanism of Intestinal Absorption of Drugs from Oil in Water Emulsions. II. Absorption from Oily Solutions." *Chemical & Pharmaceutical Bulletin* 20 (4):715–20. http://www.ncbi.nlm.nih.gov/pubmed/5045613.

Kakinoki, Sachiro, and Tetsushi Taguchi. 2007. "Antitumor Effect of an Injectable in-Situ Forming Drug Delivery System Composed of a Novel Tissue Adhesive Containing Doxorubicin Hydrochloride." *European Journal of Pharmaceutics and Biopharmaceutics* 67 (3):676–81. https://doi.org/10.1016/j.ejpb.2007.03.020.

Karim, Reatul, Claudio Palazzo, Julie Laloy, Anne-Sophie Delvigne, Stéphanie Vanslambrouck, Christine Jerome, Elise Lepeltier et al. 2017. "Development and Evaluation of Injectable Nanosized Drug Delivery Systems for Apigenin." *International Journal of Pharmaceutics* 532 (2):757–68. https://doi.org/10.1016/j.ijpharm.2017.04.064.

Kim, H, and D J Burgess. 2002. "Effect of Drug Stability on the Analysis of Release Data from Controlled Release Microspheres." *Journal of Microencapsulation* 19 (5):631–40. https://doi.org/10.1080/02652040210140698.

Konan, Yvette N, Robert Gurny, and Eric Allémann. 2002. "Preparation and Characterization of Sterile and Freeze-Dried Sub-200 Nm Nanoparticles." *International Journal of Pharmaceutics* 233 (1–2):239–52. http://www.ncbi.nlm.nih.gov/pubmed/11897428.

Kostanski, J W, and P P DeLuca. 2000. "A Novel in Vitro Release Technique for Peptide Containing Biodegradable Microspheres." *AAPS PharmSciTech* 1 (1):E4. https://doi.org/10.1208/pt010104.

Kunjachan, Sijumon, Felix Gremse, Benjamin Theek, Patrick Koczera, Robert Pola, Michal Pechar, Tomas Etrych et al. 2013. "Noninvasive Optical Imaging of Nanomedicine Biodistribution." *ACS Nano* 7 (1). Europe PMC Funders:252–62. https://doi.org/10.1021/nn303955n.

Kunjachan, Sijumon, Jabadurai Jayapaul, Marianne E Mertens, Gert Storm, Fabian Kiessling, and Twan Lammers. 2012. "Theranostic Systems and Strategies for Monitoring Nanomedicine-Mediated Drug Targeting." *Current Pharmaceutical Biotechnology* 13 (4):609–22. http://www.ncbi.nlm.nih.gov/pubmed/22214503.

Lacasse, F X, P Hildgen, J Pérodin, E Escher, N C Phillips, and J N McMullen. 1997. "Improved Activity of a New Angiotensin Receptor Antagonist by an Injectable Spray-Dried Polymer Microsphere Preparation." *Pharmaceutical Research* 14 (7):887–91. https://doi.org/10.1023/A:1012147700014.

Lammers, Twan, Silvio Aime, Wim E. Hennink, Gert Storm, and Fabian Kiessling. 2011. "Theranostic Nanomedicine." *Doi.org* 44 (10). American Chemical Society (ACS):1029–38. https://doi.org/10.1021/ar200019c.

Larsen, Claus, Susan Weng Larsen, Henrik Jensen, Anan Yaghmur, and Jesper Ostergaard. 2009a. "Role of in Vitro Release Models in Formulation Development and Quality Control of Parenteral Depots." *Expert Opinion on Drug Delivery* 6 (12):1283–95. https://doi.org/10.1517/17425240903307431.

—. 2009b. "Role of in Vitro Release Models in Formulation Development and Quality Control of Parenteral Depots." *Expert Opinion on Drug Delivery* 6 (12):1283–95. https://doi.org/10.1517/17425240903307431.

Larsen, D B, K Fredholt, and C Larsen. 2001. "Addition of Hydrogen Bond Donating Excipients to Oil Solution: Effect on in Vitro Drug Release Rate and Viscosity." *European Journal of Pharmaceutical Sciences: Official Journal of the European Federation for Pharmaceutical Sciences* 13 (4):403–10. http://www.ncbi.nlm.nih .gov/pubmed/11408155.

Larsen, Dorrit Bjerg, Stig Joergensen, Niels Vidiendal Olsen, Steen Honoré Hansen, and Claus Larsen. 2002. "In Vivo Release of Bupivacaine from Subcutaneously Administered Oily Solution. Comparison with in Vitro Release." *Journal of Controlled Release: Official Journal of the Controlled Release Society* 81 (1–2):145–54. http://www.ncbi.nlm.nih.gov/pubmed/11992687.

Larsen, Dorrit Bjerg, Henrik Parshad, Karin Fredholt, and Claus Larsen. 2002. "Characteristics of Drug Substances in Oily Solutions. Drug Release Rate, Partitioning and Solubility." *International Journal of Pharmaceutics* 232 (1):107–17. https://doi.org/10.1016/S0378-5173(01)00904-8.

Larsen, Dorrit Høj, Karin Fredholt, and Claus Larsen. 2000. "Assessment of Rate of Drug Release from Oil Vehicle Using a Rotating Dialysis Cell." *European Journal of Pharmaceutical Sciences* 11 (3):223–29. https://doi.org/10.1016/S0928-0987(00)00105-6.

Larsen, Susan Weng, Anna Buus Frost, Jesper Østergaard, Hermann Marcher, and Claus Larsen. 2008. "On the Mechanism of Drug Release from Oil Suspensions in Vitro Using Local Anesthetics as Model Drug Compounds." *European Journal of Pharmaceutical Sciences: Official Journal of the European Federation for Pharmaceutical Sciences* 34 (1):37–44. https://doi.org/10.1016/j.ejps.2008.02.005.

Larsen, Susan Weng, Jesper Østergaard, Henrik Friberg-Johansen, Marit N B Jessen, and Claus Larsen. 2006. "In Vitro Assessment of Drug Release Rates from Oil Depot Formulations Intended for Intra-Articular Administration." *European Journal of Pharmaceutical Sciences: Official Journal of the European Federation for Pharmaceutical Sciences* 29 (5):348–54. https://doi.org/10.1016/j.ejps.2006.07.002.

Latha, M S, A V Lal, T V Kumary, R Sreekumar, and A Jayakrishnan. 2000. "Progesterone Release from Glutaraldehyde Cross-Linked Casein Microspheres: In Vitro Studies and in Vivo Response in Rabbits." *Contraception* 61 (5):329–34. http://www.ncbi.nlm.nih.gov/pubmed/10906504.

Lee, H, S K Williams, S D Allison, and T J Anchordoquy. 2001. "Analysis of Self-Assembled Cationic Lipid-DNA Gene Carrier Complexes Using Flow Field-Flow Fractionation and Light Scattering." *Analytical Chemistry* 73 (4):837–43. http://www.ncbi.nlm.nih.gov/pubmed/11248901.

Levy, M Y, and S Benita. 1990. "Drug Release from Submicronized O/w Emulsion: A New in Vitro Kinetic Evaluation Model." *International Journal of Pharmaceutics* 66 (1):29–37. https://doi.org/10.1016/0378-5173(90)90381-D.

Li, Yanhua, Na Li, Wei Pan, Zhengze Yu, Limin Yang, and Bo Tang. 2017. "Hollow Mesoporous Silica Nanoparticles with Tunable Structures for Controlled Drug Delivery." *ACS Applied Materials & Interfaces* 9 (3):2123–29. https://doi.org/10.1021/acsami.6b13876.

Liao, Yu-Te, Chih-Hung Lee, Si-Tan Chen, Jui-Yang Lai, and Kevin C.-W. Wu. 2017. "Gelatin-Functionalized Mesoporous Silica Nanoparticles with Sustained Release Properties for Intracameral Pharmacotherapy of Glaucoma." *J. Mater. Chem. B* 5 (34):7008–13. https://doi.org/10.1039/C7TB01217A.

Liggins, R T, and H M Burt. 2001. "Paclitaxel Loaded poly(L-Lactic Acid) Microspheres: Properties of Microspheres Made with Low Molecular Weight Polymers." *International Journal of Pharmaceutics* 222 (1):19–33. http://www.ncbi.nlm.nih.gov/pubmed/11404029.

Liu, Fang-I, J H Kuo, K C Sung, and Oliver Y P Hu. 2003. "Biodegradable Polymeric Microspheres for Nalbuphine Prodrug Controlled Delivery: In Vitro Characterization and in Vivo Pharmacokinetic Studies." *International Journal of Pharmaceutics* 257 (1):23–31. https://doi.org/10.1016/S0378-5173(03)00110-8.

Looney, Terry. 1996. "USP Apparatus 4 (Flow- Through Method) Primer." *Dissolution Technology*, 10–12. https://doi.org/10.14227/DT030496P10.

Lootvoet, G, E Beyssac, G K Shiu, J-M. Aiache, and W A Ritschel. 1992. "Study on the Release of Indomethacin from Suppositories: In Vitro-in Vivo Correlation." *International Journal of Pharmaceutics* 85 (1):113–20. https://doi.org/10.1016/0378-5173(92)90140-W.

Luo, J P, J W Hubbard, and K K Midha. 1998. "The Roles of Depot Injection Sites and Proximal Lymph Nodes in the Presystemic Absorption of Fluphenazine Decanoate and Fluphenazine: Ex Vivo Experiments in Rats." *Pharmaceutical Research* 15 (9):1485–89. http://www.ncbi.nlm.nih.gov/pubmed/9755905.

Maestrelli, F, P Mura, and M J Alonso. 2004. "Formulation and Characterization of Triclosan Sub-Micron Emulsions and Nanocapsules." *Journal of Microencapsulation* 21 (8):857–64. https://doi.org/10.1080/02652040400015411.

Magalhaes, N S Santos, H Fessi, F Puisieux, S Benita, and M Seiller. 1995. "An in Vitro Release Kinetic Examination and Comparative Evaluation between Submicron Emulsion and Polylactic Acid Nanocapsules of Clofibride." *Journal of Microencapsulation* 12 (2):195–205. https://doi.org/10.3109/02652049509015290.

Maksimenko, O, E Pavlov, E Toushov, A Molin, Y Stukalov, T Prudskova, V Feldman, J Kreuter, and S Gelperina. 2008. "Radiation Sterilisation of Doxorubicin Bound to Poly(butyl Cyanoacrylate) Nanoparticles." *International Journal of Pharmaceutics* 356 (1–2):325–32. https://doi.org/10.1016/j.ijpharm.2008.01.010.

Martens, Thomas F, Karen Peynshaert, Thaís Leite Nascimento, Elias Fattal, Marcus Karlstetter, Thomas Langmann, Serge Picaud, et al. 2017. "Effect of Hyaluronic Acid-Binding to Lipoplexes on Intravitreal Drug Delivery for Retinal Gene Therapy." *European Journal of Pharmaceutical Sciences* 103 (November):27–35. https://doi.org/10.1016/j.ejps.2017.02.027.

Martinez, M N, M J Rathbone, D Burgess, and M Huynh. 2010. "Breakout Session Summary from AAPS/CRS Joint Workshop on Critical Variables in the in Vitro and in Vivo Performance of Parenteral Sustained Release Products." *Journal of Controlled Release* 142 (1):2–7. doi:10.1016/j.jconrel.2009.09.028.

Martinez, Marilyn N. 2011. "Factors Influencing the Use and Interpretation of Animal Models in the Development of Parenteral Drug Delivery Systems." *The AAPS Journal* 13 (4):632–49. https://doi.org/10.1208/s12248-011-9303-8.

Martinez, Marilyn, Michael Rathbone, Diane Burgess, and Mai Huynh. 2008. "In Vitro and in Vivo Considerations Associated with Parenteral Sustained Release Products: A Review Based upon Information Presented and Points Expressed at the 2007 Controlled Release Society Annual Meeting." *Journal of Controlled Release: Official Journal of the Controlled Release Society* 129 (2):79–87. https://doi.org/10.1016/j.jconrel.2008.04.004.

Martins, Dorival, Lucas Frungillo, Maristela C Anazzetti, Patrícia S Melo, and Nelson Durán. 2010. "Antitumoral Activity of L-Ascorbic Acid-Poly- D,L-(Lactide-Co-Glycolide) Nanoparticles Containing Violacein." *International Journal of Nanomedicine* 5:77–85. http://www.ncbi.nlm.nih.gov/pubmed/20161989.

Mi, Fwu-Long, Shin-Shing Shyu, Yi-Mei Lin, Yu-Bey Wu, Chih-Kang Peng, and Yi-Hung Tsai. 2003. "Chitin/PLGA Blend Microspheres as a Biodegradable Drug Delivery System: A New Delivery System for Protein." *Biomaterials* 24 (27):5023–36. http://www.ncbi.nlm.nih.gov/pubmed/14559016.

Minto, C F, C Howe, S Wishart, A J Conway, and D J Handelsman. 1997. "Pharmacokinetics and Pharmacodynamics of Nandrolone Esters in Oil Vehicle: Effects of Ester, Injection Site and Injection Volume." *The Journal of Pharmacology and Experimental Therapeutics* 281 (1):93–102. http://www.ncbi.nlm.nih.gov/pubmed/9103484.

Montasser, I, H Fessi, and A W Coleman. 2002. "Atomic Force Microscopy Imaging of Novel Type of Polymeric Colloidal Nanostructures." *European Journal of Pharmaceutics and Biopharmaceutics: Official Journal of Arbeitsgemeinschaft Fur Pharmazeutische Verfahrenstechnik e.V* 54 (3):281–84. http://www.ncbi.nlm.nih.gov/pubmed/12445557.

Mora, Laura, Karin Y Chumbimuni-Torres, Corbin Clawson, Lucas Hernandez, Liangfang Zhang, and Joseph Wang. 2009. "Real-Time Electrochemical Monitoring of Drug Release from Therapeutic Nanoparticles." *Journal of Controlled Release* 140 (1):69–73. doi:10.1016/j.jconrel.2009.08.002.

Murty, Santos B, Jack Goodman, B C Thanoo, and Patrick P DeLuca. 2003. "Identification of Chemically Modified Peptide from poly(D,L-Lactide-Co-Glycolide) Microspheres under in Vitro Release Conditions." *AAPS PharmSciTech* 4 (4):392–405. https://doi.org/10.1208/pt040450.

Negrin, C M, A Delgado, M Llabrés, and C Evora. 2001. "In Vivo-in Vitro Study of Biodegradable Methadone Delivery Systems." *Biomaterials* 22 (6):563–70. https://doi.org/10.1016/S0142-9612(00)00214-3.

Nimesh, Surendra. 2013. "Tools and Techniques for Physico-Chemical Characterization of Nanoparticles," November. https://doi.org/10.1533/9781908818645.43.

Nordling-David, Mirjam M, Roni Yaffe, David Guez, Hadar Meirow, David Last, Etty Grad, Sharona Salomon, et al. 2017. "Liposomal Temozolomide Drug Delivery Using Convection Enhanced Delivery." *Journal of Controlled Release* 261 (November):138–46. https://doi.org/10.1016/j.jconrel.2017.06.028.

Ogunjimi, A T, S M Melo, C G Vargas-Rechia, F S Emery, and R F Lopez. 2017. "Hydrophilic Polymeric Nanoparticles Prepared from Delonix Galactomannan with Low Cytotoxicity for Ocular Drug Delivery." *Carbohydrate Polymers* 157 (November):1065–75. https://doi.org/10.1016/j.carbpol.2016.10.076.

Ohnishi, Naozumi, Eiichi Yamamoto, Hiromasa Tomida, Kenji Hyodo, Hiroshi Ishihara, Hiroshi Kikuchi, Kohei Tahara, and Hirofumi Takeuchi. 2013. "Rapid Determination of the Encapsulation Efficiency of a Liposome Formulation Using Column-Switching HPLC." *International Journal of Pharmaceutics* 441 (1–2):67–74. https://doi.org/10.1016/j.ijpharm.2012.12.019.

Olerile, Livesey David, Yongjun Liu, Bo Zhang, Tianqi Wang, Shengjun Mu, Jing Zhang, Lesego Selotlegeng, and Na Zhang. 2017. "Near-Infrared Mediated Quantum Dots and Paclitaxel Co-Loaded Nanostructured Lipid Carriers for Cancer Theragnostic." *Colloids and Surfaces B: Biointerfaces* 150 (November):121–30. https://doi.org/10.1016/j.colsurfb.2016.11.032.

Ozer, A Y, and Talsma H. 1989. "Preparation and Stability of Liposomes Containing 5-Fluorouracil." *International Journal of Pharmaceutics* 55 (2–3). Elsevier:185–91. https://doi.org/10.1016/0378-5173(89)90040-9.

Packhaeuser, C B, J Schnieders, C G Oster, and T Kissel. 2004. "In Situ Forming Parenteral Drug Delivery Systems: An Overview." *European Journal of Pharmaceutics and Biopharmaceutics* 58 (2). Elsevier:445–55. https://doi.org/10.1016/j.ejpb.2004.03.003.

Parent, Marianne, Ariane Boudier, François Dupuis, Cécile Nouvel, Anne Sapin, Isabelle Lartaud, Jean-Luc Six, Pierre Leroy, and Philippe Maincent. 2013. "Are in Situ Formulations the Keys for the Therapeutic Future of S-Nitrosothiols?" *European Journal of Pharmaceutics and Biopharmaceutics* 85 (3, Part A):640–49. https://doi.org/10.1016/j.ejpb.2013.08.005.

Park, Eun Ji, Dong Hee Na, and Kang Choon Lee. 2007. "In Vitro Release Study of Mono-PEGylated Growth Hormone-Releasing Peptide-6 from PLGA Microspheres." *International Journal of Pharmaceutics* 343 (1):281–83. https://doi.org/10.1016/j.ijpharm.2007.06.005.

Park, T G, H Lee Yong, and Y Nam Sung. 1998. "A New Preparation Method for Protein Loaded poly(D, L-Lactic-Co-Glycolic Acid) Microspheres and Protein Release Mechanism Study." *Journal of Controlled Release: Official Journal of the Controlled Release Society* 55 (2–3):181–91. https://doi.org/10.1016/S0168-3659(98)00050-9.

Pedersen, Brian Thoning, Susan Weng Larsen, Jesper Østergaard, and Claus Larsen. 2008. "In Vitro Assessment of Lidocaine Release from Aqueous and Oil Solutions and from Preformed and in Situ Formed Aqueous and Oil Suspensions. Parenteral Depots for Intra-Articular Administration." *Drug Delivery* 15 (1):23–30. https://doi.org/10.1080/10717540701828657.

Pilaniya, Urmila, Kapil Khatri, and U K Patil. 2011. "Depot Based Drug Delivery System for the Management of Depression." Current Drug Delivery. http://www.eurekaselect.com/74791/article.

Pistolozzi, Marco, and Carlo Bertucci. 2008. "Species-Dependent Stereoselective Drug Binding to Albumin: A Circular Dichroism Study." *Chirality* 20 (3–4):552–58. https://doi.org/10.1002/chir.20521.

Plourde, François, Aude Motulsky, Anne-Claude Couffin-Hoarau, Didier Hoarau, Huy Ong, and Jean-Christophe Leroux. 2005. "First Report on the Efficacy of L-Alanine-Based in Situ-Forming Implants for the Long-Term Parenteral Delivery of Drugs." *Journal of Controlled Release* 108 (2–3):433–41. https://doi.org/10.1016/j.jconrel.2005.08.016.

Plum, A, H Agerso, and L Andersen. 2000. "Pharmacokinetics of the Rapid-Acting Insulin Analog, Insulin Aspart, in Rats, Dogs, and Pigs, and Pharmacodynamics of Insulin Aspart in Pigs." *Drug Metabolism and Disposition: The Biological Fate of Chemicals* 28 (2):155–60. http://www.ncbi.nlm.nih.gov/pubmed/10640512.

Porter, C J, G A Edwards, and S A Charman. 2001. "Lymphatic Transport of Proteins after S.c. Injection: Implications of Animal Model Selection." *Advanced Drug Delivery Reviews* 50 (1–2):157–71. http://www.ncbi.nlm.nih.gov/pubmed/11489338.

Pourhassan, Houman, Gael Clergeaud, Anders E Hansen, Ragnhild G Østrem, Frederikke P Fliedner, Fredrik Melander, Ole L Nielsen, Ciara K O'Sullivan, Andreas Kjær, and Thomas L Andresen. 2017. "Revisiting the Use of sPLA 2 -Sensitive Liposomes in Cancer Therapy." *Journal of Controlled Release* 261 (November):163–73. https://doi.org/10.1016/j.jconrel.2017.06.024.

Prettyman, Julie. 2005. "Subcutaneous or Intramuscular? Confronting a Parenteral Administration Dilemma." *Medsurg Nursing: Official Journal of the Academy of Medical-Surgical Nurses* 14 (2):93–8; quiz 99. http://www.ncbi.nlm.nih.gov/pubmed/15916264.

Qi, Feng, Jie Wu, Tingyuan Yang, Guanghui Ma, and Zhiguo Su. 2014. "Mechanism Studies for Monodisperse Exenatide-Loaded PLGA Microspheres Prepared by Different Methods Based on SPG Membrane Emulsification." *Acta Biomaterialia* 10. https://doi.org/10.1016/j.actbio.2014.06.018.

Rana, Sudha, Amit Tyagi, Nabo Kumar Chaudhury, and Rakesh Kumar Sharma. 2015. "In Vivo Imaging Techniques of the Nanocarriers Used for Targeted Drug Delivery." In, 667–86. Springer, Cham. https://doi.org/10.1007/978-3-319-11355-5_21.

Rawat, Archana, Upkar Bhardwaj, and Diane J Burgess. 2012. "Comparison of in Vitro-in Vivo Release of Risperdal(®) Consta(®) Microspheres." International Journal of Pharmaceutics 434 (1–2):115–21. doi:10.1016/j.ijpharm.2012.05.006.

Rawat, Archana, and Diane J Burgess. 2011. "USP Apparatus 4 Method for in Vitro Release Testing of Protein Loaded Microspheres." *International Journal of Pharmaceutics* 409 (1–2):178–84. https://doi.org/10.1016/j.ijpharm.2011.02.057.

Rawat, Archana, Upkar Bhardwaj, and Diane J Burgess. 2012. "Comparison of in Vitro-in Vivo Release of Risperdal(®) Consta(®) Microspheres." *International Journal of Pharmaceutics* 434 (1–2):115–21. https://doi.org/10.1016/j.ijpharm.2012.05.006.

Rawat, Archana, Erika Stippler, Vinod P Shah, and Diane J Burgess. 2011. "Validation of USP Apparatus 4 Method for Microsphere in Vitro Release Testing Using Risperdal Consta." *International Journal of Pharmaceutics* 420 (2):198–205. https://doi.org/10.1016/j.ijpharm.2011.08.035.

Ricci, Maurizio, Stefano Giovagnoli, Paolo Blasi, Aurelie Schoubben, Luana Perioli, and Carlo Rossi. 2006. "Development of Liposomal Capreomycin Sulfate Formulations: Effects of Formulation Variables on Peptide Encapsulation." *International Journal of Pharmaceutics* 311 (1–2):172–81. https://doi.org/10.1016/j.ijpharm.2005.12.031.

Rosenblatt, Karin M, Dionysios Douroumis, and Heike Bunjes. 2007. "Drug Release from Differently Structured Monoolein/poloxamer Nanodispersions Studied with Differential Pulse Polarography and Ultrafiltration at Low Pressure." *Journal of Pharmaceutical Sciences* 96 (6):1564–75. doi:10.1002/jps.20808.

Sakurai, F, R Inoue, Y Nishino, A Okuda, O Matsumoto, T Taga, F Yamashita, Y Takakura, and M Hashida. 2000. "Effect of DNA/liposome Mixing Ratio on the Physicochemical Characteristics, Cellular Uptake and Intracellular Trafficking of Plasmid DNA/cationic Liposome Complexes and Subsequent Gene Expression." *Journal of Controlled Release: Official Journal of the Controlled Release Society* 66 (2–3):255–69. http://www.ncbi.nlm.nih.gov med/10742585.

Sasaki, H, Y Takakura, M Hashida, T Kimura, and H Sezaki. 1984. "Antitumor Activity of Lipophilic Prodrugs of Mitomycin C Entrapped in Liposome or O/W Emulsion." *Journal of Pharmacobio-Dynamics* 7 (2): 120–30. http://www.ncbi.nlm.nih.gov/pubmed/6427444.

Schaefer, M J, and J Singh. 2002. "Effect of Tricaprin on the Physical Characteristics and in Vitro Release of Etoposide from PLGA Microspheres." *Biomaterials* 23 (16):3465–71. https://doi.org/10.1016/S0142-9612(02)00053-4.

Schliecker, Gesine, Carsten Schmidt, Stefan Fuchs, Andreas Ehinger, Jürgen Sandow, and Thomas Kissel. 2004. "In Vitro and in Vivo Correlation of Buserelin Release from Biodegradable Implants Using Statistical Moment Analysis." *Journal of Controlled Release* 94 (1):25–37. https://doi.org/10.1016/j.jconrel.2003.09.003.

Schultz, null, null Møllgaard, null Frokjaer, and null Larsen. 1997. "Rotating Dialysis Cell as in Vitro Release Method for Oily Parenteral Depot Solutions." *International Journal of Pharmaceutics* 157 (2):163–69. http://www.ncbi.nlm.nih.gov/pubmed/10477813.

Seidlitz, Anne, and Werner Weitschies. 2012. "In-Vitro Dissolution Methods for Controlled Release Parenterals and Their Applicability to Drug-Eluting Stent Testing." The Journal of Pharmacy and Pharmacology 64 (7):969–85. doi:10.1111/j.2042-7158.2011.01439.x.

Shen, Jie, and Diane J Burgess. 2012a. "Accelerated in Vitro Release Testing of Implantable PLGA microsphere/PVA Hydrogel Composite Coatings." *International Journal of Pharmaceutics* 422 (1–2):341–48. https://doi.org/10.1016/j.ijpharm.2011.10.020.

Shen, Jie, and Diane J. Burgess. 2012b. "Accelerated in-Vitro Release Testing Methods for Extended-Release Parenteral Dosage Forms." *Journal of Pharmacy and Pharmacology* 64 (7):986–96. https://doi.org/10.1111/j.2042-7158.2012.01482.x.

Shen, Jie, and Diane J. Burgess. 2013. "In Vitro Dissolution Testing Strategies for Nanoparticulate Drug Delivery Systems: Recent Developments and Challenges." *Drug Delivery and Translational Research* 3 (5). NIH Public Access:409–15. https://doi.org/10.1007/s13346-013-0129-z.

Shukla, Vasundhara, Manu Dalela, Manika Vij, Ralph Weichselbaum, Surender Kharbanda, Munia Ganguli, Donald Kufe, and Harpal Singh. 2017. "Systemic Delivery of the Tumor Necrosis Factor Gene to Tumors by a Novel Dual DNA-Nanocomplex in a Nanoparticle System." *Nanomedicine: Nanotechnology, Biology and Medicine* 13 (5):1833–39. https://doi.org/10.1016/j.nano.2017.03.004.

Söderberg, Lars, Henrik Dyhre, Bodil Roth, and Sven Björkman. 2002. "In-Vitro Release of Bupivacaine from Injectable Lipid Formulations Investigated by a Single Drop Technique—Relation to Duration of Action in-Vivo." *The Journal of Pharmacy and Pharmacology* 54 (6):747–55. http://www.ncbi.nlm.nih.gov/pubmed/12078990.

Söderberg, Lars, Henrik Dyhre, Bodil Roth, and Sven Björkman. 2006a. "The "inverted Cup" — A Novel in Vitro Release Technique for Drugs in Lipid Formulations." *Journal of Controlled Release: Official Journal of the Controlled Release Society* 113 (1):80–88. https://doi.org/10.1016/j.jconrel.2006.03.015.

Söderberg, Lars, Henrik Dyhre, Bodil Roth, and Sven Björkman. 2006b. "Ultralong Peripheral Nerve Block by Lidocaine:prilocaine 1:1 Mixture in a Lipid Depot Formulation: Comparison of in Vitro, in Vivo, and Effect Kinetics." *Anesthesiology* 104 (1):110–21. http://www.ncbi.nlm.nih.gov/pubmed/16394697.

Sun, Fengying, Cheng Sui, Lesheng Teng, Ximing Liu, Lirong Teng, Qingfan Meng, and Youxin Li. 2010. "Studies on the Preparation, Characterization and Pharmacological Evaluation of Tolterodine PLGA Microspheres." *International Journal of Pharmaceutics* 397 (1–2):44–49. https://doi.org/10.1016/j.ijpharm.2010.06.042.

Tai, Lingyu, Chang Liu, Kuan Jiang, Xishan Chen, Linglin Feng, Weisan Pan, Gang Wei, and Weiyue Lu. 2017. "A Novel Penetratin-Modified Complex for Noninvasive Intraocular Delivery of Antisense Oligonucleotides." *International Journal of Pharmaceutics* 529 (1–2):347–56. https://doi.org/10.1016/j.ijpharm.2017.06.090.

Tan, Jeremy P K, Chew H Goh, and Kam C Tam. 2007. "Comparative Drug Release Studies of Two Cationic Drugs from pH-Responsive Nanogels." European *Journal of Pharmaceutical Sciences* 32 (4):340–48. doi:10.1016/j.ejps.2007.08.010.

Vandelli, M A, F Rivasi, P Guerra, F Forni, and R Arletti. 2001. "Gelatin Microspheres Crosslinked with D,L-Glyceraldehyde as a Potential Drug Delivery System: Preparation, Characterisation, in Vitro and in Vivo Studies." *International Journal of Pharmaceutics* 215 (1–2):175–84. http://www.ncbi.nlm.nih.gov/pubmed/11250103.

Vargas de Oliveira, Erika Cristina, Zumira Aparecida Carneiro, Sérgio de Albuquerque, and Juliana Maldonado Marchetti. 2017. "Development and Evaluation of a Nanoemulsion Containing Ursolic Acid: A Promising Trypanocidal Agent: Nanoemulsion with Ursolic Acid Against T. Cruzi." *AAPS PharmSciTech* 18 (7):2551–60. https://doi.org/10.1208/s12249-017-0736-y.

Voisine, J M, B S Zolnik, and D J Burgess. 2008. "In Situ Fiber Optic Method for Long-Term in Vitro Release Testing of Microspheres." *International Journal of Pharmaceutics* 356 (1–2):206–11. https://doi.org/10.1016/j.ijpharm.2008.01.017.

Vuorimaa-Laukkanen, Elina, Ekaterina S Lisitsyna, Tiia-Maaria Ketola, Emmanuelle Morin-Pickardat, Huamin Liang, Martina Hanzlíková, and Marjo Yliperttula. 2017. "Difference in the Core-Shell Dynamics of Polyethyleneimine and Poly (L -Lysine) DNA Polyplexes." *European Journal of Pharmaceutical Sciences* 103 (November):122–27. https://doi.org/10.1016/j.ejps.2017.03.025.

Wagenaar, B W, and B W Müller. 1994. "Piroxicam Release from Spray-Dried Biodegradable Microspheres." *Biomaterials* 15 (1):49–54. http://www.ncbi.nlm.nih.gov/pubmed/8161657.

Wallace, Stephanie J, Jian Li, Roger L Nation, and Ben J Boyd. 2012. "Drug Release from Nanomedicines: Selection of Appropriate Encapsulation and Release Methodology." *Drug Delivery and Translational Research* 2 (4). NIH Public Access:284–92. https://doi.org/10.1007/s13346-012-0064-4.

Wang, Dongkai, Liwen Kong, Jing Wang, Xiaoling He, Xiang Li, and Yuting Xiao. 2009. "Polymyxin E Sulfate-Loaded Liposome for Intravenous Use: Preparation, Lyophilization, and Toxicity Assessment in Vivo." *PDA Journal of Pharmaceutical Science and Technology* 63 (2):159–67. Accessed November 29, 2017. http://www.ncbi.nlm.nih.gov/pubmed/19634354.

Wang, Juan, Barbara M Wang, and Steven P Schwendeman. 2002. "Characterization of the Initial Burst Release of a Model Peptide from poly(D,L-Lactide-Co-Glycolide) Microspheres." *Journal of Controlled Release: Official Journal of the Controlled Release Society* 82 (2–3):289–307. http://www.ncbi.nlm.nih.gov/pubmed/12175744.

Wang, Liwei, Subbu Venkatraman, and Lothar Kleiner. 2004. "Drug Release from Injectable Depots: Two Different in Vitro Mechanisms." *Journal of Controlled Release* 99 (2):207–16. https://doi.org/10.1016/j.jconrel.2004.06.021.

Washington, C. 1990. "Drug Release from Microdisperse Systems: A Critical Review." *International Journal of Pharmaceutics* 58 (1). Elsevier:1–12. https://doi.org/10.1016/0378-5173(90)90280-H.

Watson, P D. 1998. "Analysis of the Paired-Tracer Method of Determining Cell Uptake." *The American Journal of Physiology* 275 (2 Pt 1):E366-371. http://www.ncbi.nlm.nih.gov/pubmed/9688641.

Wehbe, Moe, Armaan Malhotra, Malathi Anantha, Jeroen Roosendaal, Ada W Y Leung, David Plackett, Katarina Edwards, Roger Gilabert-Oriol, and Marcel B Bally. 2017. "A Simple Passive Equilibration Method for Loading Carboplatin into Pre-Formed Liposomes Incubated with Ethanol as a Temperature Dependent Permeability Enhancer." *Journal of Controlled Release* 252 (November):50–61. https://doi.org/10.1016/j.jconrel.2017.03.010.

Wei, Guobao, Glenda J Pettway, Laurie K McCauley, and Peter X Ma. 2004. "The Release Profiles and Bioactivity of Parathyroid Hormone from Poly(lactic-Co-Glycolic Acid) Microspheres." *Biomaterials* 25 (2):345–52. http://www.ncbi.nlm.nih.gov/pubmed/14585722.

Weng Larsen, Susan, and Claus Larsen. 2009. "Critical Factors Influencing the In Vivo Performance of Long-Acting Lipophilic Solutions—Impact on In Vitro Release Method Design." *The AAPS Journal* 11 (4):762–70. https://doi.org/10.1208/s12248-009-9153-9.

Woo, B H, J W Kostanski, S Gebrekidan, B A Dani, B C Thanoo, and P P DeLuca. 2001. "Preparation, Characterization and in Vivo Evaluation of 120-Day poly(D, L-Lactide) Leuprolide Microspheres." *Journal of Controlled Release: Official Journal of the Controlled Release Society* 75 (3):307–15. http://www.ncbi.nlm.nih.gov/pubmed/11489318.

Wu, Miaojing, Yanghua Fan, Shigang Lv, Bing Xiao, Minhua Ye, and Xingen Zhu. 2016. "Vincristine and Temozolomide Combined Chemotherapy for the Treatment of Glioma: A Comparison of Solid Lipid Nanoparticles and Nanostructured Lipid Carriers for Dual Drugs Delivery." *Drug Delivery* 23 (8):2720–25. https://doi.org/10.3109/10717544.2015.1058434.

Xu, Qiaobing, Michinao Hashimoto, Tram T Dang, Todd Hoare, Daniel S Kohane, George M Whitesides, Robert Langer, and Daniel G Anderson. 2009. "Preparation of Monodisperse Biodegradable Polymer Microparticles Using a Microfluidic Flow-Focusing Device for Controlled Drug Delivery." *Small (Weinheim an Der Bergstrasse, Germany)* 5 (13):1575–81. https://doi.org/10.1002/smll.200801855.

Xu, Xiaoming, Mansoor A. Khan, and Diane J. Burgess. 2012a. "A Two-Stage Reverse Dialysis in Vitro Dissolution Testing Method for Passive Targeted Liposomes." *International Journal of Pharmaceutics* 426 (1–2):211–18. https://doi.org/10.1016/j.ijpharm.2012.01.030.

Xu, Xiaoming, Mansoor A Khan, and Diane J Burgess. 2012b. "A Two-Stage Reverse Dialysis in Vitro Dissolution Testing Method for Passive Targeted Liposomes." *International Journal of Pharmaceutics* 426 (1–2):211–18. https://doi.org/10.1016/j.ijpharm.2012.01.030.

Yang, Yi-Yan, Hui-Hui Chia, and Tai-Shung Chung. 2000. "Effect of Preparation Temperature on the Characteristics and Release Profiles of PLGA Microspheres Containing Protein Fabricated by Double-Emulsion Solvent Extraction/evaporation Method." *Journal of Controlled Release* 69 (1). Elsevier:81–96. https://doi.org/10.1016/S0168-3659(00)00291-1.

Yen, S Y, K C Sung, J J Wang, and O Hu Yoa-Pu. 2001. "Controlled Release of Nalbuphine Propionate from Biodegradable Microspheres: In Vitro and in Vivo Studies." *International Journal of Pharmaceutics* 220 (1–2):91–99. https://doi.org/10.1016/S0378-5173(01)00649-4.

Yue, Peng-Fei, Xiu-Yun Lu, Zeng-Zhu Zhang, Hai-Long Yuan, Wei-Feng Zhu, Qin Zheng, and Ming Yang. 2009. "The Study on the Entrapment Efficiency and in Vitro Release of Puerarin Submicron Emulsion." *AAPS PharmSciTech* 10 (2). Springer:376–83. https://doi.org/10.1208/s12249-009-9216-3.

Zackrisson, G, G Östling, B Skagerberg, and T Anfält. 1995. "ACcelerated Dissolution Rate Analysis (ACDRA) for Controlled Release Drugs. Application to Roxiam®." Journal of Pharmaceutical and Biomedical Analysis, Papers from the fifth international symposium on pharmaceutical and biomedical analysis, 13 (4):377–83. doi:10.1016/0731-7085(95)01293-T.

Zauner, W, N A Farrow, and A M Haines. 2001. "In Vitro Uptake of Polystyrene Microspheres: Effect of Particle Size, Cell Line and Cell Density." *Journal of Controlled Release: Official Journal of the Controlled Release Society* 71 (1):39–51. http://www.ncbi.nlm.nih.gov/pubmed/11245907.

Zheng, Yongli, Yan Wu, Wuli Yang, Changchun Wang, Shoukuan Fu, and Xizhong Shen. 2006. "Preparation, Characterization, and Drug Release in Vitro of Chitosan-Glycyrrhetic Acid Nanoparticles." *Journal of Pharmaceutical Sciences* 95 (1):181–91. https://doi.org/10.1002/jps.20399.

Zolnik, Banu S, Pauline E Leary, and Diane J Burgess. 2006. "Elevated Temperature Accelerated Release Testing of PLGA Microspheres." *Journal of Controlled Release* 112 (3):293–300. doi:10.1016/j.jconrel.2006.02.015.

Zolnik, Banu S, and Diane J. Burgess. 2007. "Effect of Acidic pH on PLGA Microsphere Degradation and Release." *Journal of Controlled Release* 122 (3):338–44. doi:10.1016/j.jconrel.2007.05.034.

Zuidema, J, Farisha Kadir, H A C Titulaer, and Christien Oussoren. 1994. "Release and Absorption Rates of Intramuscularly and Subcutaneously Injected Pharmaceuticals (II)." *International Journal of Pharmaceutics* 105:189–207. https://doi.org/10.1016/0378-5173(94)90103-1.

Appendix: Characterization Parameter and Common Characterization Tools

Parameter	Characterization Tools
ADME	Pharmacokinetic studies
Aerodynamic performance of the inhalation formulations	Cascade impactor, Twin stage impinge, Next generation impactor
Aerosol droplet size	Laser diffraction technique
Assay	Chromatography or spectroscopy
Bacterial endotoxins/pyrogen testing	LAL test
Biodistribution in animal models	Radiolabeling gamma scintigraphy
Buffer capacity of the cationic polymers/lipids as gene delivery vectors	Acid base titration method
Carrier–drug interaction	Differential scanning calorimetry, FTIR
Catanionic Polymer/Lipid –DNA/siRNA complexation	Gel electrophoresis
Cell cycle analysis, cell apoptosis and necrosis	Flow cytometry
Charge determination	Laser doppler anemometry, Zeta potentiometer
Composition structure	Mass spectroscopy
Conformation change of protein–metallic NP conjugate	Surface enhanced Raman spectroscopy
Content uniformity	HPLC or UV spectroscopy
Crystallinity	Powder X-Ray diffractometry
Detection of a specific DNA sequence in DNA samples	Southern blotting
Detection of mitochondrial dysfunction	Oxygen consumption assessment (via polarographic technique), Measurement of ATPase activity (via luciferin–luciferase reaction), Membrane potential (via fluorescent probe analysis) and Morphology (via electron microscopy)
Detection of specific proteins in tissue homogenate or extract	Western blotting
Determination of depth of penetration in ocular tissues	Optical coherence tomography

(Continued)

Parameter	Characterization Tools
Determination of fracture force	Texture analyser
Determination of the insertion force in ocular tissues	Texture analyser
Drug permeation or diffusion	Franz diffusion cell, Excised tissue
Drug Solubility study	Shake flask method
Drug-excipient compatibility	FTIR, differential scanning calorimetry
Elasticity and tensile strength	Texture analyser
Elemental analysis	Inductive coupled plasma optical emission spectrometry
Emulsion/microemulsion droplet size	Microscopic method and Dynamic light scattering particle size technique
Entrapment efficiency	Centrifugation technique, dialysis membrane technique, gel exclusion chromatography
Gel strength	Texture analyser
Hemocompatibility of the formulations	Erythrocyte aggregation method and hemolysis study
Hydrodynamic dimension binding kinetics	Fluorescence correlation spectroscopy
Hydrodynamic size and size distribution (indirect analysis)	Raman scattering
In-vitro cell cytotoxicity study	MTT Assay
In-vitro permeation studies, cell uptake studies, and transfection studies	Different in-vitro cell culture (cell lines)
Lipid peroxidation	Thiobarbituric acid reactive substances assay
Measurement of mass change	Quartz crystal microbalance surface method
Micelle CMC determination	Conductometry, spectroflurimetry
Moisture content determination	Thermo gravimetric analyzer
Molecular weight	Gel permeation chromatography Mass spectroscopy
Morphology and surrounding properties of the nanoparticles	Atomic force microscope
Mucoadhesion	Texture analyzer
Nanocarriers stability in-vivo	Electrolyte induced flocculation method
Nanoparticle dispersion stability	Critical flocculation temperature
Oxidative stress	Fluorescent dichlorodihydrofluoroscein assay
Particle size and distribution	Atomic force microscopy, Laser diffractometry, Photon correlation spectroscopy, Scanning electron microscopy, Transmission electron microscopy
Phospholipid content	Bartlett assay, Stewart assay
Qualitative cell uptake study	Confocal Microscopy

(*Continued*)

Parameter	Characterization Tools
Qualitatively detection of gene expression	Reverse transcription polymerase chain reaction
Quantitative cell uptake study	Flow Cytometry (FACS Analysis)
Quantum dot binding efficiency	Spectrofluorimeter
Release profile	In-vitro release characteristics under physiologic and sink conditions
Residual solvent or volatile impurities	Gas chromatography
Shape	Small-angle X-ray scattering
Size and shape of nanomaterials	Near-field scanning optical microscopy
Size/size distribution	Small-angle X-ray scattering
Skin irritation and sensitization test	Draize Test (Intradermal technique), Freund's Complete adjuvant test (Intradermal technique), Buckler's test, Open epicutaneous test
Spreadability	Texture analyzer
Structural, chemical, and electronic properties	Tip-enhanced Raman spectroscopy
Structure	Small-angle X-ray scattering
Structure and conformational change of biomolecules and thermal stability	Circular dichroism
Surface chemical analysis	X-Ray photoelectron spectroscopy
Surface hydrophobicity	Rose Bengal(dye) binding, Water contact angle measurement, X-ray photoelectron spectroscopy
Surface plasmon response of metal nanoparticles (silver, gold)	Ultraviolet-Visible (UV-Vis) Spectroscopy
Surface properties	Mass spectroscopy
Thickness of ocular insert	Screw gauge
To study gene expression by detection of RNA (or isolated mRNA)	Northern blotting
Viscosity	Brookfield viscometer Cone and plate viscometer Cup and bob viscometer
Water content	Volumetric/coulometric Karl Fischer, near-IR
Zone of inhibition for antibiotics	Agar plate method

Index

E

F

G

H

I

U

X

Y

Z